Catalysis –
Science and Technology

Volume 5

EDITORS:

Dr. J. R. Anderson (Melbourne/Australien)
Prof. Dr. M. Boudart (Stanford/USA)

CONTRIBUTORS:

Dr. P. Gallezot, Dr. J. B. Peri,
Dr. J. R. Rostrup-Nielsen,
Dr. K. C. Taylor

CATALYSIS – Science and Technology

Volume 5

With 122 Figures

Akademie-Verlag · Berlin
1984

Die
Originalausgabe erscheint
im Springer-Verlag Berlin · Heidelberg
New York · Tokyo

Vertrieb
ausschließlich für alle Staaten mit Ausnahme
der sozialistischen Länder:
Springer-Verlag Berlin · Heidelberg
New York · Tokyo

Vertrieb
für die sozialistischen Länder:
Akademie-Verlag Berlin

Erschienen im
Akademie-Verlag, DDR-1086 Berlin, Leipziger Straße 3—4

Lizenznummer: 200 · 100/497/84
Printed in the German Democratic Republic
Gesamtherstellung: VEB Druckerei „Thomas Müntzer", 5820 Bad Langensalza
LSV 1215
Bestellnummer: 763 387 6 (6844)
13200

General Preface to Series

In one form or another catalytic science reaches across almost the entire field of reaction chemistry, while catalytic technology is a cornerstone of much of modern chemical industry. The field of catalysis is now so wide and detailed, and its ramifications are so numerous, that the production of a thorough treatment of the entire subject is well beyond the capability of any single author. Nevertheless, the need is obvious for a comprehensive reference work on catalysis which is thoroughly up-to-date, and which covers the subject in depth at both a scientific and at a technological level. In these circumstances, a multi-author approach, despite its well-known drawbacks, seems to be the only one available.

In general terms, the scope of *Catalysis: Science and Technology* is limited to topics which are, to some extent at least, relevant to industrial processes. The whole of heterogeneous catalysis falls within its scope, but only biocatalytic processes which have significance outside of biology are included. Ancillary subjects such as surface science, materials properties, and other fields of catalysis are given adequate treatment, but not to the extent of obscuring the central theme.

Catalysis: Science and Technology thus has a rather different emphasis from normal review publications in the field of catalysis: here we concentrate more on important established material, although at the same time providing a systematic presentation of relevant data. The opportunity is also taken, where possible, to relate specific details of a particular topic in catalysis to established principles in chemistry, physics, and engineering, and to place some of the more important features into a historical perspective.

Because the field of catalysis is one where current activity is enormous and because various topics in catalysis reach a degree of maturity at different points in time, it is not expedient to impose a preconceived ordered structure upon *Catalysis: Science and Technology* with each volume devoted to a particular subject area. Instead, each topic is dealt with when it is most appropriate to do so. It will be sufficient if the entire subject has been properly covered by the time the last volume in the series appears. Nevertheless, the Editors will try to organize the subject matter so as to minimize unnecessary duplication between chapters, and to impose a reasonable uniformity of style and approach. Ultimately, these aspects of the presentation of this work must remain the responsibility of the Editors, rather than of individual authors.

The Editors would like to take this opportunity to give their sincere thanks to all the authors whose labors make this reference work possible. However, we all stand in debt to the numerous scientists and engineers whose efforts have built the discipline of catalysts into what it is today: we can do no more than dedicate these volumes to them.

Preface

Catalytic steam reforming has grown during the last two or three decades into one of the world's great catalytic processes. It is of major economic significance since the products from it form the feed for a number of other major processes. Nevertheless, catalytic steam reforming is a relatively difficult technology. It operates at high temperatures where problems of the maintenance of materials integrity and of catalyst stability and activity are severe, the establishment of high thermal efficiency of the plant is economically vital, and reactor operation is strongly influenced by mass and heat transport effects. The process is the subject of a thorough review by Dr. J. R. Rostrup-Nielsen who discusses both the basic catalytic chemistry and the way in which this is interrelated with reactor and plant design.

The use of catalytic converters for the purification of automotive exhaust gases is a relatively new technology which was brought into existence by social pressures for the preservation of acceptable environmental conditions. The majority of catalytic practitioners have been able to watch the growth of this technology from its inception to its current state of sophistication. Automotive catalytic converter technology is now in a mature state, and the chapter in this volume by Dr. K. C. Taylor provides a review which covers both the process chemistry and the most important converter design factors.

Catalyst characterization is essential if reaction chemistry and catalyst performance are to be adequately understood. The final two chapters in this volume address two aspects of the characterization problem. Infrared spectroscopy is one of the most important methods for obtaining information about the nature of adsorbed species on practical and model catalysts, but it is also relevant to the nature of the catalyst material itself. It is a mature technique which is also in the process of elaboration via the introduction of newer instrumental methods. The chapter by Dr. J. B. Peri provides a thorough account of this subject.

X-ray methods provide what is probably the most important collection of related methods for determining the structure of solids. Again, although a number of these techniques are mature and well-established, they have recently been complemented by newer methods which are of particular relevance to situations where long range order is minimal or absent. The final chapter in this volume by Dr. P. Gallezot gives a comprehensive account of the application of X-ray methods to the characterization of catalyst materials.

Contents

List of Contributors

Dr. Pierre Gallezot
Institut de Recherches sur la Catalyse
2, Avenue Albert Einstein
F-69626 Villeurbanne-Cedex, France

Dr. John B. Peri
Amoco Oil Company
Research and Development Dept.
Amoco Research Center
P.O. Box 400
Naperville, IL 60566, USA

Dr. Jens R. Rostrup-Nielsen
Haldor Topsøe A/S
Nymollevej 55
DK-2800 Lyngby, Denmark

Dr. Kathleen C. Taylor
Physical Chemistry Dept.
General Motors Research
Laboratories
Warren, Michigan 48090, USA

Chapter 1

Catalytic Steam Reforming

Jens R. Rostrup-Nielsen

Haldor Topsøe A/S, DK-2800 Lyngby, Denmark

Contents

1. Introduction

A. Reforming Reactions

1. Terminology

The steam reforming process converts hydrocarbons into mixtures of hydrogen, carbon monoxide, carbon dioxide, and methane

$$C_nH_m + n\,H_2O \rightarrow n\,CO + \left(n + \frac{m}{2}\right) H_2 \;(-\Delta H^0_{298} < 0) \quad (1)$$

$$CO + H_2O \leftrightarrows CO_2 + H_2 \quad (-\Delta H^0_{298} = 41.2\ \text{kJ mol}^{-1}) \quad (2)$$

$$CO + 3\,H_2 \leftrightarrows CH_4 + H_2O \quad (-\Delta H^0_{298} = 206.2\ \text{kJ mol}^{-1}) \quad (3)$$

The expression *reforming* is misleading since it is used also for the well-knowm process for improvement of the octane number of gasoline [1]. In the gas industry, reforming has generally been used for "the changing by heat treatment of a hydrocarbon with high heating value into a gaseous mixture of lower heating value" [2].

Reaction (1) involves the decomposition of the hydrocarbon by means of oxygen atoms, and Padovani [3] has suggested to term reaction (1) "oxygenolysis" corresponding to the related "pyrolysis" and "hydrogenolysis" of hydrocarbons using heat and hydrogen, respectively, to split the hydrocarbon. However, as the term "reforming" is very common for reaction 1, it will be used in this treatise.

There are several processes based on gasification of hydrocarbons in the presence of steam [3, 4]. As indicated in Table 1, some of these processes make use of additional of air. Cyclic reformers operate alternating with steam/hydrocarbons and air for decoking and heating. Other processes operate

Table 1. Characteristics of "Oxygenolysis" processes for gasification of hydrocarbons

Operation	Oxidizing agent	Heating	Active catalyst component	Examples
1. continuous	H_2O	external autothermic	Ni Ni	Topsøe, ICI, Kellogg, etc. British Gas Council (CRG), Lurgi/BASF (Recatro, Gassynthan). J.G.C., (MRG).
2. continuous	H_2O + air	external autothermic	Ni Ni	Didier, Otto, etc. Topsøe — SBA
3. cyclic	H_2O alt. air	internal internal	Ni Lime	Onia, Gegi, etc. Segas
4. continuous	$O_2(H_2O)$	autothermic	none	Shell, Texaco

continuously with the addition of air. Finally, some noncatalytic processes operate with oxygen and steam. The present treatise is restricted to processes and catalysts for continuous reforming with no addition of air or oxygen.

A general and detailed description of steam reforming was given by Bridger [5, 6], Rostrup-Nielsen [7], and Jockel [4], whereas reviews by Ross [8], Trimm [9], Van Hook [10] and Bartholomew [11] have dealt with specific catalytic problems. This treatise is no review. The scope has been to give an integrated treatment of steam reforming illustrating the close connection between catalyst properties and reformer design principles. The emphasis will be on topics critical for industrial operation.

2. The Equilibria

2.1. Gas Compositions

The number of independent reactions representing the complete stoichiometry of the steam reforming system is given by equations (1) to (3). Equation (1) should be written for each higher hydrocarbon ($n > 1$) present in the system. As will be shown the break-down on nickel catalysts of the higher hydrocarbons proceeds normally *via* (1) directly to C_1-components with no intermediate products. This is in contrast to the steam dealkylation of toluene to benzene over noble metal catalysts [12] and reactions over nonmetal catalysts [13].

Reaction (1) is the reverse of the Fischer-Tropsch synthesis [14], which is performed below 620 K. However, at the higher temperatures of interest in reforming processes, the endothermic reaction (1) can be considered as being irreversible, if the hydrocarbon is not methane.

As an example, consider steam reforming of *n*-heptane at 773 K, 3 MPa (= 30 atm) and $H_2O/C = 4$ mol/atom. The equilibrium constant for reaction (1) (K_p^1 (773 K)) is 6.92×10^{-3}, and the molar fraction of *n*-heptane is calculated to be 1.7×10^{-22} after equilibration of reactions (1–3).

It is misleading [8] to carry out such calculations including only reaction (1). For ethane K_p^1 (773 K) for reaction (1) is 6.58×10^{-4}. Equilibrium calculations at 773 K and 3 MPa result in a molar fraction of ethane of 0.07 when considering reaction (1) alone, whereas the correct value 3.7×10^{-6} is obtained when considering all three reactions.

Reaction (2), the shift reaction, and reaction (3), the methanation reaction, are reversible at reforming temperatures. It is evident from the principle of Le Chatelier that at the higher temperatures less methane and more carbon monoxide are present in the equilibrium gas, and that the methane content increases with pressure and decreases with increasing ratio of steam to carbon. This is illustrated in Figure 1 and 2. Several simplified methods for the estimation of equilibrium conditions have been published [15, 16, 17]. The

Figure 1. Equilibrium compositions steam reforming of methane. Various H_2O/CH_4 as parameter $y(H_2O)$ is calculated from $y(CH_4$, dry) and $y(CH_4$, wet) obtained from the curves. Then, $y(CO)$, $y(CO_2)$, and $y(H_2)$ can be calculated from three linear equations expressing the O/C, H/C and the sum of molar fractions. The result should be considered a rough estimate ▶

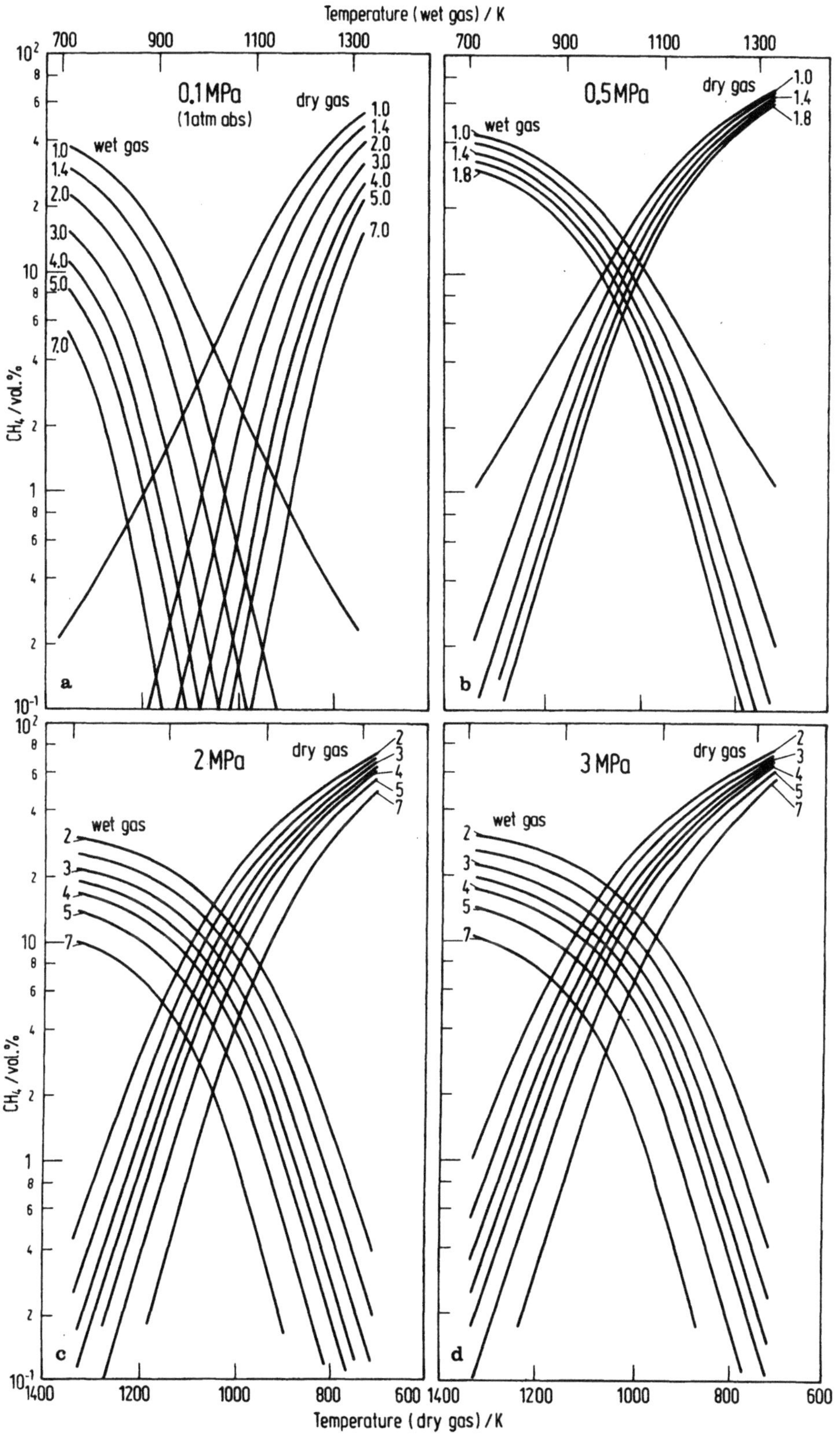
Temperature (wet gas) / K
700
900
1100
1300
0.1 MPa
(1atm abs)
dry gas
wet gas
1.0
1.4
2.0
3.0
4.0
5.0
7.0
a
0.5 MPa
dry gas
wet gas
1.0
1.4
1.8
b
2 MPa
dry gas
wet gas
2
3
4
5
7
c
3 MPa
dry gas
wet gas
2
3
4
5
7
d
CH₄ / vol.%
1400
1200
1000
800
600
Temperature (dry gas) / K

equilibrium calculations in this treatise are based on the computer methods for thermodynamic calculations developed by Kjær [18]. Some values of the equilibrium constants for reactions (2) and (3) are listed in Table 2.

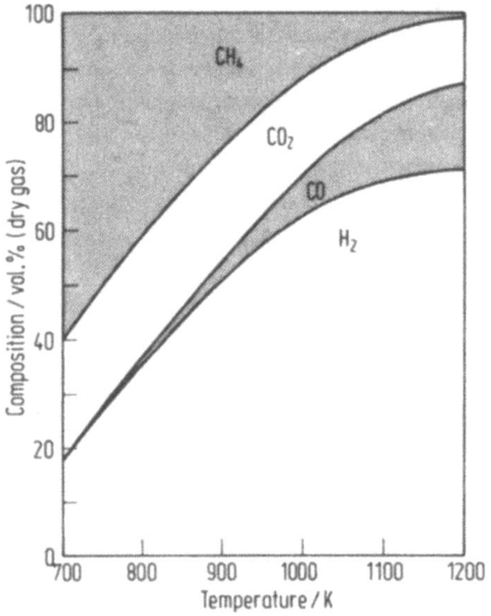

Figure 2. Equilibrium compositions, dry gas. Steam reforming of *n*-heptane. p = 3 MPa, H_2O/C = 4 mol/atom. (Reproduced with permission from ref. [7])

The approximate product gas composition can be estimated from thermodynamic calculations because in most cases it will be close to that of the equilibrated gas. In industrial practice the "approach to equilibrium" for a given reaction is expressed by a temperature difference defined as[1]

$$\Delta T(\text{approach}) = T(Q_R) - T(\text{exit catalyst}) \qquad (4)$$

This implies that ΔT(approach) is positive for exothermic reactions and negative for endothermic reactions. $T(Q_R)$ is the temperature at which the reaction quotient, Q_R, calculated from the product gas is equal to the equilibrium constant K_p. This measure of the affinity can be misleading when comparing different reactions since it does not consider that the temperature dependence of K_p (*i.e.* the heat of reaction) varies from reaction to reaction.

The various gas compositions which can be obtained through reactions (1) to (3) have resulted in the use of the reforming process as the essential step in the preparation of gas for several purposes. In practice, the pressure is often determined by the overall lay-out of the process. This leaves the steam to carbon ratio and the catalyst exit temperature as the major parameters determining the gas composition. Table 3 shows typical combinations of these parameters and resulting product gas analyses for the more important applications of steam reforming.

---

[1] A comprehensive list of symbols is given at the end of this chapter.

Table 2. Equilibrium constants[a, b, c]

Temperature/K	K_p^1 $CO + H_2O = CO_2 + H_2$ (reaction 2)	K_p^1 $CH_4 + H_2O = CO + 3H_2$ (reverse of reaction 3)	K_p^1 [a] $2CO = C + CO_2$ (reaction 5)	K_p^1 [a] $CH_4 = C + 2H_2$ (reaction 6)
273	4.555×10^5	0.066×10^{-29}	5.678×10^{24}	0.820×10^{-11}
373	3.587×10^3	0.269×10^{-19}	8.082×10^{15}	0.606×10^{-7}
473	2.299×10^2	0.487×10^{-13}	6.047×10^{10}	1.283×10^{-5}
498	1.380×10^2	0.747×10^{-12}	0.661×10^{10}	3.582×10^{-5}
523	8.737×10	0.892×10^{-11}	8.941×10^8	0.914×10^{-4}
548	5.780×10	0.858×10^{-10}	1.451×10^8	2.153×10^{-4}
573	3.973×10	0.682×10^{-9}	2.758×10^7	4.736×10^{-4}
598	2.823×10	4.592×10^{-9}	6.028×10^6	0.981×10^{-3}
623	2.066×10	2.669×10^{-8}	1.489×10^6	1.924×10^{-3}
648	1.552×10	1.361×10^{-7}	4.101×10^5	3.597×10^{-3}
673	1.192×10	0.618×10^{-6}	1.244×10^5	6.445×10^{-3}
698	9.352	2.528×10^{-6}	4.111×10^4	1.111×10^{-2}
723	7.470	0.941×10^{-5}	1.468×10^4	1.849×10^{-2}
748	6.066	3.222×10^{-5}	5.620×10^3	2.984×10^{-2}
773	4.999	1.021×10^{-4}	2.291×10^3	4.678×10^{-2}
798	4.176	3.016×10^{-4}	0.989×10^3	7.144×10^{-2}
823	3.530	0.837×10^{-3}	4.495×10^2	1.066×10^{-1}
848	3.018	2.189×10^{-3}	2.143×10^2	1.554×10^{-1}
873	2.606	5.429×10^{-3}	1.067×10^2	2.223×10^{-1}
898	2.270	1.282×10^{-2}	5.523×10	3.120×10^{-1}
923	1.995	2.968×10^{-2}	2.967×10	4.303×10^{-1}
948	1.767	6.267×10^{-2}	1.647×10	5.843×10^{-1}
973	1.541	1.305×10^{-1}	9.438	7.817×10^{-1}
998	1.415	2.623×10^{-1}	5.564	1.031
1023	1.278	5.098×10^{-1}	3.368	1.344
1048	1.160	0.961	2.090	1.731
1073	1.059	1.759	1.327	2.203
1098	9.717×10^{-1}	3.133	8.606×10^{-1}	2.775
1123	8.953×10^{-1}	5.444	5.695×10^{-1}	3.462
1148	8.284×10^{-1}	9.236	3.838×10^{-1}	4.280
1173	7.694×10^{-1}	1.533×10	2.633×10^{-1}	5.245
1198	7.172×10^{-1}	2.491×10	1.836×10^{-1}	6.374
1223	6.708×10^{-1}	3.969×10	1.300×10^{-1}	7.689
1248	6.294×10^{-1}	6.209×10	0.934×10^{-1}	9.207
1273	5.923×10^{-1}	0.955×10^2	6.794×10^{-2}	1.095×10
f'	1.000	97.402	0.10133	9.869

[a] graphite basis
[b] partial pressures in MPa
[c] K_p^1 values should be multiplied by f' if partial pressures are in atm. $K_p = f' \cdot K_p^1$

Table 3. Calculated product gas compositions[a]

Final Product	1	2	3	4	5	6	7	8	9
	NH_3	NH_3	H_2	H_2	MeOH	Town gas	SNG[b]	Oxo-alcohols	Reducing gas
Feedstock	CH_4	CH_4	naphtha	CH_4	CH_4	naphtha	naphtha	CH_4	CH_4
p_{exit}/MPa	3.3	3.3	2.7	2.7	1.7	2.4	3.5	1.7	0.5
T_{exit}/K	1073	1103	1073	1123	1123	948	788	1138	1223
H_2O/C_nH_m mol/C atom	3.7	2.5	4.5	2.5	3.0	2.4	1.6	1.8[b]	1.15
H_2O/vol%	44.30	34.17	49.62	31.09	32.20	45.58	47.70	25.21	4.28
H_2	39.12	44.74	34.60	48.59	50.28	25.62	8.25	28.05	70.92
CO	5.04	7.59	5.33	9.22	9.53	3.25	0.43	25.91	22.44
CO_2	6.00	5.49	8.03	5.24	5.42	10.14	11.18	19.71	0.90
CH_4	5.54	8.01	2.42	5.86	2.57	15.41	32.44	1.12	1.46

[a] $\Delta T_R = -10$ K, $\Delta T_s = 0$ K
[b] $\Delta T_R = 0$ K
1 MPa = 10 atm. abs.

Table 4. Enthalpy of formation[a]

Temperature K	Enthalpy of formation/J mol^{-1}								
	H_2	H_2O	N_2	CO	CO_2	CH_4	C_2H_6	n-C_4H_{10}	n-C_7H_{16}
298	0	−242175	0	−110684	−394080	−74955	−84789	−126329	−188090
573	8045	−232622	8070	−102572	−382385	−63144	−65528	− 89993	−126304
623	9507	−230786	9583	−101042	−380022	−60516	−61061	− 81592	−112058
673	10978	−228926	11112	− 99492	−377605	−57761	−56335	− 72722	− 97032
723	12453	−227036	12658	− 97929	−375128	−54852	−51374	− 63416	− 81282
773	13928	−225113	14221	− 96345	−372606	−51826	−46182	− 53703	− 64857
823	15411	−223165	15796	− 94749	−370033	−48675	−40777	− 43610	− 47808
873	16903	−221187	17384	− 93136	−367414	−45403	−35175	− 33164	− 30177
923	18398	−219180	18989	− 91506	−364762	−42018	−29389	− 22391	− 12009
973	19907	−217148	20611	− 89863	−362068	−38523	−23422	− 11313	+ 6654
1023	21419	−215 082	22241	− 88208	−359344	−34924	−17292	+ 50	25781
1073	22940	−212991	23891	− 86536	−356592	−31224	−11011	11678	45332
1123	24670	−210875	25551	− 84852	−353809	−27436	− 4579	23548	65276
1173	26012	−208730	27223	− 83155	−351002	−23556	+ 1986	35645	85585
1223	27562	−206560	28911	− 81446	−348170	−19597	8682	47951	106234
1273	29125	−204360	30608	− 79724	−345320	−15562	15499	60454	127201

[a] ideal gases

2.2. Heat of Reaction

Although the irreversible reaction (1) is strongly endothermic, the over-all heat of reactions (1) to (3) may be positive, zero, or negative, depending on the process conditions.

At low steam to carbon ratios and at low catalyst exit temperatures, the over-all heat of reaction is positive. This is reflected by the high contents of methane in the product gas. An example is the adiabatic gasification processes for substitute natural gas (SNG), (Table 3, case 7) which operate with an adiabatic temperature increase of 50–60 K. However, for the production of town gas, synthesis gases, and hydrogen with lower methane contents in the reformer effluent, the over-all reaction becomes endothermic. In order to supply the heat for the endothermic reaction, the catalyst is loaded into a number of tubes placed inside a furnace. This reactor is called a tubular reformer. This version of the process is the most important for industrial applications.

The required heat input, the reformer duty Q, is the enthalpy difference between the exit and the inlet gas and it can easily be calculated from enthalpy tables such as Table 4. The duty consists of heat of reaction as well as the heat for raising the temperature to the level of the reformer exit. For typical ammonia plant conditions (Table 3, case 1), the over-all enthalpy change amounts to approximately 214 kJ mol^{-1} CH_4 (feed) with 63% and 37% used for the chemical reaction and the temperature increase, respectively.

The duty varies only little with the type of hydrocarbon feedstock at otherwise given reaction conditions [19]. It is slightly higher (4–5%) for methane than for liquid hydrocarbons.

2.3. Side Reactions

Reactions (1) to (3) may be accompanied by reactions forming carbon

$$2\,CO \leftrightarrows C + CO_2 \quad (-\Delta H^0_{298} = 172.4 \text{ kJ mol}^{-1}) \tag{5}$$

$$CH_4 \leftrightarrows C + 2\,H_2 \quad (-\Delta H^0_{298} = -74.9 \text{ kJ mol}^{-1}) \tag{6}$$

$$C_nH_m \rightarrow \text{“carbonaceous deposits”} + xH_2 \tag{7}$$

At high temperatures (above about 920 K), the hydrocarbons may react in parallel to reaction (1) by thermal cracking [20, 21] (“steam cracking”) into olefins which may easily form coke *via* reaction (7). Reactions (5) and (6) are reversible, whereas (7) is irreversible for $n > 1$. Equilibrium constants for reactions 5 and 6 are listed in Table 2. The risk of carbon formation must be eliminated in industrial operations since carbon causes severe operational trouble.

B. Historical and Future Aspects

1. Early Work

The catalytic interaction between hydrocarbons and metals was already observed in 1817 by Davy [22] during his famous experiments with the wire-gauze safety-lamp. Davy also observed that "a thin film of carbonaceous matter destroys the igniting power of platinum, and a slight coating of sulphuret deprives palladium of this property". Thus, the disturbing actions of carbon and sulfur (causing severe problems in steam reforming) were recognized even before the concept catalysis was introduced by Berzelius in 1836.

A process for conversion of hydrocarbons into hydrogen in the presence of steam was described by Tessie du Motay and Maréchal in 1868 [23]. The hydrocarbons and steam were passed over calcium oxide resulting in the formation of calcium carbonate and hydrogen. The application of nickel for this process was claimed in 1889 by Mond [24].

At the same time, Lang [25] studied the homogeneous reaction between steam and methane. The experiments, which were performed at molar ratio H_2O/CH_4 of unity, resulted in very small conversions even at 1220–1320 K. Moreover, the reaction was accompanied by formation of coke.

Although some industrial interest was reflected by patents of Dieffenbach and Moldenhauer in 1909 [26], and by BASF (Mittasch and Schneider) in 1912 [27], Sabatier did not mention steam reforming in his book published 1920 [28] which, among other topics, summarized his comprehensive studies of reactions on nickel catalysts.

The first detailed study of the catalytic reaction between steam and methane to be published is apparently that of Neumann and Jacob [29] from 1924. The experiments resulted in gas mixtures close to the equilibria of reactions (2) and (3). Shortly after, an increasing interest developed in utilizing the reforming reactions for industrial conversion of natural gas or methane-rich gases into synthesis gas or hydrogen [30, 31]. This resulted in numerous patents issued around 1930 [32] among which, one described a process where the catalyst was placed in externally heated tubes of alloy steel. A broad range of catalyst compositions was claimed as for example [33] "catalysts consisting of iron, nickel or cobalt activated by the addition of other metals or metallic compounds. As activating agents, metals whose oxides are reducible with difficulty, or compounds thereof, are especially useful, *e.g.* chromium, vanadium, and compounds of alkali, alkaline earth, and earth metals, such as potassium, magnesium, and aluminum" or more simply [34] "a substance comprising a metal of the iron group with an activating addition of a non-reducible oxide of a metal from groups 2 to 6 of the periodic system".

2. *Industrial Developments*

The first industrial steam reformer was installed at Baton Rouge by Standard Oil of New Jersey [35] and commissioned in 1930. Six years later a steam reformer was commissioned at ICI, Billingham [36]. The reforming process was adopted mainly in the U.S. where natural gas was abundantly available as feedstock.

During the fifties light distillate naphthas became an economical feedstock for steam reforming in Europe. At the same time metallurgical developments made it possible to design reformers for operation at elevated pressures. This improved the energy efficiency of the overall process [37], because higher pressure facilitates the heat recovery and partly results in savings in compression energy in ammonia and methanol plants. In 1962, two tubular reformers operating at around 1.5 MPa (= 15 atm) and using "high molecular weight hydrocarbons" as feed were commissioned by ICI [36]. Less than five years later, a Topsøe reformer was operating at 4 MPa (= 40 atm).

Another route for naphtha reforming was followed by the British Gas Council where Dent *et al.* [38, 39] in 1957 described a process for adiabatic gasification of naphtha. The first plant based on these principles (the CRG process) was commissioned in 1964 [40]. Similar processes were developed by Lurgi [41] and JGC [42].

3. Present Trends

Today steam reforming is a principal process for production of hydrogen and synthesis gases. The most important alternatives are partial oxidation of fuel oil and coal gasification. However, capital costs of a fuel oil based ammonia plant are approximately 1.5 times and for a coal based plant approximately twice that of an ammonia plant using steam reforming of natural gas [43]. Moreover, the energy consumption for the two alternatives is larger (approximately 20% and approximately 50%, respectively) than for steam reforming. Therefore, the use of alternatives to steam reforming can be justified only in the case of an attractive price difference between heavy oil fraction or coal and the hydrocarbon feedstock for steam reforming.

Today, more than 80% of the world ammonia production [44] is based on steam reforming of hydrocarbons. Natural gas, which alone accounts for 70% (up from 64% in 1964), is the preferred feedstock in almost all new plants [44]. This is not surprising in view of the high thermal efficiency (c.f. Table 5). Forecasts of the future use of fertilizers mainly in the developing countries indicate a substantial growth in steam reforming.

Table 5. Energy consumption for various applications of steam reforming of natural gas

Product	Energy consumption/GJ t^{-1} practical	Energy consumption/GJ t^{-1} theoretical	Thermal efficiency %	Estimated world production 1980 10^6 t g^-
Ammonia[a]	27–31	20.9	72 (66)[a]	80
Hydrogen	178	100	63	3
Methanol	27–31[b]	17.5	60	12
Gasoline	ca. 74	ca. 40	ca. 54	0
Sponge iron *via* direct reduction	10–125	7.1	60–70	30

[a] N_2 added with air. Product as liquid ammonia at 240 K. For $3.54\ CH_4 + 4.93\ H_2O_{liq} + 4\ N_2 + 1.07\ O_2 \rightleftharpoons 3.54\ CO_2 + 8\ NH_{3\,liq}$, $\Delta H^0_{298} = -0.41$ GJ per t NH_3

[b] excluding distillation (*ca.* 1 GJ t^{-1})

In the petrochemical industry steam reforming is used for the production of methanol and for synthesis gases for oxo-synthesis.

Oil refineries are using more and more hydrogen [45] mainly because of increased demands for desulfurization and hydrocracking. The available amounts of by-product hydrogen (from catalytic reforming etc.) have not been sufficient to cover the needs, and in particular in Japan and the U.S., the gap is being filled by steam reforming or by partial oxidation of heavy feedstock. The requirements for hydrogen will further increase with the processing of more heavy and hydrodeficient feedstocks such as tar sands and coal.

Reforming of natural gas has found a new application in the manufacture of reducing gas for steel production [46, 47]. Plants of this type are being constructed in particular in countries, where cheap natural gas is available.

The choice of feedstock for steam reforming is influenced by regional availabilities of hydrocarbons. In Japan and India, naphtha is still an important feedstock for steam reformers, but elsewhere natural gas is dominating. After the discovery of natural gas in Western Europe, many European town gas units, which were previously based on naphtha, were closed or converted into using natural gas; ammonia and methanol plants are also using mainly natural gas as feedstock. In the U.S., natural gas is still easily available, although some industrial plants based on natural gas have been forced to look for alternative feedstocks [48].

4. Future Aspects

The huge amounts of natural gas being flared jointly with oil production at many locations represent a challenge to chemical technology. It has been considered to convert the gas into transportable energy carriers, when pipelines to consumers cannot be established. Large scale conversion of natural gas *via* steam reforming into fuel methanol was suggested after the oil crisis in 1973 [49]. This would soon dwarf the present methanol production for petrochemical use. However, so far this solution has not been feasible compared to transport of liquified natural gas (LNG).

Many natural gas resources are located off-shore or at coastal areas with difficult access. Several of these resources are of minor size which cannot justify the construction of a pipeline nor a LNG-plant. In this situation it can be attractive to operate barge mounted ammonia or methanol plants, which can be transferred to a new location when the field is exhausted.

Sufficiently high energy prices may justify the use of fuel methanol. If so, this would probably also lead to the introduction of natural gas based gasoline, manufactured by the further conversion of methanol into gasoline *via* the Mobil MTG process [50].

The steam reforming process is an important element in compact fuel cell systems based on natural gas or liquid hydrocarbons [51, 52]. The reforming step produces the hydrogen for the fuel cell and excess gas is used as fuel for the reforming process. Units with a capacity up to 24 Mw are under construction.

A special application of the steam reforming process is the German ADAM/EVA system [53] in which it is foreseen to use hot helium from a high temperature gas cooled nuclear reactor as heat source for the reforming reaction. The produced carbon monoxide and hydrogen is transported over long distances to cities where the reaction heat is recovered [54] by the methanation reaction (3) and utilized for the production of electricity and hot water for district heating. The methane is recycled to the reformer. The helium heated reforming system may also be used for the manufacture of hydrogen for coal liquifaction [53]. A similar reactor system has been studied in Japan for the utilization of nuclear energy in production of reducing gas for steel production [55].

The endo-thermicity of the steam reforming reactions is also utilized in explorative studies on the conversion of solar energy into chemical energy [56]. The solar energy may be transferred *via* a sodium heat pipe [57]. Another future application may be steam reforming of gasoline or methanol [58] for combustion engines using the hot exhaust gas as heating medium.

2. Characteristics of Steam Reforming Process

A. Process Schemes

1. Energy Converter

The tubular reformer is an energy converter, since most of the energy input to many processes is added *via* the reformer with the hydrocarbon feed and the fuel (often the same hydrocarbon). The energy is transferred into hot, steam-containing synthesis gas, and hot flue gas. The synthesis gas is subsequently processed further and converted into products by mainly exothermic reactions releasing more heat. The heat of the process streams and of the flue gas must be recovered to achieve high energy efficiency and this requires tight integration of the reformer with other parts of the process. This is illustrated by two examples, an ammonia plant and a plant for direct reduction of iron ore.

2. Ammonia Plant

2.1. Conventional Technology

Figure 3 shows a simplified process diagram for a typical ammonia plant based on natural gas [59]. The natural gas at *ca.* 3.5 MPa (= 35 atm) is purified for sulfur compounds by absorption over zinc oxide (or previously by adsorption over active carbon). If organic sulfur compounds (mercaptans, thiophenes etc.) are present in the feed they are hydrogenated over a sulfided Co—Mo catalyst (620–670 K) before absorption of the hydrogen sulfide in zinc oxide. Natural gas is mixed with steam (H_2O/C = 3.7–4.0, (*c.f.* Table 3, case 1) and after preheating (670–770 K) passed to the tubular (primary) reformer in which the gas is equilibrated at *ca.* 1060 K. Air is added to a secondary, autothermic reformer to supply the

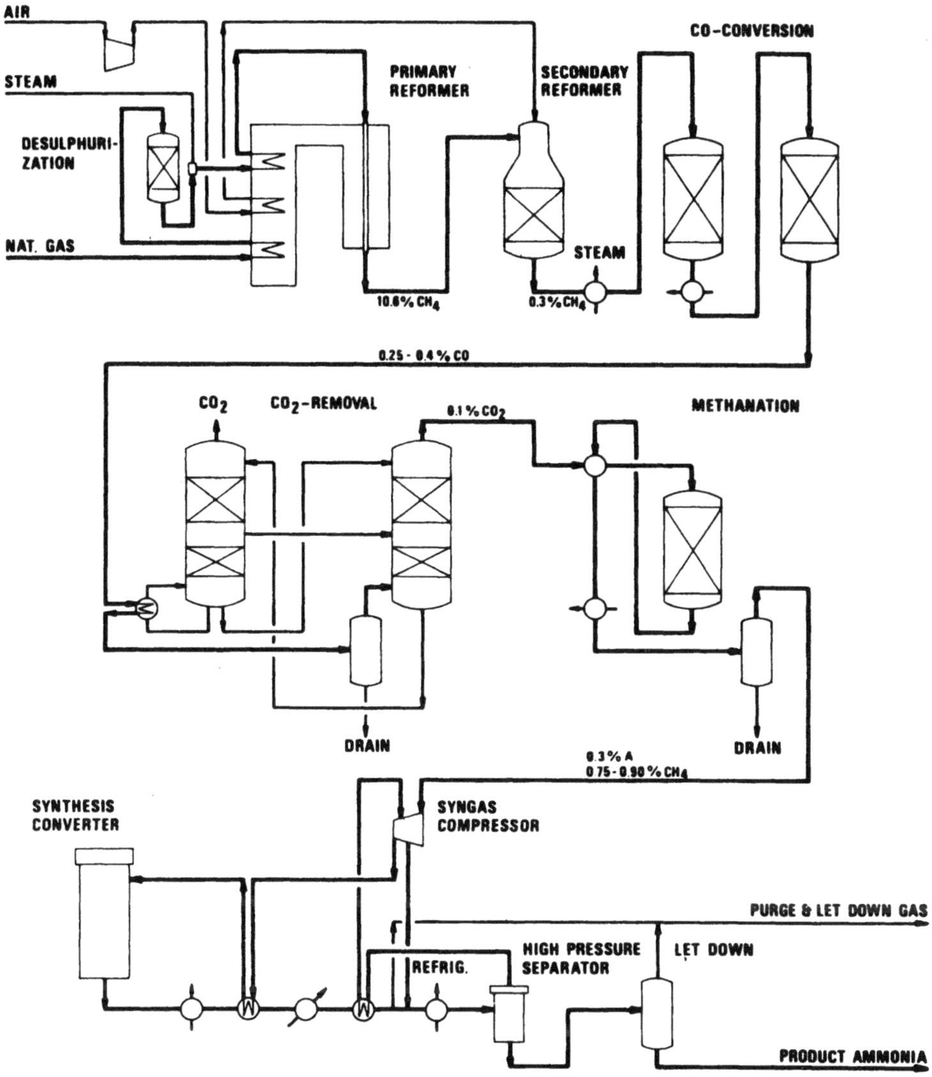

Figure 3. Process diagram for natural gas based ammonia plant. Reformer conditions: $p_{H_2O}/p_{CH_4} = 3.7$, $T_{exit} = 1073$ K, $p_{exit} = 3.3$ MPa

nitrogen required for the ammonia synthesis. The oxygen and excess methane from the primary reformer are converted into more synthesis gas over a nickel catalyst.

The heat produced by the reactions in the secondary reformer is nearly 40% of the fired duty in the primary reformer. The sensible heat of the effluent from the secondary reformer (*ca.* 1270 K) is used in a steam boiler to generate high pressure steam (11 MPa). This steam flow (corresponding roughly to the required amount of process steam) plus steam from an auxiliary boiler is expanded to *ca.* 3.8 MPa in the turbine driver for the synthesis gas compressor (*ca.* 1300 kWh per t NH_3) and used partly as process steam, partly as motive steam for other compressors.

Carbon monoxide is converted in a high temperature shift reactor (650–710 K, iron oxide catalyst) and subsequently in a low temperature shift reactor (490–505 K, Cu-catalyst), after which carbon dioxide is removed by chemical wash (potassium carbonate (Benfield, Vetrocoke), amines such as MEA, DEA, etc.). The heat required for regeneration of the solution (CO_2 desorption) is obtained from condensation of excess steam in the process gas leaving the shift system. Remaining traces of carbon oxides are removed by methanation before compression to *ca.* 27 MPa. The synthesis gas (H_2/N_2 = 3) finally enters the ammonia synthesis loop.

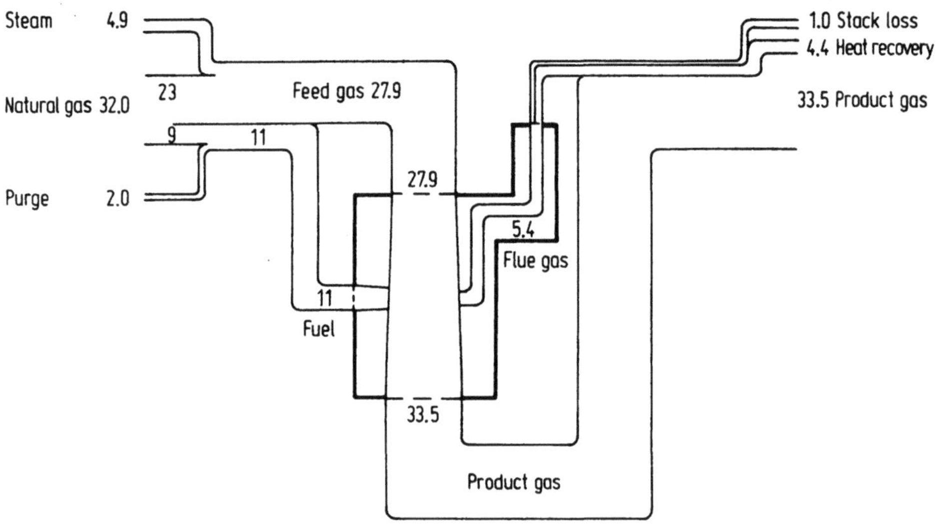

Figure 4. Energy flow diagram of tubular reformer. Conventional ammonia plant (*c.f.* Figure 3). The numbers give GJ (liq. NH_3 at 240 K)$^{-1}$

In the primary reformer, natural gas is added as process feed and usually as fuel as well. This amounts to nearly 95% of the total energy consumption of the plant. Fuel natural gas is supplemented by purge gas, recycled from the ammonia loop. The fired duty amounts to *ca.* 50% of heat content in the process natural gas as indicated in Figure 4. About half the fired duty is transferred through the reformer tubes (maximum tube wall temperature *ca.* 1170 K) and adsorbed by the process (60% for reaction, 40% for temperature increase). The other half of the fired duty leaves the reformer as hot flue gas (*ca.* 1300 K) and more than half is recovered in the waste heat channel for preheat of the process streams and boiler feed water as well as for superheating of steam. The stack gas leaves the plant at *ca.* 470 K. The thermal efficiency of the reformer amounts to *ca.* 80% referred to the fired duty and *ca.* 95% referred to the total energy flow.

2.2. Low Energy Technology

Energy costs have placed emphasis on improving the efficiency of ammonia plants and modern designs show energy consumptions which are *ca.* 20% less than that for conventional plants (Table 5). One trend has been to decrease the synthesis pressure thus reducing the need for steam for compression. Another trend has been to decrease the steam-to-carbon ratio in the reformer feed from 3.7–4.0 in the conventional lay-out (Table 3, case 1) to 2.5 to 3.0 (Table 3, case 2).

It is true that the excess process steam in the conventional plant (Figure 3) is utilized for the regeneration of the solvent in the CO_2-absorption section, but this represents a degradation of energy since medium pressure steam is used where low pressure steam would do. However, the required amount of steam can be significantly reduced by using a physical wash (Selexol, Purisol etc.) instead of a chemical wash.

A lower steam-to-carbon ratio results in a smaller duty and size of the reformer. It also means a higher potential for carbon formation (reactions (5) to (7)) in the reformer and, therefore, more precise design and better catalysts are required. In addition, the lower steam-to-dry gas ratio necessitates modification of the shift system [59].

2.3. Plants with Similar Reforming Conditions

In hydrogen plants the reformer is followed by the same process steps as in ammonia plants (Figure 3) with the exception of secondary reforming. In recent years the CO_2-wash and methanation have often been replaced by a pressure swing adsorption system using molecular sieves. This has allowed a higher methane leakage from the reformer and hence a reduction of the steam-to-carbon ratio as shown in Table 3 (cases 3 and 4).

Front ends of methanol plants are even more simple than hydrogen plants because the shift and CO_2-removal systems are eliminated. The synthesis purge gas is often utilized as fuel in the reformer. There has been a similar tendency to use a lower steam-to-carbon ratio by accepting higher contents of methane in the purge gas (Table 3, case 5).

Reformers in oxo-synthesis plants usually operate with a substantial recycle of carbon dioxide to reduce the over-all hydrogen-to-carbon ratio (Table 3, case 8).

3. Reducing Gas Plant

In plants for direct reduction of iron ore on the basis of natural gas, the reducing gas is manufactured by the reforming process. The mixture of mainly carbon monoxide and hydrogen reduces the iron ore without forming molten iron. A number of processes [46, 47] have been introduced which are characterized by extensive integration of the reformer and the reduction furnace (Hyl, Midrex, Nippon Steel, Purofer, etc.).

The most economical solutions involve minimum excess water in the reformer effluent gas, and sending it directly to a shaft furnace in which the gas reacts in counter-current with a moving bed of iron ore (*c.f.* Figure 5).

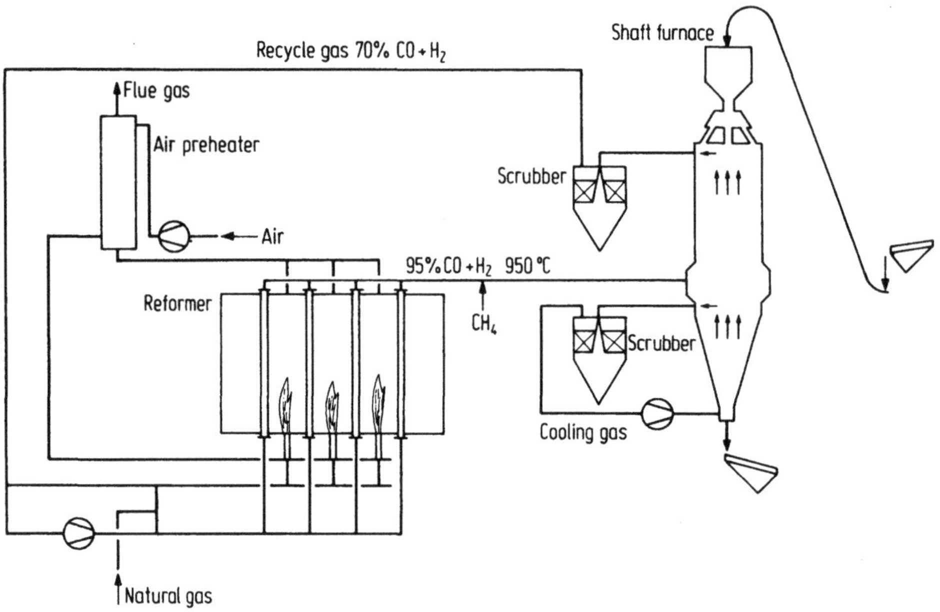

Figure 5. Flowsheet of the Midrex process for direct reduction of iron ore. (Reproduced with permission from ref. [61])

This operation requires a nearly stoichiometric atomic ratio of oxygen to carbon in the reformed feed to ensure maximum reduction potential of the gas, which is essential for the efficient utilization of the shaft furnace [60]. The operation involves high reformer exit temperature (1120–1220 K) and low pressure (0.1–0.5 MPa) to obtain maximum methane conversion (*c.f.* Table 3, case 9).

For thermodynamic and stoichiometric reasons the consumption of hydrogen and carbon monoxide for the reduction of haematite (Fe_2O_3) is restricted to only about half the introduced gas [60]. In practice the gas utilization amounts to 30–40% [46] and this situation makes the effective use of the unconverted gas in the effluent (top gas) from the shaft furnace decisive for the process economy. Two different principles have been applied to solve the problem. The top gas is either recycled to the shaft furnace inlet after removal of product water (Armco) and carbon dioxide (Nippon Steel, Purofer) or the gas is recycled to the reformer inlet (Midrex). For both solutions, part of the top gas is used as fuel for the reformer.

Figure 5 shows a simplified diagram of the Midrex-process, where advantage is taken of the fact that carbon dioxide may to a large extent replace steam in the reforming process. The carbon dioxide in the top gas is used as oxidant in the reforming process, which operates at an over-all stoichiometry close to

$$CO_2 + CH_4 \leftrightarrows 2\,CO + 2\,H_2 \; (-\Delta H^0_{298} = 247.4 \text{ kJ mol}^{-1}) \qquad (8)$$

The Midrex process was described in some detail by Töpfer [61]. The feed gas (*ca.* 17 GJ per t Fe) enters an up-flow reformer at *ca.* 0.24 MPa and is converted into a reducing gas (*ca.* 21 GJ per t Fe) at *ca.* 1220 K containing less than 5% of the oxidizing components, steam and carbon dioxide. The hot gas is passed to the shaft furnace, and after passing this, the cooled and dust-scrubbed top gas (*ca.* 13 GJ per t Fe) contains 18–20% carbon dioxide. One third of the top gas is used with natural gas as fuel in the reformer, whereas the remaining two thirds are passed to the reformer inlet after saturation with water. The hot flue gas from the reformer is utilized for preheating of feed gas and of combustion air. With this scheme the overall consumption of natural gas amounts to less than 11 GJ per t sponge iron. Similar energy consumptions are achieved in process schemes with recycle of CO_2-scrubbed top gas to the shaft furnace inlet [62] (Nippon Steel).

There may be a risk in designing plants too highly integrated. The improved economy is paid for by higher sensitivity towards upsets. Operating problems in one part of the plant can result in serious consequences for the rest of the plant. Reformer autonomy cannot be achieved in a flow scheme as Figure 5 since any disturbance in the shaft furnace operation is transferred to the reformer.

Figure 6. Primary reformer of 1250 MTPD ammonia plant. Anic, Manfredonia. Design Topsøe, Contractor Snam Progetti. The Reformer has two furnace chambers with eight rows of burners on the side walls. The inlet hairpins are apparent at the top. The waste heat boiler (exit sec. reformer) is seen to the left of the reformer furnace, and the vessel at the far left contains the high and low temperature shift reactors. The syngas compressors are installed in the construction to the right

B. The Tubular Reformer

1. Furnace Types

Apart from a few examples (heat exchange reformers [52, 53, 55, 63] the reformer catalyst tubes are heated up in a fired furnace. Figure 6 shows a photograph of a reformer for a 1250 MTPD ammonia plant, and four representative furnace types characterized by the placement of burners are shown in Figure 7.

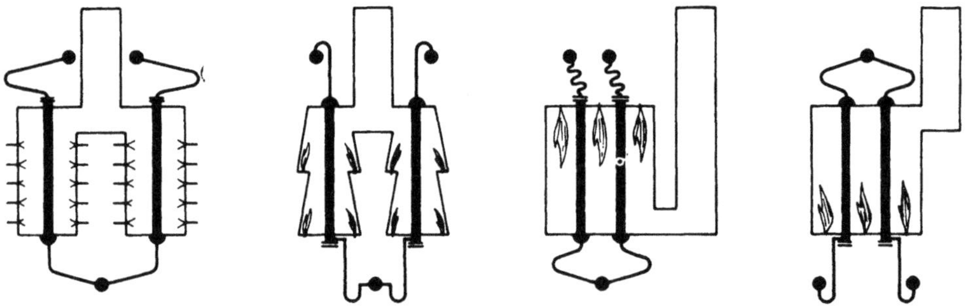

Figure 7. Typical configurations of reformer furnace

In the side-wall fired furnace (Topsøe, Selas) radiant type burners are placed in serveral rows on the two-side walls of the furnace chamber. The tubes are placed in single rows in the furnace chamber. In another type, the burners are placed in terraces at the side walls (Foster Wheeler). A common type uses burners placed at the top of the furnace (Kellogg, ICI etc.). This arrangement has several rows of tubes in the same furnace box separated by burner rows in the furnace ceiling. Burners may also be placed in the bottom of the furnace (Exxon).

Top and bottom furnaces use long flame burners. While the side-wall radiant type furnace is very flexible in adjusting firing profiles, the top-fired furnace may provide a more compact design with a smaller number of burners [64]. Figure 8 illustrates how different firing arrangements result in different temperature and heat flux profiles.

2. Tube Design

A typical reformer may contain typically between 40 and 400 tubes. The internal diameter is in the range 70–160 mm with a tube thickness 10–20 mm. The heated length is 6–12 m depending on furnace type. The tubes are made from high alloy nickel chromium steel (HK40: Cr 25%, Ni 20%, Co 4%; IN519: Cr 24%, Ni 24%, Nb 1.5%, Co 3%). The tubes are manufactured

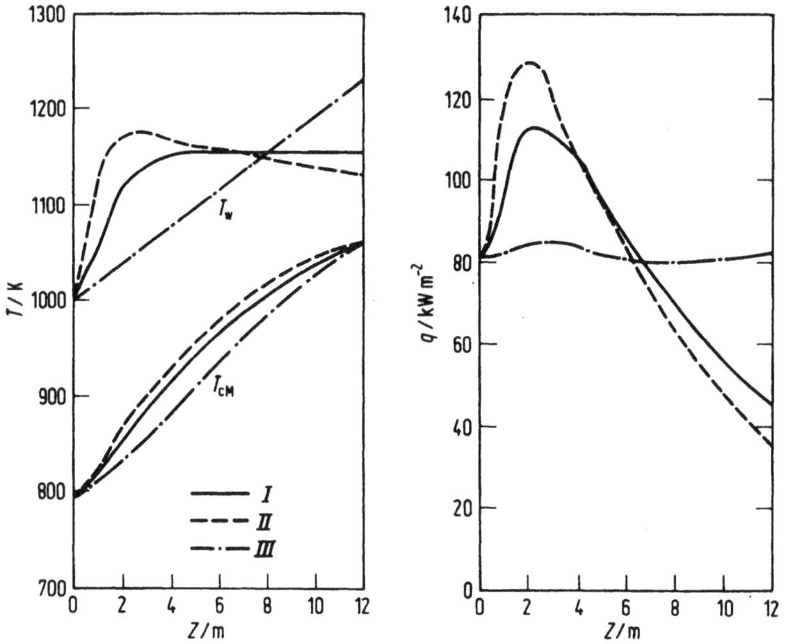

Figure 8. Temperature and heat flux profiles. Impact of firing arrangement for given operating conditions and tube geometry. Calculations based on one dimensional pseudo-homogeneous model for catalyst tube $p_{H_2O}/p_{CH_4} = 3.9$, $p_{exit} = 3.1$ MPa, $q_{av} = 81$ kw m^{-2}, $d_t = 122$ mm, $Z = 12$ m
I: heat input concentrated at top (top fired furnace)
II: constant T_{wall} (side wall fired furnace)
III: constant q.
Practical designs will seldom maintain tube geometry when changing firing arrangement

by centrifugal casting in *ca.* 6 m sections, which are welded together to the required tube length. The tubes are supported outside the furnace chamber either in the floor or in the ceiling. An example of the design is shown in Figure 9.

The high alloy reformer tubes are expensive and account for a large part of the reformer costs. The reliability of the tubes is also important because tube failures could result in long down-periods for retubing and hence loss of production. The mechanical design is critical to ensure plant reliability at minimum costs.

At high operating temperature ($T_w > 1100$ K) the tube metal creeps under stress. Above the Hüttig temperature, (one third of the melting point of the metal, T_m, K), the total strain becomes dependent on time [65]. The creep rate, which is almost constant with time, increases nearly linearly with the applied stress and exponentially with temperature. The creep eventually results in rupture along grain boundaries of the metal.

Experimental data on time for rupture, t_R and the applied stress, σ, are usually correlated with the wall temperature (T_w, K) in plots of $\log_{10} \sigma$

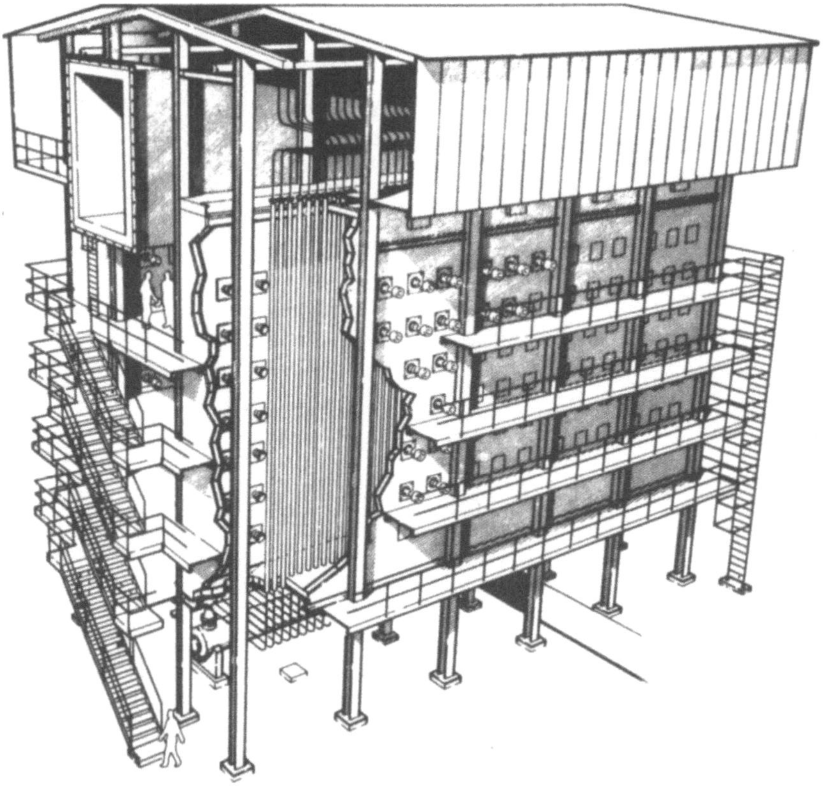

Figure 9. Reformer furnace. Tube and burner arrangement. Topsøe design. In the non-fixed end of the tubes thermal expansions of the tubes are absorbed by hairpins. The preheated gas (673–773 K) is distributed to the inlet hairpins (low alloy steel or stainless steel), and the exit gas (973—1173 K) passes *via* high alloy hairpins (incoloy etc.) to a bricklined collector system

and P_{LM} in Larson-Miller plots [66, 67], where P_{LM}, is the Larson-Miller parameter given by

$$P_{LM} = \frac{T_w}{100}(\log_{10} t_R + C_{LM}) . \tag{9}$$

Recommended values for C_{LM} for HK40 vary between 9.4 and 66. The allowable stress for tube life of 100,000 hours is found by linear extrapolation. The experimental data show significant scatter [66] and normal design practice [68] has been based on 75–90% of the mean value. A 90% limit implies a design for less than 2% risk for tube failure within 100,000 hours. Knowing the allowable stress, σ, the tube thickness, t_t, can be calculated from the given inner tube diameter, d_t, and design pressure, p

$$t_t = \frac{pd_t}{a\sigma + bp} \tag{10}$$

in which a and b are constants.

For a given tube, the life is very sensitive to small variations in maximum tube wall temperature, $T_{w,max}$ and to a smaller extent to variations in operating pressure. As shown in Figure 10, even a slight increase of $T_{w,max}$ of 10 K could result in reduction of the tube life by 30%. The number of shut downs affects the tube life, as well. This emphasizes the need for stable operation [65, 69].

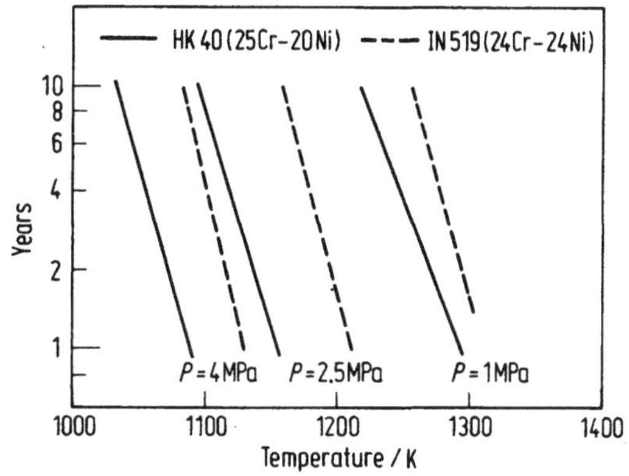

Figure 10. Tube life and maximum tube wall temparature. Given tube dimensions d_t = 100 mm, t_t = 5.6 mm. Less than 2% risk for tube failure within indicated tube life. The curves should be used for relative comparisons. Practical design involves more parameters

The tube life is also influenced by thermal stresses caused by the temperature difference over the tube wall. This limits the maximum heat flux to be applied for given T_w and d_t [70].

In conclusion, the tube wall temperature profile and the heat flux pattern are highly important parameters of the reformer. Today tubular reformers operate at an average heat flux, q_{av}, (referred to the internal surface) of 45–90 kWm^{-2} (40,000–75,000 kcal m^{-2} h^{-1}).

Top fired furnaces have a tendency for forming a maximum in the T_w-profile about one third of the tube length from the inlet [64, 71] of *ca.* 10–40 K higher than the average T_w, whereas it is possible in a side-wall fired furnace to approach a constant tube wall temperature. Since the design must be based on $T_{w,max}$, it is desirable to keep this value as low as possible. Figure 8 illustrates this situation for different firing arrangements for a given tube. It is shown that both firing arrangements are accompanied by a peak in the heat flux profile being largest for the top fired furnace. Thus, for given tube geometry the latter arrangement results in higher thermal stresses. However, it should be emphasized that commercial design compensates for this by proper selection the optimum tube dimensions.

An almost constant heat flux down the tube can be obtained by a linear T_w-profile, but, as shown in Figure 8, this results in a very high tube wall temperature at the exit of the tubes and hence a non-economical design.

The tube geometry has a complex influence on the reformer design. Increasing the length of the tubes is more economical than increasing the number of tubes, because more tubes mean enlargement of the complex inlet and outlet systems. However, the tube length is limited by the risk of tube bending and by restrictions in pressure drop across the catalyst.

For given tube length, feedstock flow and reformer duty, Q, the number of tubes, n, is determined by the selected tube diameter, d_t, the average heat flux, q_{av}, and the space velocity, SV. These parameters are interrelated. It can be shown by a total heat balance and by assuming constant inlet and outlet conditions that, independent of n

$$q_{av} \simeq d_t SV . \tag{11}$$

This means that only two of these three parameters can be selected freely.

The tube diameter (for given q_{av} and tube length) affects the cost of the reformer by the amount of tube metal determined by equation (10), and its influences on size of the reformer box. Thus, the amount of steel increases proportional to the tube diameter whereas the length of the tube row (for given pitch, usually $\lesssim$ *ca.* $2d_t$) remains constant.

In conclusion, industrial reformer design is a compromise involving many parameters.

C. Reformer Models

1. Simulation of Furnace Chamber

Tubular steam reforming is a complex interaction of heat transfer and coupled chemical reactions. The heat released by the burners is transferred *via* radiation and convection to the reformer tubes. The heat passes through the tube walls by conduction and is transferred to the catalyst bed by convection and radiation at the same time as a network of chemical reactions create temperature and concentration gradients in the radial direction of the tube and around and within the porous catalyst particles.

An ideal model should be able to simulate the reformer performance on the basis of the individual burner duties, the feed stream characteristics, catalyst properties and reformer geometry.

Simplified models like the well-stirred furnace model for the heat transfer in the fire box for a side-wall fired reformer were presented by Singh *et al.* [72] assuming constant view factors in radiation calculations, and by Roesler [73] for a top fired reformer being considered as a multistream heat exchanger coupled by radiation beams. Both methods include only radiation and they contain implicit strong limitations on reactor geometry. More promising attempts are based on the Hottel zone method [74] by which the furnace chamber is divided into zones with individual view factors (or rather radiation exchange factors). This model can be refined to include convection as well as radiation. It has been used by Vercammen and Froment [75] for an ethane cracking furnace.

Simulations of the process gas side in tubular reformers are generally uncoupled from the furnace box by assuming an outer tube wall temperature profile or a heat flux profile. These profiles are established or checked by feedback from measurements in industrial plants and monotube pilot plants. It should be pointed out, however, that measurements of tube wall temperatures are difficult. Pyrometric methods involve complex corrections

[76] because of reflections from furnace walls and flames. The correction is largest at the coldest position of the tube at the reformer inlet, where reaction conditions at the same time are most complex. Thermocouples welded into the tube wall give more exact information but their life is very limited. Another uncertainty is caused by shadowing effects in the tube row [77]. The extent of this distortion increases with decreasing tube pitch [4, 77].

2. One-Dimensional Model for Catalyst Tube

Calculation methods [72, 78, 79, 80, 81] for the simulation of the reformer tube are normally based on a one-dimensional pseudo-homogeneous model [82], which does not account explicitly for the presence of the catalyst and which assumes that concentration and temperature gradients in the bed occur only in the axial direction. The conservation equation for mass, heat and momentum can be written [82] for a segment of the tube $\mathrm{d}Z$

$$-u_\mathrm{s}\frac{\mathrm{d}C}{\mathrm{d}Z} = r_\mathrm{v} \tag{12}$$

$$u_\mathrm{s}\varrho_\mathrm{g}c_\mathrm{p}\frac{\mathrm{d}T}{\mathrm{d}Z} = (-\Delta H)\,r_\mathrm{v} + 4\frac{U}{d_\mathrm{T}}(T_\mathrm{w} - T_\mathrm{CM}) \tag{13}$$

$$-\frac{\mathrm{d}p}{\mathrm{d}Z} = \frac{2f\varrho_\mathrm{g}u_\mathrm{s}^2}{gd_\mathrm{p,v}}\,. \tag{14}$$

Equation (12) deals with the conversion profile, (here written for a single reaction only), (13) with the heat balance, and (14) with the pressure drop in the segment. Equation (13) may be written alternatively as the duty transferred per unit volume of catalyst

$$\frac{4}{\pi d_\mathrm{t}^2}\frac{\mathrm{d}Q}{\mathrm{d}Z} = \frac{4}{d_\mathrm{t}}\,q = \frac{4}{d_\mathrm{t}}\,U(T_\mathrm{w} - T_\mathrm{CM}) = \Delta H\,r_\mathrm{v} + u_\mathrm{s}\varrho_\mathrm{g}c_\mathrm{p}\frac{\mathrm{d}T}{\mathrm{d}Z}\,. \tag{15}$$

The two terms on the right hand side represent the heat absorbed by the reforming reactions and for raising the gas temperature, respectively. The overall heat transfer coefficient, U, is composed of the thermal conductivity of the tube, the wall film coefficient, α_W, and an effective thermal conductivity of the catalyst bed, λ_er, in series. The relative significance of the three resistances depends on operating conditions but the main resistance is normally concentrated in the film at the tube wall.

λ_er involves terms [18] for the heat transfer by convection, radiation, conduction through the catalyst and by a static (molecular) contribution as well, of which the convection (turbulence) part, $\lambda_\mathrm{r,t}$, is dominating. $\lambda_\mathrm{r,t}$ and the turbulence part of α_W are influenced by the Reynolds number $Re = \dfrac{d_\mathrm{p,v}G_\mathrm{M}}{\mu}$ which means that the heat transfer is improved by increasing the catalyst particle size and the mass velocity. In ammonia plant reformers Re is in the range 7,500–10,000 and the typical values of U are 300–500 W m^{-2} K^{-1}.

In the one-dimensional model, reformer simulations are not very sensitive to the selection of the expression for the reaction rate r_v, which in many cases [72, 78, 79, 80] is expressed as a first order reaction with respect to the hydrocarbon including rate constants fitted from feedback from industrial reformers or monotube pilot plants.

Figures 8 and 11 show results from reformer simulations on the basis of a one-dimensional model. Figure 8 has been discussed above (section 2.B). Figure 11 compares the methane concentration profiles in an ammonia plant reformer for different catalyst activities for a fixed catalyst temperature profile. It is evident that this curve is only slightly influenced by the catalyst activity. Apart from the inlet of the reformer, the gas composition is not far from the equilibrium composition at the mean catalyst temperature [83] (*c.f.* Figure 31, see later). This explains the limited influence of reaction kinetics on the models.

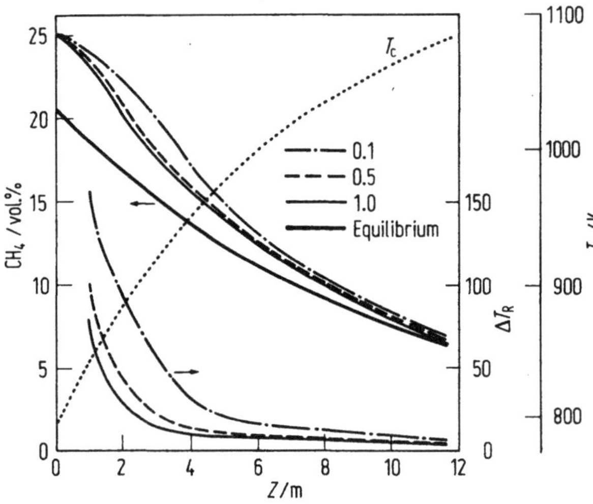

Figure 11. Conversion profiles in tubular reformer. $p_{H_2O}/p_{CH_4} = 3.0$, $p_{exit} = 3.5$ MPa, $T_{exit} = 1083$ K. Calculations based on one-dimensional heterogeneous model accounting for interfacial and intra particle gradients.
Three levels of intrinsic activity of catalyst. Fixed temperature profile (corresponds to Figure 31)

A kinetic analysis [83] shows that the effectiveness factor, η, for the reforming reactions decreases from the tube inlet. Typical values are below 0.1 (*c.f.* section IV.D).

3. *Two-Dimensional Model for Catalyst Tube*

The one-dimensional pseudo-homogeneous model is adequate for studying reformers at non-critical conditions and for simulation of the over-all performance. It is, however, insufficient for reformers of tight design or reformers operating close to carbon limits. For these cases a more detailed analysis of the local phenomena in the reformer is required.

The radial temperature and concentration profiles are included in two-dimensional pseudo-homogeneous models, whereas the gradients in and around the catalyst pellets are neglected.

Froment [84] introduced a two-dimensional model using an effective transport concept to describe the mass and heat flux in a radial direction in terms of an effective radial diffusivity D_{er} and an effective thermal conductivity λ_{er}. This means that equations (12) and (13) are replaced by [82, 84]

$$u_s \frac{\partial C}{\partial Z} = \varepsilon D_{er} \left(\frac{\partial^2 C}{\partial R^2} + \frac{1}{R} \frac{\partial C}{\partial R} \right) - r_v \tag{16}$$

$$u_s \varrho_g c_P \frac{\partial T}{\partial Z} = \lambda_{er} \left(\frac{\partial^2 T}{\partial R^2} + \frac{1}{R} \frac{\partial T}{\partial R} \right) + (-\Delta H)\, r_v \tag{17}$$

D_{er} is calculated from an expression for the Peclet number for radial mass transport, $Pe_{m,r}$

$$Pe_{m,r} = \frac{G_M d_{p,v}}{\varepsilon \varrho_g D_{er}}. \tag{18}$$

λ_{er} is assumed to be constant in the catalyst bed and the resistance to heat transfer close to the tube wall is modelled by introducing a wall coefficient α_w defined by

$$\alpha_w (T_w - T_{CW}) = \lambda_{er} \left(\frac{\partial T}{\partial R} \right)_{R=R_t} \tag{19}$$

λ_{er} consists of a static and a turbulent part of which the turbulent part $\lambda_{r,t}$, is by far dominating at the high Re in industrial reformers. $\lambda_{r,t}$ is calculated from a correlation of the Peclet number for radial heat transfer, $Pe_{h,r}$

$$\frac{\lambda_{r,t}}{\lambda_g} = \frac{Re\,Pr}{Pe_{h,r}} \quad \text{or} \quad \lambda_{r,t} = \frac{G_M d_{p,v} c_p}{Pe_{h,r}} \tag{20}$$

using the analogy of heat and mass transfer.

α_w consists of a radiation and convection (turbulent) part. The turbulent part $\alpha_{w,t}$ is normally estimated from the Nusselt number Nu by use of a correlation of the following type

$$Nu = k[Re^x] \times [Pr^{0.33}], \qquad 0.5 < x < 1.0$$
$$Nu = \frac{\alpha_{w,t} d_{p,v}}{\lambda_g}. \tag{21}$$

The two-dimensional model was applied for a steam reformer by Christiansen *et al.* [85, 86] using the method by Villadsen and Michelsen [87] for the numerical solution of the partial differential equations. Froment *et al.* [88] used the Crank-Nicholson procedure for solution of a similar model for steam cracking of hydrocarbons.

The usefulness of these more sophisticated models is not better than the accuracy by which the transport parameters are known. The formulas for heat transfer have been established mainly from data at *Re* lower than typical for steam reforming and often with the use of spheres instead of rings. The most critical parameters is the Peclet number for radial heat transport $Pe_{h,r}$. Correlations in literature [18, 82, 89] show dependency on the tube-to-particle diameter ratio, but data are scarce at the high *Re* in tubular reformers. The constants to be used in equation (21) are also being discussed [82, 89].

Christiansen and Tøttrup analyzed data from a monotube steam reforming pilot plant [86] to provide a more accurate calculation basis. Figure 12 shows measured radial temperature profiles in the upper part of the reformer. Other data are shown in Figure 51 (see later).

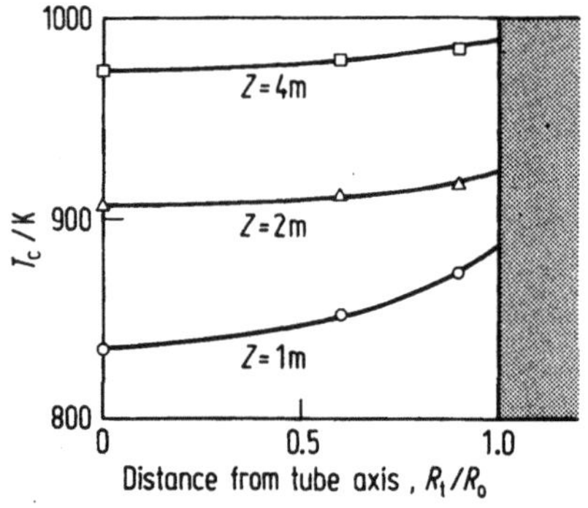

Figure 12. Radial temperature profiles. Measured data from monotube pilot plant. $H_2O/C = 2.4$. $p_{ex} = 3.3$ MPa, $d_t = 84$ mm, $Z = 12$ m, $q_{av} = 83$ kw m^{-2}, $Re_{inlet} = 7250$. (Reproduced with permission from ref. [86])

The radial temperature gradient is significant and it is evident that the risk of break-down of hydrocarbon into carbon is higher close to the tube wall. It will be illustrated how the two-dimensional model can be a useful tool to predict conditions for carbon-free operation (*c.f.* section V.B.1).

D. The Role of the Catalyst in Steam Reforming

1. Operator's Requirements

The operator's requirements to the reformer are:
- *full conversion* which means a close approach to equilibrium at the reformer exit,
- *low tube wall temperatures* to ensure long life,
- *constant pressure drop* to maintain full process flow equally distributed through all reformer tubes.

To meet the requirements, the catalyst must have sufficient activity, resistance to carbon formation and mechanical strength.

2. Approach to Equilibrium

As indicated above (Figure 11) most commercial catalysts will ensure the product gas to be close to the thermodynamic equilibrium of reactions (1) and (3) at temperatures in the region of 1070 K. In reformers operating at relatively low exit temperatures, *i.e.* 920–970 K (town gas) differences in catalyst activity will appear and a low catalyst activity may significantly affect the approach to equilibrium. The same is true in the case of severe poisoning by sulfur.

The methane content of the exit gas is often higher than expected from equilibrium calculations at the catalyst exit temperature. The approach to equilibrium of the methane reforming reaction, ΔT_R, may typically be in the range of —10 to —20 °K as calculated by equation (4). This deviation can normally be ascribed to variations in catalyst exit temperatures from tube to tube. The measured exit temperature from the reformer is the arithmetic mean of the individual tube exit temperatures, whereas the temperature approach in equation (4) includes the equilibrium constant.

For a single tube the approach to equilibrium can be expressed by a simplified expression for a one-dimensional model [83, 90] by

$$-\Delta T_R \sim C_{CH_4} - C_{CH_4,eq} = -\frac{u_s}{\varrho_B}\frac{1}{k_v}\frac{d(C_{CH_4,eq})}{dT}\frac{dT}{dZ} \tag{22}$$

$d(C_{CH_4,eq})/dT$ can be obtained from an equilibrium table, whereas dT/dZ represents the slope of the temperature profile. Equation (22) illustrates that the approach to equilibrium depends not only on catalyst activity (k_v) but also on load (u_s) and the slope of the temperature profile close to the exit of the reformer.

3. Tube Wall Temperature

It is evident from equation (15) that a lack of catalyst activity will cause overheating of the gas and of the tube. For given local heat flux q (or $T_w - T_{CM}$) a higher catalyst activity can be utilized to absorb the same reaction heat at a lower catalyst temperature, T_{CM}, and consequently at a lower tube wall temperature T_w (*c.f.* section IV.D.3) which could mean a substantial increase of the tube life (*c.f.* section II.B.2).

4. Constant Pressure Drop

The catalyst is loaded carefully into the tubes by adjusting the variance of pressure drop in individual tubes to be within *ca.* ±10%. This is necessary to secure equal distribution of flow in the tubes. Unequal gas distribution may cause local overheating and result in shorter tube life. The catalyst material should be stable under process conditions and at the conditions used during start-up and shut-down of the plant. In particular, resistance to conditions during upsets may become critical. Degradation of catalyst may cause partial or total blockage of some tubes resulting in dovelopment of "hot spots", "hot bands" or totally hot tubes [91], and the uneven flow distribution may cause a self-accelerating situation with further overheating of the hot tubes.

Coking can cause similar problems, if the catalyst tubes become blocked and the catalyst breaks down. The catalyst can eliminate carbon formation firstly by a high activity (by maintaining low reaction temperatures and minimum amounts of unreacted hydrocarbon) and secondly by high selectivity for gasification reactions.

5. Catalyst Properties and Reformer Design

High catalyst activity is important for decreasing tube wall temperatures and for minimizing the approach to equilibrium and the risk of carbon formation. The reformer design is equally important. The shape of the temperature profile may affect the approach to equilibrium (equation (22)). The average heat flux and the furnace design are the main factors determining the flux distribution and the external tube wall temperature (Figure 8, equations (11) and (13)). The local heat flux determines the radial temperature gradient in the catalyst bed (equation (17)) and hence the risk for carbon formation. Therefore, the role of the catalyst and the role of the reformer design are closely connected.

3. Chemical and Physical Properties of Steam Reforming Catalysts

A. Chemistry of Reforming Catalysts

1. Composition and Stability

1.1. Catalyst Formulation

The metals of group VIII of the periodic system are active for the steam reforming reaction. However, apart from a few exceptions [12, 13], nickel appears to be the active metal in industrial catalysts. Iron will be oxidized and cobalt may hardly be stable in a metallic state at the H_2O/H_2 ratios for typical steam reforming conditions (*c.f.* Table 3 and Figure 3). The noble metals are too expensive for general commercial use.

A great number of oxides have been proposed as promoters for the catalysts to improve either the activity or the ability to prevent formation of coke. Although it is true that certain components have a specific effect, there has been a typical trend to relate the function of catalysts for reforming of, for instance, naphtha to the presence of certain crystal phases. Examples are the presence of alkali in a certain volatile state [20], the formation of nickel uranate to ensure small nickel crystals, and the action of urania as an "electron promoter" [92], as well as a certain preparation method to form an intermediate product (a "precursor") of nickel magnesium oxide [93]. Hence, correlations of properties of the individual components proposed by Dowden *et al.* [94, 95] appear insufficient for the "design" of a catalyst for steam reforming.

In addition to the catalytic function, the mechanical properties of the catalyst are critical. The temperature level and the steam partial pressure restrict the choice of support material for the catalyst. Most industrial catalysts for tubular reforming are based on ceramic oxides or oxides stabilized by a hydraulic cement. Typical ceramic supports are α-alumina,

Figure 13. Ceramic support for steam reforming catalyst

Table 6. Composition of typical industrial steam reforming catalysts[a]

Producer	Catalyst	Feedstock	Components/wt %						
			NiO	Al_2O_3	MgO	$MgAl_2O_4$	CaO	SiO_2	K_2O
BGC	CRG B	naphtha	79	20–21					0.75–3.3
ICI	46-1	naphtha	22	26	11		13	16	7
ICI	57-3	natural gas	12	78	—		10	(0.1)	
Topsøe	RKS-1	natural gas	15			85		(0.1)	
Topsøe	R67	natural gas/ LPG	15			85		(0.1)	
Topsøe	RKNR	naphtha	34	12	54				
UCI	C-11-9	natural gas	11–20[b]						
UCI	G-56	light hydrocarbons	15–25[b]						
UCI	G-90	light hydrocarbons	7–15[b]						

[a] With the permission of British Gas Corp., Imperial Chemical Industries Ltd. and United Catalysts Inc.

[b] balance consists of unspecified refractory

magnesia, magnesium aluminium spinel, and zirconia fired at temperatures well above 1270 K, (Figure 13). The cement-type catalysts are normally stabilized by a silica-free binder such as calcium aluminate. Examples of composition of industrial catalysts are listed in Table 6.

1.2. Stability of Support

High area supports such as γ-alumina, chromia, etc. can be used for catalysts for low temperature adiabatic reforming [40, 41], but these supports suffer from substantial sintering and weakening at temperatures above 770 K [96, 97]. The deterioration is strongly accelerated by the high steam partial pressure [97, 98, 99, 100] and stability tests at atmospheric pressure can therefore be misleading. Stabilization methods applied in, for instance, auto exhaust catalysts [101] may become ineffective.

Catalysts stabilized with a cement may show shrinkage and decrease in strength after exposure to high temperatures [102, 103]. Therefore, there has been a trend towards a greater use of ceramic based catalysts.

Silica being volatile (as $Si(OH)_4$) at high temperatures in high pressure steam [5] is now excluded from catalysts for steam reforming, if it is not combined with alkali (Table 6). For the same reason silica-free materials are applied for the brick-lined exit gas collector. Silica would be slowly removed from the catalyst (or brickwork) and deposited in boilers, heat exchangers and catalytic reactors down-stream of the reformer.

Alkali used as promoter to eliminate carbon formation will escape slowly from the catalyst. The alkali loss is enhanced by high temperatures but may to some extent be controlled by addition of acidic components (silica) [5, 20].

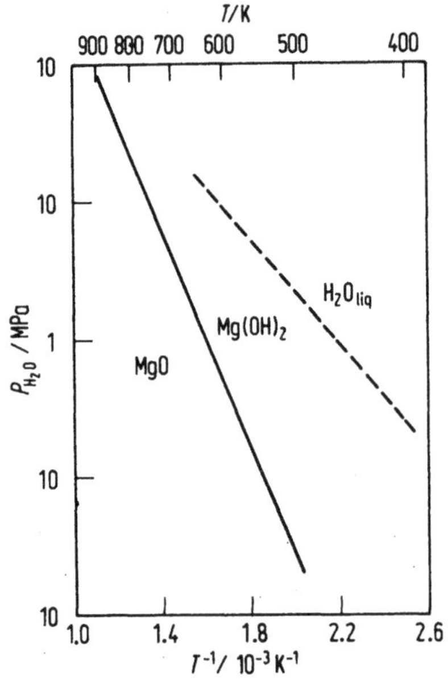

Figure 14. Equilibrium steam pressure for $Mg(OH)_2 = MgO + H_2O$. Dotted curve: pressure of saturated steam. (Data: Barin and Knacke [104])

The volatized alkali may deposit in colder parts of the plant, where the resulting hydroxyl ions will strongly promote stress corrosion in stainless steel. Moreover, alkali may react with some catalyst support materials, such as alumina, resulting in a decrease of the mechanical strength.

While resistant to high temperature steaming, catalysts based on magnesia are sensitive to steaming at low temperatures because of the risk of hydration

$$MgO + H_2O \leftrightarrows Mg(OH)_2 \quad (-\Delta H^0_{298} = 81.3 \text{ kJ mol}^{-1}) \,. \tag{23}$$

The reaction may result in break-down of the catalyst because it involves an expansion of the molecular volume. The equilibrium constant for reaction (23) as a function of temperature is plotted in Figure 14. It is seen that for pressures typical for tubular reformers hydration cannot take place at temperatures above 620 K. Kinetic studies of the hydration reaction [105, 106] have shown that magnesia reacts *via* liquid phase reaction in which water condenses in the internal pores. Therefore, the rate depends on the relative humidity of the atmosphere, and in practice hydration is a problem only when the magnesia-based catalyst is exposed to liquid water or is operated close to the condensation temperature (dotted line in Figure 14).

1.3. Reactions Between Nickel and Support

At steaming conditions at high temperature the nickel in the catalyst may react with the support, even with the nearly inactive α-alumina

$$\begin{array}{l} NiO + Al_2O_3 \rightarrow NiAl_2O_4 \\ (-\Delta H^0_{298} = 5.6 \text{ kJ mol}^{-1} \text{ for } \alpha\text{-}Al_2O_3) \end{array} \tag{24}$$

The formation of the blue-colored nickel aluminium spinel may start at temperatures above about 970 K [103, 107, 108, 109], but a less well-defined interaction between nickel oxide and η- or γ-alumina is apparent already at lower temperatures [101, 108, 110, 111]. It is possible to form a "surface spinel" [111] below 870 K, which may hardly be identified by X-ray methods alone.

A similar trend [103, 112] is observed for the reaction between nickel and magnesium oxide

$$x\text{NiO} + (1-x)\,\text{MgO} \rightarrow (\text{Ni}_x, \text{Mg}_{1-x})\,\text{O} \,. \tag{25}$$

High temperatures favor the formation of the green, nearly ideal [113], solid solution of nickel and magnesium oxide, whereas less well-crystallized structures are found at low temperatures. Other reactions may lead to the formation of mixed spinels [109].

2. *Activation*

2.1. Equilibrium

The catalyst in industrial plants can be activated by various reducing agents such as hydrogen, ammonia, methanol, and hydrocarbons, added to steam. The reaction with hydrogen

$$NiO + H_2 \leftrightarrows Ni + H_2O \qquad (-\Delta H^0_{298} = 1.2 \text{ kJ mol}^{-1}) \tag{26}$$

is nearly thermoneutral and, accordingly, the equilibrium constant, $K_p = p_{H_2O}/p_{H_2}$ varies little with temperature. As indicated by the upper curve in Figure 15, metallic nickel will be stable with approximately 0.3 and 0.6 vol.% hydrogen in steam at 673 K and 1073 K, respectively.

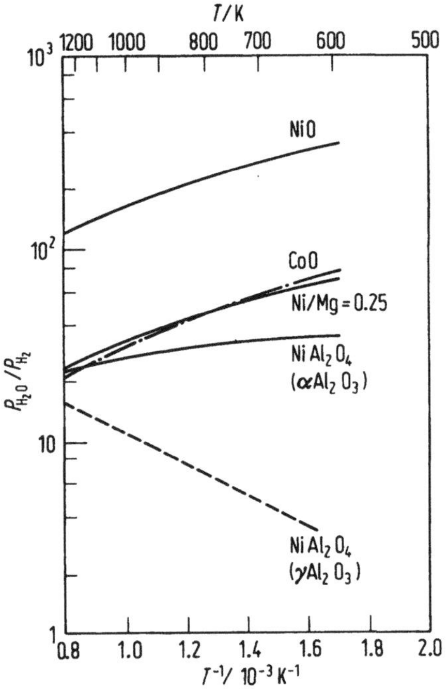

Figure 15. Equilibrium constants for catalyst reduction. (Data: Barin and Knacke [104], Fricke *et al.* [116])

In practice K_p may be lower, as the free energy of nickel oxide decreases due to interaction with the support material. As an example, the free energy of nickel oxide dissolved in magnesia (reaction (25)) can be expressed by:

$$G = G^0 + RT \ln x_{NiO} \tag{27}$$

where x_{NiO} is the mole fraction of nickel oxide in the ideal solid solution [113]. For a given p_{H_2O}/p_{H_2} there will be a certain x_{NiO} at equilibrium. The resulting effect on K_p is illustrated in Figure 15.

An analogous situation exists when nickel oxide and alumina interact [114, 115]. Figure 15 shows equilibrium data for

$$NiAl_2O_4 + H_2 \leftrightarrows Ni + Al_2O_3 + H_2O\,. \tag{28}$$

The alumina phase formed during reduction is typically γ- (or δ) alumina [115, 116], which may subsequently be transformed into α-alumina. The formation of metastable phases causes the free energy of reaction to be smaller than calculated for α-Al_2O_3. This means a lower p_{H_2O}/p_{H_2} at equilibrium. A similar effect results from the formation of small nickel crystals [116].

For practical purposes support interaction and crystal size effects mean that reduced catalysts will be oxidized when exposed to steam containing 2–10 vol.% H_2 (*i.e.* $p_{H_2O}/p_{H_2} = 10–50$) [7, 117] or to mixtures of steam and reducing agents resulting in an equivalent oxidation potential. The actual equilibrium ratio depends on the structure of the individual catalysts.

2.2. Rate

The reduction of pure nickel oxide by hydrogen starts at temperatures 470–520 K depending on the calcination temperature [118, 119].

Supported catalysts require higher temperatures to show a reasonable reduction rate. This may be ascribed to interaction with the support as indicated above. The activation of alumina-supported catalysts is reported to be difficult [108, 114, 120, 121, 122, 123, 124]. The reducibility may depend on the degree of aggregation of the nickel oxide. A fine distribution will result in stronger support interaction [123]. The calcination process of the catalyst may also influence the support interaction, and hence the reduction properties [108, 124]. If the reduced catalyst is reoxidized at low temperatures, the activation of the reoxidized catalyst may proceed more easily than with the fresh catalyst [108, 123]. The addition of small amounts (*ca.* 0.5% on Ni basis) of platinum, palladium or copper [119, 125] to the catalyst may enhance the activation rate probably by providing sites for the dissociation of hydrogen.

When the formation of nickel aluminium spinel has taken place, temperatures above 1070 K may be required [123, 126] for complete reduction. Magnesia dissolved in the nickel oxide phase, even in small amounts, drastically influences the reduction rate of nickel oxide [127, 128] and for practical purposes magnesia-based catalysts should be activated at temperatures around 1020–1120 K. The diffusion of nickel ions in magnesia [129] has a high activation energy (180 kJ mol^{-1}) probably because the nickel ion must diffuse from a preferred octahedral position through a tetrahedral position in the magnesia lattice. Counterdiffusion of nickel and aluminium appears to be rate determining for reduction of nickel present in spinel phases [130].

2.3. Industrial Activation

Activation in industrial plants is usually carried out by means of the feed stream of steam and hydrocarbons at a high steam-to-carbon ratio (5–10) and at a low pressure. The hydrocarbons crack thermally and the resulting hydrogen or carbon [131] acts as an initiator for the reduction process. As soon as metallic nickel is available, the steam reforming process will produce sufficient hydrogen for a quick reduction of the catalyst.

The temperature at which the activation is "ignited" depends on the reactivity of the hydrocarbon and on the state of the catalyst. Often the activation starts in the hotter part of the tube and the activation zone moves backwards to the colder inlet of the tube. The mechanism probably involves back diffusion of hydrogen from one activated pellet to the neighboring inactivated pellet.

Due to the thermo-stability of methane the reduction is more easily initiated when the natural gas contains higher hydrocarbons. This is in particular the case if the catalyst is in a state where, for thermodynamic reasons, a low p_{H_2O}/p_{H_2} is required.

The addition of small amounts of hydrogen to the feed stream (a few percent on steam basis) may have a tremendous effect on starting the activation in the inlet part of the tube. This can be done by recycling some product hydrogen or by addition of small amounts of components easily dissociated into hydrogen (methanol, ammonia etc.). Another possibility is installation of prereduced catalyst [48] at the inlet of the tube.

Even at normal operation, the high stability of methane may result in p_{H_2O}/p_{H_2} ratios over the inlet part of the catalyst bed being higher than required to maintain the catalyst in a reduced state. Consequently, many steam reformers operate with a non-activated inlet layer [103]. This might be accepted in older plants designed for mild conditions, but in modern reformers operating at high heat flux it may become critical by resulting in higher tube wall temperatures and higher risk for carbon formation. In such plants a small recycle of product hydrogen [117] will help to ensure maximum activity of the catalyst filling.

In laboratory units for steam reforming of methane, operating at differential conditions, hydrogen must be present in the feed stream to avoid oxidation of the catalyst [7, 12].

B. Physical Structure of Reforming Catalysts

1. Particle Shape

The high mass velocities in steam reforming plants (G_M = 40,000–70,000 kg m^{-2} h^{-1}) necessitate a large particle size of the catalyst to minimize the pressure drop across the catalyst bed. The maximum particle size should be less than about $^1/_5$ of the tube diameter to ensure effective packing. The pressure drop Δp can be expressed by for instance, the Max Leva equation [82, 132] for turbulent flow involving an empirical expression for the friction factor (*e.g.* equation (14))

$$-\frac{\Delta p}{\Delta Z} = 0.35 \frac{G_M^{1.9} \mu^{0.1}}{\varrho_g g} \frac{1}{d_{p,s}^{1.1}} \frac{(1-\varepsilon)^{1.1}}{\varepsilon^3} \tag{29}$$

where $\Delta p/\Delta Z$ is expressed in MPa m^{-1}, and $d_{p,s}$ ($= 6V_p/S_p$) is the equivalent particle diameter on a surface basis. $d_{p,s}$ is often expressed as $d_{p,v}$ (volume basis) multiplied by a shape factor ψ. More accurate methods [18, 82] for the calculation of the pressure drop include more complex functions, but equation (29) is sufficient for semiquantitative estimates.

In order to maximize the void fraction of the bed, ε, the reforming catalyst is usually manufactured in ring shapes. Typical dimensions are shown in Table 7 with the equivalent diameters, $d_{p,s}$ and $d_{p,v}$, and the

Table 7. Typical shapes of reforming catalysts: Influence on pressure drop and effective activity[c]

d_{ex}/	d_{in}/ mm	h/ mm	$d_{p,s}$/ mm	$d_{p,v}$/ mm	Ψ	ε d_t = 100 mm	relative ΔP d_t = 100 mm (equation (29))	relative r_{eff} (equation (30))
12	5	7	7.00	10.77	0.65	0.465	1.000	1.000
12	5	12	8.13	12.49	0.63	0.490	0.68	0.820
16	8	10	8.57	14.23	0.60	0.544	0.42	0.696
16	8	16	9.60	16.64	0.58	0.552	0.35	0.611
18	8	10	10.00	15.74	0.64	0.515	0.45	0.587
18	8	16	11.43	15.41	0.62	0.525	0.36	0.543
16	3.5[a]	16	5.57	15.99	0.35	0.602	0.42	0.933
23	10	23	15.20	24.55	0.62	0.497[b]	0.33	0.433
30	14	22	17.60	28.53	0.62	0.523[b]	0.22	0.354

[a] seven 3.5 mm holes
[b] estimated for D_t = 200 mm for same G_M

[c] $d_{p,s} = 6V_p/S_p$, $\quad 1/d_{p,s} = \frac{2}{3}\frac{1}{d_{ex} - d_{in}} + \frac{1}{3h}$, $\quad d_{p,v} = \frac{d_{p,s}}{\Psi}$

estimated porosity ε in a typical reformer tube. The relative pressure drops, Δp_r, have been calculated from equation (29).

The catalyst particle size also affects the heat transfer as is apparent from equations (18) and (19). The effective thermal conductivity, λ_{er}, is proportional to d_p, whereas, in principle, the wall coefficient α_w should be almost independent of d_p.

Equations (18) and (19) involve the equivalent sphere diameter $d_{p,v}$ on a volume basis in contrast to $d_{p,s}$ in equation (29). However, it is a question how ring-shaped particles should be treated. Equation (18) predicts λ_{rt} to be lower for rings than for cylinders with the same outer dimensions, whereas experimental data [86, 89] indicate the opposite. More experimental work is required with ringshaped particles to provide consistent formulae.

Industrial steam reforming involves operation with effectiveness factors below 0.1. This means [83, 86, 133] that the effective rate $r_{eff,v}$ (on a reactor volume basis) for the first order reversible steam reforming of methane can be expressed by

$$r_{eff,v} = \eta k_{i,v}(1 - \varepsilon)(C_{CH_4} - C_{CH_4,eq}) f(p_{H_2O}, p_{H_2})$$

$$= \frac{(C_{CH_4} - C_{CH_4,eq})(1 - \varepsilon) 6}{d_{p,s}} \sqrt{D_{eff} k_{i,v}(K_0/(1 + K_0)) f(p_{H_2O}, p_{H_2})}$$

$$= \frac{(p_{CH_4} - p_{CH_4,eq})(1 - \varepsilon) 6}{d_{p,s}} \exp((E_a + E_D)/2RT)$$

$$\times \sqrt{A_D \frac{A_{i,w}}{RT}(K_0'/(1 + K_0'))\, \varrho_p f(p_{H_2O}, p_{H_2})} \qquad (30)$$

and accordingly r_{eff} is inversely proportional to the equivalent particle size, $d_{p,s}$ (on surface basis). Table 7 shows relative activities for typical particle shapes. The selection of particle size is a compromise between pressure drop and catalyst activity.

2. Pore Structure

Equation (30) illustrates the importance of the pore structure on the effective activity. The effective diffusion coefficient D_{eff} is composed of the particle porosity, ε_p, the tortuosity factor τ, and the diffusion coefficient in the bulk and the Knudsen diffusion coefficient, D_B and D_K

$$D_{eff} = \frac{\varepsilon_p}{\tau} \cdot \frac{1}{\frac{1}{D_B} + \frac{1}{D_K}} \tag{31}$$

D_B decreases proportionally with pressure, whereas D_K is independent of pressure and inversely proportional to the pore radius R_p.

Pore volume distributions for typical reforming catalysts are shown in Figure 16. Catalyst A represents a typical high temperature ceramic catalyst with low surface area, whereas higher surface areas have been obtained in catalysts B and C by the incorporation of a stable micropore system in the ceramic support.

The resulting effective diffusion coefficients, D_{eff}, are shown in Figure 17 as function of pressure. At low pressures the low values of D_K dominate for catalysts B and C, but at the normal pressures for steam reforming ($\geqq$2 MPa = 20 atm), D_{eff} depends mainly on the porosity and tortuosity

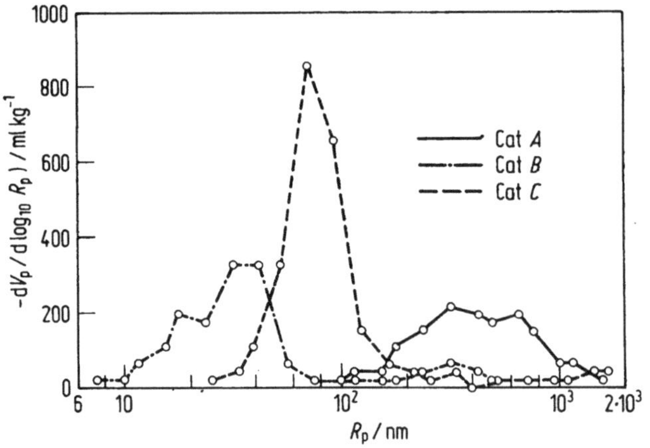

Figure 16. Pore volume distributions of typical reforming catalysts.
Cat. *A* ($Ni/MgAl_2O_4$) $S_{BE} \sim 1\ m^2\ g^{-1}$
Cat. *B* ($Ni/MgAl_3O_4$) $S_{BET} \sim 13\ m^2\ g^{-1}$
Cat. *C* (Ni/MgO) $S_{BET} \sim 17\ m^2\ g^{-1}$
Catalysts were sintered 500 h at 3 MPa, 1073 K
$p_{H_2O}/p_{H_2} = 1$, prior to measurements. (Corresponds to Figure 17)

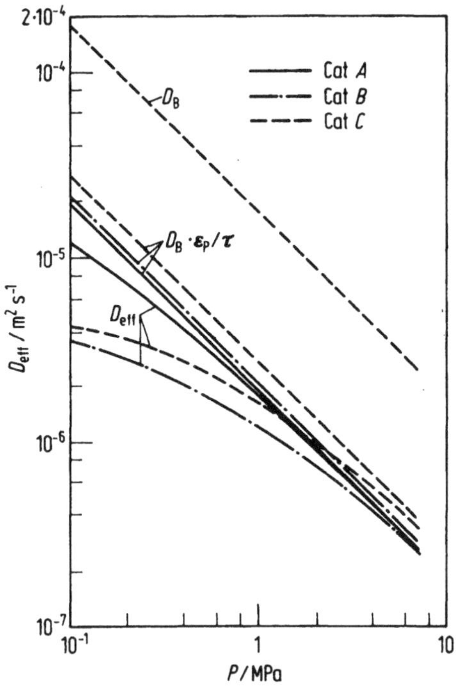

Figure 17. Effective diffusion coefficients of typical reforming catalysts. Catalysts *A*, *B* and *C* correspond to Figure 16

factor. This illustrates how low pressure tests on large catalyst particles can be misleading in evaluating the activity of reforming catalysts.

At reforming conditions there is a tendency for the volume of the micro pore system (and thereby the total surface area) to decrease in the hottest part of the tube [99, 103]. This may have a positive effect on D_{eff}, but for some catalysts it may be accompanied by shrinkage and a decrease in porosity, counteracting this effect. Therefore, estimates of diffusivities should be made on stabilized catalysts.

The tortuosity factor may vary from catalyst to catalyst. For an A-type catalyst (Figure 16) τ was estimated to be *ca.* 3 [134].

C. Nickel Surface

1. Dispersion and Crystal Shape

The activity of a steam reforming catalyst is related to the nickel surface. Figure 18 shows an electron micrograph of a low area ceramic type catalyst. The nickel crystals appear with nearly ideal six-fold symmetry of the crystals in the size of 15–150 nm. This corresponds to a low dispersion of less than 0.5%. On catalysts with higher stable surface area (for instance, catalyst *C* in Figure 16) it is possible to obtain nickel crystals in the range *ca.* 20–50 nm corresponding to a dispersion of *ca.* 5–2%. Still these figures remain low in comparison with catalysts for low temperature service [135] with nickel crystals below 5 nm and dispersion of $\geq 10\%$.

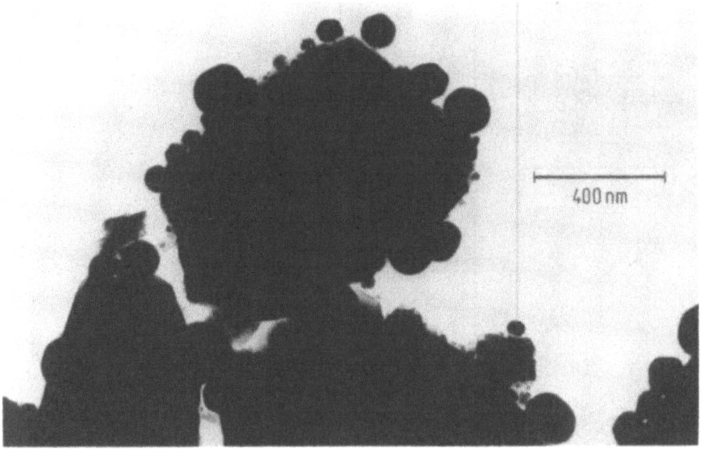

Figure 18. Electron micrograph of ceramic reforming catalyst. Corresponds to catalyst *A* in Figures 16 and 17. The nickel crystals appear with nearly ideal six fold symmetry in sizes up to *ca.* 150 nm (*c.f.* Figure 35)

For a low temperature catalyst (Ni/γ-Al_2O_3) Shephard [136] found various particle shapes. Apparently, the exposure to high temperatures and the small dispersion tend to stabilize more ideally shaped particles. Electron diffraction on individual crystals [137] of catalysts A (Figure 18) revealed the presence of (100) and (110) faces. The hexagonal appearance of the nickel crystals in Figure 18 may represent a cubo-octahedral cluster as discussed by Yacaman *et al.* [138] for supported rhodium.

The nickel cluster may not be a single crystal. As demonstrated by Smith *et al.* [139] multiple twinning is very likely within metal particles as small as 15 nm. The particles may be composed of a number of tetrahedra, which, however, are not completely space filling. Therefore, some lattice distortion is required and these dislocations might play a role in the catalytic reaction [139]. The observed metal "crystal" may be a poly-particle of one of such multiply twinned crystals [139].

Although little is known about the number and size of active sites on real, inhomogeneous metal surfaces, it is useful to refer observed activity data to unit surface area in terms of the turn over frequency, N, or specific activity, r_s. This approach requires the determination of the nickel surface area.

2. *Measurement of Nickel Surface Area*

2.1. General

Several physical methods are available for estimation of the nickel surface area. Some are based on the determination of the crystallite size from x-ray line broadening or, directly, by observation in an electron microscope [140]. These methods remain inaccurate as the distribution of crystal sizes and shapes are to be accounted for.

Crystal size determination by x-ray-analysis may be biased by lattice defects or the existence of multiple twinning [139] as discussed above. This will result in the prediction of smaller crystals than those derived from chemisorption data [109].

The methods most commonly applied are those based on chemisorption of gases, which are selectively retained by the metal. Chemisorption of various gases has been proposed for the determination of the surface area of nickel. One of them is oxygen [141, 142, 143] which however could involve a severe reconstruction of the nickel surface and multilayer oxidation [144]. Carbon monoxide is sometimes applied [109, 145, 146], but the evaluation of results could be impeded by the different ways in which carbon monoxide may chemisorb on nickel [147], by the risk of forming of nickel carbonyl [144], and by the possibility that carbon monoxide may adsorb on the support as well [148].

2.2. Chemisorption of Hydrogen

Hydrogen has been most widely used, probably owing to a simple monolayer criterion, H/Ni = 1/1, proposed by Beeck [141, 149] around 1950. Beeck and Ritchie [149] compared hydrogen chemisorption capacities (0.01 Pa, 77 K) of randomly oriented nickel films with BET-areas obtained with rare gas adsorption. The surface area of a hydrogen site was calculated to be 6.2×10^{-2} nm^2, which is close to the mean area of a nickel site, 6.5×10^{-2} nm^2, when assuming an equal distribution of the (100), (110) and (111) planes (6.21×10^{-2}, 8.77×10^{-2}, and 5.35×10^{-2} nm^2, respectively). A site density [150] of 1.54×10^{19} atoms m^{-2} (6.5×10^{-2} $m,^2$ per site) has since been used uncritically in most studies of the H_2/Ni system.

More recently, Bartholomew *et al.* [144] found consistent measurements on nickel powder when comparing with BET-area (N_2, A) when assuming a nickel site area of 6.77×10^{-2} nm^2.

The evaluation of hydrogen capacities on the basis of rare gas adsorption has been questioned [151] because the area being active for rare gas adsorption may be much less than that available for hydrogen adsorption in view of the different sizes of the atoms.

The H_2/Ni system is complicated because of the existence of different hydrogen states [152], which are reflected by hysteresis phenomena in isobars [141] or by varying amounts of reversible and irreversible hydrogen up-takes [137, 153]. It has been suggested [154, 155] that part of the hydrogen is present in subsurface sites. This may explain why hydrogen uptakes were found to increase with pressure up to 10 MPa [156].

Selwood's [142] magnetic studies of supported catalysts showed constant H/Ni stoichiometry over the whole range of surface coverage, but the arguments for H/Ni = 1/1 are still based on a mean site area of 6.5×10^{-2} nm^2. For supported catalysts, like the well-crystallized reforming catalysts, an equal distribution of surface planes appears unlikely (*c.f.* section III.C.1) and the mean site area of 6.5×10^{-2} nm^2 remains arbitrary. A mean area of 5.35×10^{-2} nm^3 ((111)-plane) or 5.78×10^{-2} nm^2 (mean of (111) and

(100) planes) might as well be justified, which means an uncertainty of 10 to 20%.

Until now agreement has not been reached on chemisorption conditions characterizing a "saturation" layer of hydrogen. This is reflected by the broad variety of methods described in literature [7, 157]. However, Bartholomew *et al.* [153] obtained chemisorption data for a series of supported catalysts that were roughly in agreement with corresponding results from x-ray and electron microscopy (assuming H/Ni = 1/1 and mean nickel site area 6.77×10^{-2} nm^2). Typical isotherms [153] at 298 K are shown in Figure 19. As shown in Table 8 chemisorption at 298 K gives results comparable to data obtained at 201 K as used in previous studies of steam reforming catalysts [7, 158]. Bartholomew *et al.* [153] reported that *ca.* 40% of the adsorbed hydrogen could be removed by evacuation. A similar result was obtained in the measurements at 201 K.

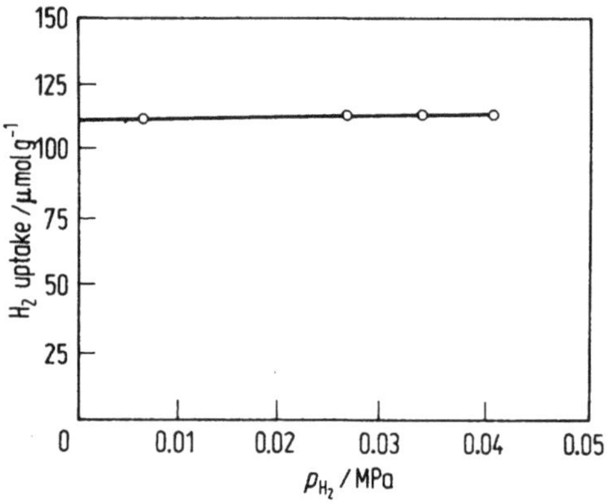

Figure 19. Hydrogen isotherms on Ni/SiO_2 catalyst. Procedure for hydrogen capacity [153]: After reduction, sample is evacuated at 673 K for 1–2 hours until *ca.* 10^{-3} Pa. H_2 uptake is measured at room temperature using 45 min. for equilibration at p_{H_2} in the range 0–70 kPa. The hydrogen capacity is determined by extrapolation to zero p_{H_2}. (Reproduced with permission from ref. [153])

Table 8. Determination of Nickel Surface Area by various methods[a] [137]

	Total adsorption capacity/ 10^{-6} g atom per g	Reversible adsorption/ %	Estimated area[b]/ m^2 g^{-1}	d_{Ni}/nm
H_2 · chemisorption, room temp.	4.68	39	0.18	270
H_2 · chemisorption, 201 K	4.78	25	0.19	260
H_2S · chemisorption, 773 K	3.13	—	0.23	240
x-ray				100–120

[a] catalyst: $Ni/Mg_2Al_2O_4$
[b] obtained using 6.5×10^{-2} nm^2 per adsorbed H and 1 m^2 equivalent to 440 μg S per g Ni

2.3. Chemisorption of Hydrogen Sulfide

Another method [7, 158] for determination of the nickel surface area is based on chemisorption of hydrogen sulfide.

Hydrogen sulfide is the stable sulfur compound at conditions for tubular reforming. It is the most severe poison for the reaction (*c.f.* section VI). The adsorption of hydrogen sulfide on nickel is rapid even below room temperature [159]. At temperatures of industrial interest hydrogen sulfide is chemisorbed dissociatively on nickel [158, 160, 161, 162]

$$Ni + H_2S \leftrightarrows Ni{-}S + H_2 \qquad (32)$$

in which Ni represents an ensemble of nickel atoms on the surface. Stable saturation uptakes of sulfur are observed [158] at p_{H_2S}/p_{H_2} from $1\text{–}10 \times 10^{-6}$ up to $100\text{–}1000 \times 10^{-6}$ above which formation of bulk nickel sulfide takes place [163]. This means that the saturation layer is stable at p_{H_2S}/p_{H_2} ratios several magnitudes below the ratios required for formation of bulk sulfide. Adsorption isotherms are discussed in section VI.

LEED studies [161, 164, 165, 166] have demonstrated that the chemisorption of hydrogen sulfide at low temperatures results in the occupation of sites of high coordination [167]. At higher temperatures and above a certain coverage, islands of well-ordered structure are formed. It is still discussed whether sulfur just forms an adlayer on the surface [167, 168] or whether the process involves reconstruction of the surface (disruptive adsorption [169], two-dimensional sulfide [165] or a penetration of sulfur into subsurface sites [170]. The Ni—S bonds are assumed to be stronger than the Ni—Ni bonds to the underlying metal [161]. Surface diffusion studies have even indicated that the surface phase may melt [171].

A similar behavior has been observed for other systems including the chemisorption of hydrogen sulfide on other metals [162, 172, 173] (Ag, Fe, Cu, Co, Ru; etc.). The bond strength of sulfur on group, the metal was found [173] to decrease in the order Ni, Co, Ru, Fe.

The composition of the saturation layer (the two dimensional sulfide) has been determined by radiochemical analysis [161] and related to Auger spectroscopy [162, 174]. The sulfur content is close to 44.5×10^{-9} g S cm^{-2} nickel [161]. This result, which does not vary significantly from face to face, corresponds to *ca.* 0.5 sulfur atom per nickel atom (S/Ni = 0.5)

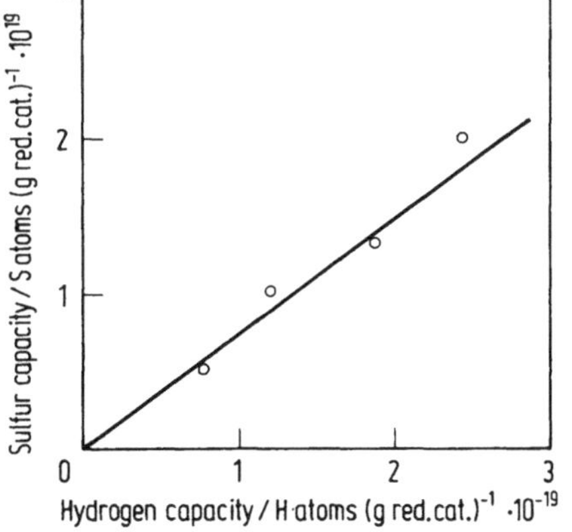

Figure 20. Correlation of hydrogen capacity (201 K) and sulfur capacity (823 K). Procedure for sulfur capacity s_0 [7, 158]: After reduction, sample is exposed to flow of p_{H_2S}/p_{H_2} = 10 to 20×10^{-6} at 773–823 K. (At lower temperature, gas stream should be saturated with steam to eliminate adsorption on the support [176]). After equilibration (1 to 2 days as estimated from flow, sample size, particle size and expected s_0), s_0 is determined by chemical analysis. The method cannot be applied for catalysts containing Ca or Ba, and only with difficulty for pyrophoric catalysts. (Reproduced with permission from ref. [7, 158])

on the (100) surface, which is in accordance with the geometry of the c (2×2) structure observed by LEED. On the (111) surface the structure is more complex [165, 166, 168]. Complete saturation corresponds to a lower S/Ni of *ca.* 0.4 because of the higher density of the (111) plane.

The sulfur capacity correlates with the hydrogen capacity [158] as shown in Figure 20. The slope, *i.e.* the atomic ratio S/H, is determined to 0.74. This result was confirmed by Oliphant *et al.* [175]. It implies an apparent discrepancy when assuming a H/Ni = 1.0 and a S/Ni = 0.5 – 0.4 depending on the surface plane.

In conclusion, more knowledge is required for the description of the site density on real nickel catalysts.

Nickel areas listed in this chapter have been determined by this procedure if not otherwise specified. The nickel area is calculated by using $s_0 = 440$ wt. ppm equivalent to 1 $m^2\ g^{-1}$, s_0 being the sulfur capacity of the catalyst (μg S per g Ni). A mean nickel particle diameter can thus be estimated from

$$d_{Ni} = 3\times 10^3\,\frac{X_{Ni}}{s_0} \qquad (33)$$

where d_{Ni} is given in nm, and where X_{Ni} is the weight % nickel in the reduced state. The corresponding dispersion (%) of nickel can be calculated from

$$D = 0.034\times\frac{s_0}{X_{Ni}} = 1.01\times 10^2/d_{Ni} \qquad (34)$$

assuming spherical nickel particles and (arbitrarily) $6.5\times 10^{-2}\ nm^2$ for a nickel site.

The turnover frequency (molecules $site^{-1}\ s^{-1}$) may be calculated from

$$N = 4.78\times 10^3\times r_w/s_0 = 10.86\times r_s \qquad (35)$$

in which r_w and r_s are intrinsic rates, mol hydrocarbon per h per gram of catalyst, and per m^2 of nickel surface, respectively.

3. Factors Influencing Size of Nickel Surface

3.1. Nickel Content

The nickel surface area is generally increased with higher nickel contents in the catalyst [6, 7, 177, 178, 179, 180] but the dispersion or utilization of the nickel tends to decrease with increasing nickel content [178, 181, 182]. Accordingly, many commercial catalysts are optimized at nickel contents around 15 wt.% (*c.f.* Table 6).

Through special preparation methods [108, 135, 183] it is possible to obtain high dispersions even at high nickel contents, but because of sintering effects at high nickel loadings practice often shows an optimum in nickel area [7, 179, 184] with nickel content as indicated in Figure 26 (see later).

3.2. Activation

The activation procedure influences the size of the resulting nickel surface [7, 108, 136, 179, 180, 182].

Table 9. Influence on nickel surface area of the atmosphere during the activation[a] [7]

Experiment No.	Atmosphere during		Content of reduced Ni/ wt. %	Ni area/ $m^2\,g^{-1}$
	heating	activation		
1	H_2	H_2	23.9	6.8
2	N_2	H_2	24.4	5.4
3	H_2O	H_2	24.6	4.6
4	H_2	$H_2O/H_2 = 3$	22.6	3.7
5	H_2O	$H_2O/H_2 = 3$	23.0	2.7

[a] magnesia-based catalyst; heating up period, 1 h; activation period, 16 h; activation temperature, 1123 K

The effect of steam is illustrated by the experimental results [7] for a magnesia based catalyst shown in Table 9. A temperature higher than 1020 K was required to obtain a reasonable reduction rate. The highest nickel area is attained when using dry hydrogen during heating-up as well as during the reduction. Only about one third of this area was observed when heating-up and activating in presence of steam.

This effect could be explained by the assumption that steam oxidizes the smallest nuclei of nickel or prevents their formation. Consequently, the number of nuclei and the resulting number of crystals is decreased, which means a smaller area.

A significant effect is caused by the presence of steam during heating-up. The small decrease of the area observed in the experiments including heating-up in nitrogen can be explained by the relatively large steam production caused by the high reduction rate at the abrupt addition of hydrogen. Contrary to this, heating in dry hydrogen may result in stabilization of the nuclei by the slow reduction rate at low temperatures.

Other investigations [136, 157, 179, 181, 182] support these observations. Even the small amounts of steam formed during activation may cause a decrease of the resulting nickel area.

Direct reduction in controlled atmosphere of nickel nitrate [182] was found to be favorable compared to reduction of the oxide after decomposition.

However, these effects may hardly be utilized in activation in industrial reformers taking place in the presence of steam, and even if they were achieved they would quickly be wholly or partly lost by sintering effects.

3.3. Sintering

Sintering of the nickel crystals results in loss of surface area, and in principle recrystallization may change the nickel ensembles available, and also cause a decrease of the specific activity. Results [7, 185] from heat treatment of nickel crystals on a stable low area ceramic support are shown in Figure 21. No sintering is observed at 820 K over a period of 1000 hours, whereas the nickel surface area drops to around 40 and 25% over the same

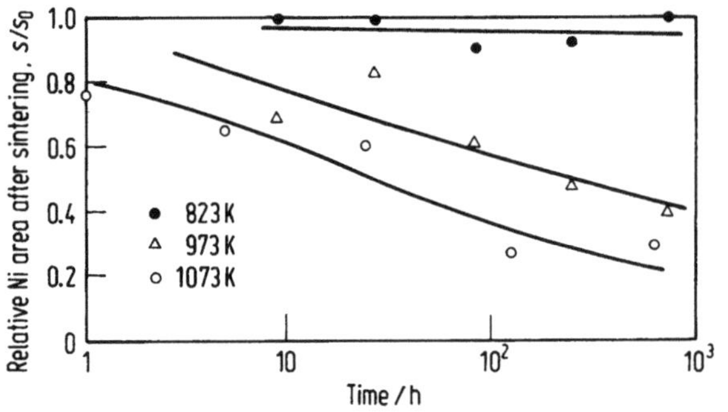

Figure 21. Sintering of nickel surface of ceramic reforming catalyst. Catalyst A (*c.f.* Figure 16) $p_{H_2O}/p_{H_2} = 3$, $p = 0.1$ MPa. Sintering of nickel crystals is apparent above the Tammann temperature (864 K). (Reproduced with permission from ref. [7, 185])

period at 970 K and 1120 K, respectively. This result corresponds to the rule of Tammann, [95] according to which sintering is expected above 0.5 times the melting point (T_m, K) of the metal, *i.e.* 864 K for nickel.

Surface diffusion may start above the Hüttig temperature [95] ($T_m/3$, K), *i.e.* 576 K for nickel and result in reorganization of the nickel crystals. Reorganization of nickel films under steam reforming conditions at 840 K was found [186] to be influenced by hydrogen, whereas steam had no effect. Sintering of nickel crystals on silica was also observed to be water insensitive [181]. The presence of carbon on the surface retarded the reorganization [186].

The growth mechanism of supported metal crystals appears to be very complex and not fully understood [187, 188]. The growth rate might be influenced by the wetting properties of the metal on the support, and by the micropores of the support. The growth is impeded when the size of the metal crystallite is of the order of magnitude of the diameter of the pore [177, 189, 190]. In general, the metal particles may hardly grow to a size larger than the pore diameter of the support. This means that a stabilized micropore system of the support may prevent sintering of the nickel crystals.

4. Activity of Steam Reforming Catalysts

A. Reactivity of Hydrocarbons

1. Thermal Reactions

Methane is a stable molecule because of its sp^3 hybrids. The high excitation energy of the carbon atom involved in the hybridization (796 kJ mol^{-1}) is compensated for by the formation of four C—H bonds each stabilizing the molecule by *ca.* 420 kJ mol^{-1} [191]. This bond strength is reflected by a similar activation energy for pyrolysis of methane. Temperatures higher than 1270 K are required for measurable conversions into ethylene and acetylene [192].

$$CH_4 \rightarrow C_2H_6 \rightarrow C_2H_4 \rightarrow C_2H_2 \rightarrow C\,. \qquad (36)$$

Likewise, thermal steam reforming of methane [193] should be carried out at *ca.* 1770 K to show feasible conversion levels.

Higher alkanes [191] have C—H bond energies in the range 350–400 kJ mol^{-1} and C—C bond energies of *ca.* 320 kJ mol^{-1}. The pyrolysis becomes significant above 920 K at temperatures as used in the production of olefins by steam cracking [21] (in which steam, however, remains nearly inactive). An acidic material may promote the cracking at temperatures above *ca.* 870 K [7, 180].

2. Interaction with Nickel Surfaces

The transition metals activate hydrocarbons at temperatures as low as 370–570 K. The activation below 670 K has been studied mainly in connection with deuterium exchange reactions and hydrogenolysis. On nickel, deuterium exchange [194] results in the formation of CH_3D and CD_4, the latter reaction giving the more dominant product

$$CH_4 + 2\,D_2 \leftrightarrows CD_4 + 2\,H_2\,. \qquad (37)$$

The activation energies of the two reactions were determined to be 100 and 135 kJ mol^{-1}, respectively [194]. Frennet *et al.* [195] concluded that the dissociative chemisorption of methane on rhodium requires a "landing site" of seven metal atoms. A similar result was obtained for nickel [196, 197]. The exchange processes involve a competition between hydrocarbon and deuterium (hydrogen) molecules, which is reflected by large negative reaction orders with respect to deuterium. Leach *et al.* [196] suggested a model with separate reaction routes for CH_3D and CD_4 meaning that formation of CD_4 involves complete dissociation of methane into adsorbed carbon and hydrogen atoms. Chahar [198] found similar results on a steam reforming catalyst ($Ni/MgAl_2O_4$) at *ca.* 570 K.

Exchange studies on ethane [196, 199] below 470 K showed mainly symmetrical incorporation of deuterium, which indicates 1, 2 attachment of ethane to the nickel surface. The simultaneous hydrogenolysis reaction resulted in CD_4.

Martin [199] found that a site of twelve adjacent nickel atoms was required for the activation of ethane. The adsorption involved complete disruption of the molecule. The breakage of the carbon-carbon bond on nickel starts at a low temperature. Thermal desorption studies of adsorbed ethane [200] showed only methane in the gas phase at temperatures above 470 K. The breakage of the carbon-carbon bond has been investigated in great detail in connection with the hydrogenolysis reaction of lower paraffins [201]

$$C_2H_6 + H_2 \leftrightarrows 2\,CH_4 \qquad (-\Delta H^0_{298} = 65\ \text{kJ mol}^{-1})\,. \qquad (38)$$

A large inhibition effect of hydrogen is observed [201, 202], which may be explained by competitive adsorption according to the model of Frennet [195]. The breakage of the carbon-carbon bond is generally assumed to be rate determining [201] although kinetics could be explained by a two-step mechanism [202] involving irreversible adsorption followed by surface reaction.

The hydrogenolysis at low temperatures of hydrocarbons higher than ethane shows that nickel attacks selectively the ends of the chains [203] by successive α-scission, in contrast to, for instance, platinum. This means that methane is the main product on nickel. The result corresponds to adsorption studies [204, 205] showing the same LEED-pattern after adsorption of methane, ethane, propane, and neopentane.

Schouten *et al.* [206] studied the interaction of methane with various surfaces of nickel by AES-LEED. On (100) and (110) planes, adsorption resulted in complete dissociation into adsorbed carbon atoms at 470–570 K as suggested by Leach *et al.* [196]. The carbon atoms were able to diffuse into the bulk nickel above *ca.* 620 K. In contrast, no adsorption was observed on the (111) plane at 570 K. Martin *et al.* [207] reported the same trend for hydrogenolysis, the (111) plane being less active than other crystal planes. The influence of surface roughness was studied by Lehwald *et al.* [208] who observed much higher rates for decomposition of hydrocarbons on the steps of a stepped (111) surface than on terrace stoms.

It must be stressed that these observations for low temperature reactions cannot be transferred directly to conditions for steam reforming. The multiatom site required for complete dissociation of the hydrocarbon may not be that required for chemical reaction, nor may it be required for reaction at the high temperatures applied in steam reforming.

The adsorption of steam on nickel probably involves complete dissociation of water into adsorbed oxygen and hydrogen atoms as observed by McNaught *et al.* [209] in exchange reactions of hydrogen with deuterium oxide. Methane, propane and *n*-hexane did not exchange with deuterium oxide [114, 209, 210] but reacted by steam reforming at temperatures about 570 K. This indicates irreversible adsorption of the hydrocarbon in presence of adsorbed water.

3. *Reactivity in Steam Reforming*

Hydrocarbons show different reactivities for the steam reforming reaction (1). The literature [211, 212, 213, 214] reports different sequences of reactivities depending on temperature and the active metal [215].

Results from gradientless tests [7, 184] on methane, ethane, and *n*-butane are shown in Table 10. Methane has a lower reactivity than the higher hydrocarbons. The apparent activation energy is higher for methane than for ethane and *n*-butane, the two latter showing nearly identical activation energies. On a molar basis and at given steam-to-carbon ratio, butane is less reactive than ethane, whereas the reactivities remain similar on a carbon atom basis.

Other reforming studies on pure hydrocarbons [7, 184] at 773 K and 3 MPa (30 atm) (representative for the inlet of a tubular reformer), showed that apart from benzene most hydrocarbons present in normal naphthas are more reactive than methane. The molar reactivity decreased with increasing molecular weight [184, 213] for a given steam-to-carbon ratio, confirming the trend in Table 10. Accordingly, the reactivity of full range

Table 10. Reactivities of various hydrocarbons[a]

Feed	H_2O/C mol/atom	$N_0/$ molec site^{-1} s^{-1}	$N_0'/$ C-atoms site^{-1} s^{-1}	activation energy/ kJ/mol^{-1}
CH_4	8	0.65	0.65	110 (1.6)[b]
C_2H_6	4	1.30	2.61	76 (0.6)
n-C_4H_{10}	2	1.50	6.00	78 (2.5)

[a] reforming experiments at atmospheric pressure under isothermal, gradientless conditions (*cf.* ref. [184]) rates calculated for $H_2O/C_nH_m = 8$, $H_2O/H_2 = 10$, 773 K, constant $p_{C_nH_m} = 0.01$ MPa: if calculations were based on fixed H_2O/C, N_0' would show less variation between hydrocarbons

[b] figures in brackets show accuracy of activation energy data

naphtha is less than that of light naphtha. This is illustrated in Figure 22, which also shows significant effect of benzene on the reactivity [216].

The limitations in converting heavy feedstocks in tubular reforming appear to be related to the desulfurization step rather than to boiling range. When desulfurized to less than 0.1 wt. ppm S, light gas oil could be completely converted into C_1 components [216]. In practice, pore condensation in the desulfurization catalyst dictates the feedstock limitation.

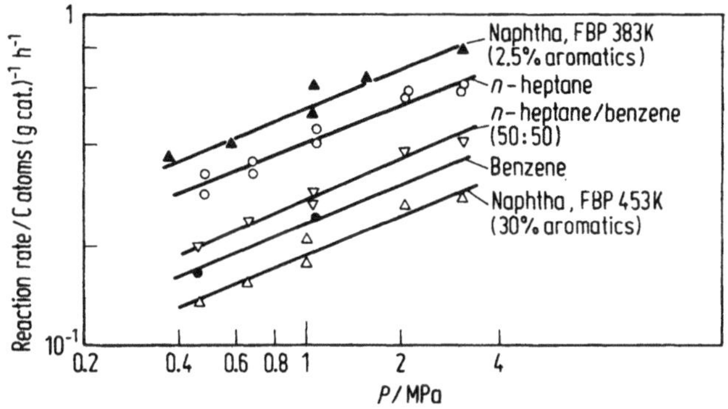

Figure 22. Steam reforming of various liquid hydrocarbons. Ni/MgO catalyst, 4.5 × 4.5 mm cylinders. $H_2O/C = 4$, $p_{H_2O}/p_{H_2} = 20$. The impact of final boiling point and content of aromatics is apparent as well as a similar overall dependency of rate on pressure for all feedstocks. (Reproduced with permission from ref. [216])

Analyses of higher hydrocarbons in the reactor effluents of tests on pure hydrocarbons [7, 184] showed compositions very close to one of the feedstock. Benzene represented an exception to this picture.

This result is in line with the observation by Traply *et al.* [213, 217], who found no hydrocarbons other than methane among the products over a wide conversion range, provided the temperature was kept below 840 K. The same was true for experiments with pure hydrocarbons at 740 K carried out by Jackson *et al.* [218].

The experiments with *n*-butane reported in Table 10 showed no higher hydrocarbons among the products. The measurements were carried out in the temperature range 670–800 K. However, at 870 K Schnell [219] detected substantial amounts of lower olefins in reforming experiments with propane and butane as shown in Figure 23. The tests were carried out at very short contact times ($\sim 2 \times 10^{-3}$ s) but still within the range applied in the experiments [184] at 770 K reported in Table 10. The olefins were claimed [20, 179, 219] to be intermediates in the reforming reaction, but at the high temperature, parallel reactions like thermal pyrolysis or cracking on the support may become significant.

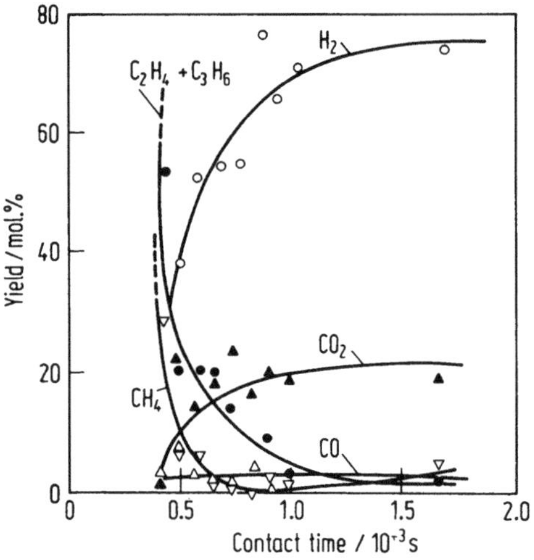

Figure 23. Steam reforming butane. Product distribution at short contact times. Akali promoted reforming catalyst $H_2O/C = 3$, 0.1 MPa, 873 K. (Reproduced with permission from ref. [219])

Whereas olefins might be intermediates at 870 K this assumption can be exluded for temperatures around 770 K, the latter being more representative for the inlet layer of a steam reformer. At this temperature level it appears reasonable to assume that the hydrocarbons are adsorbed irreversibly on the nickel surface and that only C_1-components leave the surface.

The minor amounts of ethane, propane and other higher hydrocarbons observed in the effluent of some tubular reformers can be considered as hydrogenated products from pyrolysis (steam cracking) or cracking on the support, thus indicating that the activity of the nickel catalyst has been insufficient to ensure complete conversion into C_1-species below ca. 900 K.

B. Reaction Kinetics

1. Steam Reforming of Methane

1.1. Transport Restrictions

The kinetics of steam reforming of methane have been the subject of several studies [10], and while there is general agreement on first order kinetics with respect to methane, the activation energies found are scattered from *ca.* 20 to *ca.* 160 kJ mol^{-1}. This can be explained by pore diffusion and heat transfer restrictions influencing in particular the early studies [220, 221, 222].

Pore diffusion restrictions may change the apparent activation energy but the reaction remains first order. However, reaction orders with respect to other components are changed ($\alpha/2$ for $\eta < 0.1$) [133] (*c.f.* equation (30)). This effect as well as the pressure dependency of D_{eff} (*c.f.* Figure 17) may result in misleading results concerning the influence of total pressure if the data are influenced by uncontrolled diffusion restrictions.

As an example of the impact of transport restrictions, calculations on Akers's data [220] on 3.2 mm cylinders show $\eta = 0.15$ and a temperature drop of *ca.* 12 K over the gas film (using Akers' run 233 and an estimated pore radius of 20 nm).

The temperature drop over the gas film may become very large at special reaction conditions and hence confuse results if not recognized. This is illustrated by an example [223] from a TGA apparatus where a catalyst pellet (4.5 × 4.5 mm cylinders) was hung in a thermocouple ($d = 0.1$ mm). The gas temperature, 920 K, was measured with another thermocouple situated just below the pellet. With the low flow of reactants ($G_M = 26$ kg m^{-2} h^{-1}, $p_{H_2O}/p_{CH_2} = 1.2$, $p_{H_2O}/p_{H_2} = 8$) and a methane conversion of 10%, the catalyst temperature droped to 889 K, whereas the gas temperature remained constant.

1.2. Work by Temkin Group

A series of systematic studies was performed by Temkin's group using a nickel foil as catalyst in order to eliminate the bias due to pore diffusion.

Bodrov *et al.* [224] found the following rate expression for the conversion of methane from tests at 1073 and 1173 K

$$r = \left[\frac{1.1 \times 10^9 \exp(-15.6 \times 10^3/T)}{1 + a\dfrac{p_{H_2O}}{p_{H_2}} + b p_{CO}}\right] \times p_{CH_4} \tag{39}$$

where r is expressed in mol m^{-2} h^{-1}. At 1073 K, a = 0.5, b = 20 MPa^{-1}; at 1173 K, a = 0.2, b = 0. All partial pressures are expressed in MPa.

The pre-exponential factor was found to be of the same order as the number of collisions with the surface, indicating a probability factor for reaction of close to one. The activation energy 130 kJ mol^{-1} is close to that for deuterium exchange (reaction (37)).

The rate expression (39) was derived from the following sequence

$$CH_4 + * \rightarrow CH_2 - * + H_2 \tag{40}$$

$$CH_2 - * + H_2O \rightarrow CO - * + 2\,H_2 \tag{41}$$

$$CO - * \rightarrow * + CO \tag{42}$$

$$H_2O + * \rightarrow O - * + H_2 \tag{43}$$

$$CO + O - * \rightarrow CO_2 + * \tag{44}$$

assuming methane adsorption (reaction (40)) as the rate determining step (rds), and neglible surface concentration of $CH_2 - *$.

The rate in experiments where carbon dioxide replaced steam [225] (reaction (8)) could be described by equation (39). It was assumed that steam for the steam reforming reaction (reaction (1)) was formed by reaction between carbon dioxide and hydrogen (reverse of shift gas reaction (2)). When hydrogen and steam were absent in the feed gas the rate dropped by a factor 15 to 20.

Measurements on a nickel/alumina catalyst [226] at 670–970 K resulted in the rate being retarded by hydrogen. This effect vanished at high temperatures as reflected by the reaction order α_{H_2} increasing from -1 at 670–770 K to zero at 970 K. Bodrov *et al.* explained the results by assuming the surface reaction (41) as rds and $O - *$ concentration to be small. The turn over frequency derived from the data ($N_0 = 0.65\ s^{-1}$) corresponds to that listed in Table 10.

In later studies Temkin *et al.* [227, 228] avoided the discussion of a rds by using a steady state approach. Khomenko *et al.* [227] determined the effective stoichiometric number [229] $\bar{\nu}$ for the steam reforming reaction to be one from measurements close to equilibrium. On the basis of Temkin's general kinetic identity [229, 230], the following rate expression [228] was obtained for data on nickel foil at 740–970 K

$$r = \frac{k p_{CH_4} p_{H_2O} \left[1 - \frac{Q_R}{K_p}\right]}{f(p_{H_2O}, p_{H_2}) \left[1 + K_w \frac{p_{H_2O}}{p_{H_2}}\right]} \tag{45}$$

in which $f(p_{H_2O}, p_{H_2})$ is a polynomial in p_{H_2O} and p_{H_2}. Equation (45) contains five temperature dependent constants.

In deriving the expression, the sequence (40) to (44) was used with the reaction step (41) being replaced by

$$CH_2 - * + H_2O \rightarrow CHOH - * + H_2 \tag{46}$$

$$CHOH - * \rightarrow CO - * + H_2 \tag{47}$$

or alternatively

$$CH_2 - * \rightarrow C - * + H_2 \tag{48}$$

$$C - * + H_2O \rightarrow CO - * + H_2\,. \tag{49}$$

$O - *$ was assumed to be the most abundant reaction intermediate (mari).

Agranat *et al.* [231] tested equation (45) on data obtained at high pressure (2.1–2.4 MPa) on nickel foil. The rate constant was found to vary inversely with pressure thus indicating the limitation of equation (45). However, extrapolation to 0.1 MPa resulted in a rate constant corresponding to that reported by Khomenko *et al.* [228].

1.3. Alternative Approach

Some aspects of the Temkin sequence (reactions (40) to (44)) may be questioned.

Recent exchange studies [196, 198] do not support the existence of $CH_2 - *$. Reaction steps (41), (46), and (49) involve a Rideal mechanism with gas phase steam as reactant. This may hardly be justified with adsorbed oxygen atoms as mari. Moreover, the work of Bodrov *et al.* [224, 226] involves the assumption of different rds at different temperatures.

Ross *et al.* [232] found evidence from tracer studies for different rds depending on catalyst composition. A nickel/alumina catalyst coprecipitated with sodium carbonate showed a retarding effect of steam, whereas this effect was less pronounced on an impregnated catalyst (*c.f.* p. 62). The results indicated a role of the support in the kinetics as also reported elsewhere [184, 211, 233]. The influence of products on the rates observed by Ross *et al.* [114, 232] was insignificant.

Temkin *et al.*, as mentioned, solved the rds problem by using a steady state approach, which however lead to the complicated equation (45). An alternative solution might be based on simplified two step kinetics [202] and the following sequence

$$CH_4 + n* \rightarrow CH_x - *_n + \frac{4-x}{2} H_2 \tag{50}$$

$$CH_x - *_n + O - * \rightarrow CO + \frac{x}{2} H_2 + (n+1)* \tag{51}$$

$$H_2O + * \rightarrow O - * + H_2 \tag{52}$$

$$H_2 + 2* \rightarrow 2H - * . \tag{53}$$

Reaction (50) follows the Frennet model [195] for hydrocarbon adsorption on a multi site, which however need not involve as many nickel atoms as indicated in low temperature studies. Moreover, a complete dissociation of the molecule may not be required for the reaction. Reactions (50) and (51) are considered irreversible. This may be justified sufficiently far from equilibrium. $O - *$ is assumed to be the mari. On this basis, the following rate expression is obtained

$$r = \frac{k p_{CH_4}}{\left[1 + K_H \sqrt{p_{H_2}} + K_w \frac{p_{H_2O}}{p_{H_2}}\right]^n} \tag{54}$$

This expression contains the same elements as equation (45). For $n = 2$ and $K_w \ll 1$, equation (54) represents the retarding effect of hydrogen observed by Bodrov *et al.* [226]. For $K_w \gg 1$ the retarding effect of steam observed by Ross *et al.* [232] is apparent.

2. Steam Reforming of Higher Hydrocarbons

2.1. Reaction Sequence

The reaction sequence (50) to (53) can be generalized for higher hydrocarbons. It is assumed that the hydrocarbon is chemisorbed on a dual site followed by successive α-scission of the carbon-carbon bonds. The resulting C_1-species react with adsorbed steam. This results in a reaction model represented by the following sequence [184, 7]

$$C_nH_m + 2* + \xrightarrow{k_A} C_nH_z - *_2 + \frac{m-z}{2} H_2 \tag{55}$$

$$C_nH_z - *_2 + n^* + \xrightarrow{K_H} C_{n-1}H_{z'} - *_2 + CH_x - *_n \tag{56}$$

$$CH_x - *_n + O - * \xrightarrow{k_r} CO + \frac{x}{2} H_2 + (n+1)^* \tag{57}$$

$$H_2O + * \overset{K_W}{\rightleftarrows} O - * + H_2 \tag{58}$$

$$H_2 + 2* \overset{K_H}{\rightleftarrows} 2H - * \tag{59}$$

Assuming the concentration $C_nH_z - *_2$ to be negligible and using Langmuir equations, the following rate equation [184, 234] is obtained

$$r = \frac{k_A p_{C_nH_m}}{\left[1 + \frac{nk_A}{k_r K_w}\frac{p_{H_2}}{p_{H_2O}} p_{C_nH_m} + K_w\left(\frac{p_{H_2O}}{p_{H_2}}\right) + \sqrt{K_H p_{H_2}}\right]^{2n}} \tag{60}$$

Reaction (58) consists of the following steps

$$H_2O + \text{support} = H_2O - \text{support} \tag{61}$$

$$H_2O - \text{support} + * = O - * + H_2 \,. \tag{62}$$

Reactions (61) and (62) represent the ability of the support and various promoters to enhance adsorption of steam, which is then transformed (spilled over) to the nickel surface. Strong support interaction may cause the nickel surface to be covered by water. Since steam is also adsorbed directly on the nickel surface

$$H_2O + * = O - * + H_2 \,, \tag{63}$$

K_w cannot be a true equilibrium constant without violating the principle of microscopic reversibility. K_w should be considered a constant reflecting a steady state condition as illustrated in Figure 24. With these assumptions it follows that

$$K_w = K_{61}\frac{k_{62}}{k_{-63}} \,. \tag{64}$$

From equation (60) it is possible quantitatively to explain the varying kinetic coefficients reported in the literature as listed in Table 11. The kinetic order with respect to water may become positive or negative, depending on the size of K_w and the relative rate constants for k_A, and k_r. When the second term in the denominator is dominating, the kinetic order with respect to the hydrocarbon will be lower than one. The higher k_A the lower the reaction order. A possible temperature dependence of reaction orders is evident since the relative size of the terms in the denominator may change differently with temperature.

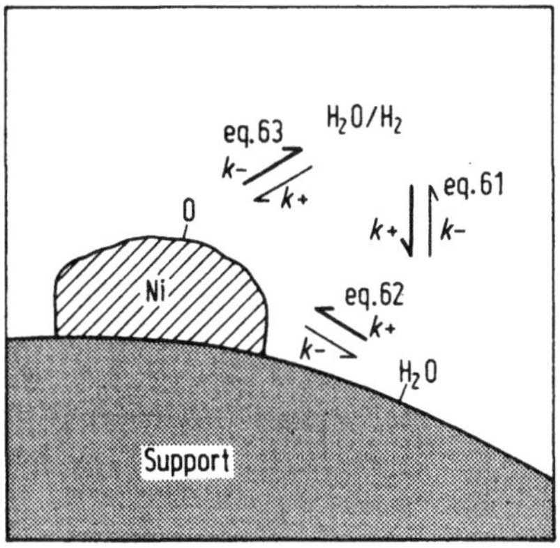

Figure 24. Interaction of steam and catalyst. Steady state condition corresponding to equation (64) for K_w

2.2. Results

For a Ni/MgO catalyst the following expression [184] was obtained for steam reforming of ethane on the basis of gradientless tests at atmospheric pressure and 773 K

$$r_{773} = \frac{20 p_{C_2H_6}}{\left[1 + 30 p_{C_2H_6} \dfrac{p_{H_2}}{p_{H_2O}} + 1.26 \times 10^{-2} \dfrac{p_{H_2O}}{p_{H_2}}\right]^2} \tag{65}$$

where r_{773} is expressed in mol m^{-2} (Ni) h^{-1}, and pressures are in MPa. Equation (64) may be simplified to a power law expression

$$r = 2.2 \times 10^5 \exp(-9100/T)\, p_{C_2H_6}^{0.54}\, p_{H_2O}^{-0.33}\, p_{H_2}^{0.2}\,. \tag{66}$$

Licka [235] and Meshenko [236] reported expressions similar to (65) with the exception of the second term in the denominator, which corresponds to assuming O — * as mari.

The reaction order with respect to steam, α_{H_2O}, decreased with increasing temperature [184] from *ca.* —0.6 at 723 K to —0.2 at 823 K.

For butane and higher hydrocarbons, $\alpha_{C_nH_m}$ tends to approach zero as k_A increases (*c.f.* Table 11). The retarding effect of benzene (aromatics) on the rate (*c.f.* Figure 22) may be explained by a large value of k_A (easy adsorption) of the unsaturated molecule.

Table 11. Results from kinetic studies of steam reforming[a]

Authors	Catalyst system	Hydrocarbon	Temperature/K	Pressure/MPa	Kinetic coefficients			Activation energy/hJ mole^{-1}
					$\alpha_{C_nH_m}$	α_{H_2O}	α_{H_2}	
Bodrov *et al.* [226]	Ni/Al_2O_3	methane	673–873	0.1	1	—	–1	152
Rostrup-Nielsen [184]	Ni/MgO	methane	723–823	0.1	1	—	—	110
Rostrup-Nielsen [184]	Ni/MgO	ethane	723–823	0.1	0.6	–0.4	0.2	76
Meshenko [236]	Ni/Cr_2O_3	ethane	573–633	0.1	1	<0		(38)
Figueiredo *et al.* [237]	$Ni/\alpha\text{-}Al_2O_3$	propylene	773–913	0.1	0.6	—	—	67
Bhatta *et al.* [238]	$Ni/\gamma\text{-}Al_2O_3$	*n*-butane	698–748	3	0	1	—	
Bhatta *et al.* [233]	$Ni/\gamma\text{-}Al_2O_3$	*n*-butane	677–764	3	0	1	—	54
	$Ni/\alpha\text{-}Al_2O_3$, UO_2 (0.3% K)			3	1	–0.6	—	
Saito *et al.* [239]	Ni/SiO_2	*n*-butane	643–723	0.1	0	1	—	
Balashova *et al.* [240]	Ni/SiO_2	cyclohexane	673–733	(0.1)	0	0–1	—	96
Kato *et al.* [241]	Ni/WO_3	*n*-heptane	723–773	0.1	0	1	—	117
Tøttrup [242][b]	Ni/MgO	*n*-heptane	723–773	0.5–3	0.1–0.3	–0.2	0.8	67

[a] supported nickel catalysts: data selected to avoid bias by transport restrictions: other data are listed in refs. [6, 10]
[b] corrected for pore diffusion

The data [216] in Figure 22 show that the overall pressure dependency is nearly identical for different liquid hydrocarbons, which may reflect common kinetic expressions. Since the effectiveness factor in these tests was below 0.1, it was possible to fit the data to a general effective rate equation [133] like equation (30) in which naphtha could be considered as one component [216], giving

$$r_{eff} = k_{eff} \exp(-E/RT) [y_{C_nH_m}]^{(\alpha_{C_nH_m}+1)/2} [y_{H_2O}]^{(\alpha_{H_2O})/2} \times [y_{H_2}]^{(\alpha_{H_2})/2} p^{\beta} \quad (67)$$

in which y is component mole fraction, and the α values are kinetic coefficients for intrinsic rates. k_{eff} depends on $d_{p,s}$, D_{eff} and k_i as shown in equation (30). β was determined to be 0.4 for the Ni/MgO catalyst in question (not identical to that forming the basis for equation (65)). For tests on n-heptane close to 773 K on a Ni/MgO catalyst, Tøttrup [216, 242] derived the intrinsic rate by back-calculation, yielding

$$r_{C_7} = k \exp(-8150/T)\, p_{C_7H_{16}}^{0.2}\, p_{H_2O}^{-0.2}\, p_{H_2}^{0.4} \quad (68)$$

where the rate is expressed in mol m^{-2} (Ni) h^{-1}. Alternatively, the data could be fitted to an expression like (60), giving

$$r_{C_7} = \frac{24 \times 10^5 \exp(-8150/T)\, p_{C_7H_{16}}}{\left[1 + 252\, p_{C_7H_{16}} \dfrac{p_{H_2}}{p_{H_2O}} + 0.08 \dfrac{p_{H_2O}}{p_{H_2}}\right]^2} \quad (69)$$

where the rate is expressed in mol m^{-2} (Ni) h^{-1} and the pressures are in MPa. As will be explained below in section 4, the kinetic expression is strongly influenced by the catalyst composition mainly *via* K_w.

3. *The Water Gas Shift Reaction*

In most kinetic studies of steam reforming the water gas shift reaction (3) has been assumed to be at equilibrium, which indeed facilitates the kinetic analysis. Experimentally, it may be difficult to follow the shift reaction at the high temperatures of normal steam reforming studies, because the shift reaction may proceed thermally in the hot sample line after the reactor. This may be the reason for conflicting results.

Some investigations [213, 226] have indicated that the shift reaction was approached from the CO_2-rich side and, accordingly, that carbon dioxide was a primary product in the steam reforming reaction, whereas Ross *et al.* [114] observed a slow approach from the CO-rich side. van Hook [10] presented strong evidence for the latter view by a correlation between the approach to the shift equilibrium and the methane conversion. The general correlation, which included data from Akers *et al.* [220], Bodrov *et al.* [224, 225], Ross *et al.* [114] and his own work showed that the surplus of carbon monoxide decreased from *ca.* 70% to zero when the methane conversion increased from 10 to 80%.

A correlation of this kind cannot replace a kinetic expression for the shift reaction, which however has not been published so far for conditions of interest for steam reforming. Allen *et al.* [243] solved the problem by proposing separate rate equations for the formation of carbon monoxide and carbon dioxide, respectively.

The shift reaction was studied by Grenoble *et al.* [244] at 573 K on a series of supported metals including nickel. The activity did not vary drastically among metals, but the support material affected the steam adsorption and hence the rate. Kinetics on nickel showed $\alpha_{H_2O} = 0.6$ and $\alpha_{CO} = -0.14$, whereas the activation energy was 78 kJ mol^{-1}. The turnover frequency at 573 K was 0.1 s^{-1}. Although the extrapolated turnover frequency at 773 K remains semiquantitative it is still a few orders of magnitude larger than those estimated in Table 10 for methane reforming. Therefore, the results of Grenoble *et al.* [244] strongly support the view that the shift reaction is very fast at reforming conditions. However, when the reforming reaction is influenced by pore diffusion the effective shift rate becomes limited by the diffusion rate of methane into the catalyst particle. This may explain the data reported by van Hook [10].

C. Catalyst Structure and Activity

1. *Nickel Crystal Size and Surface Topography*

Activities of a number of different catalysts were compared [184] on the basis of the turn over frequency, N_0, for steam reforming of ethane at atmospheric pressure (H_2O/C = 4 mol/atom; p_{H_2O}/p_{H_2} = 10, 773 K). This is a convenient test reaction, and the data obtained at these conditions

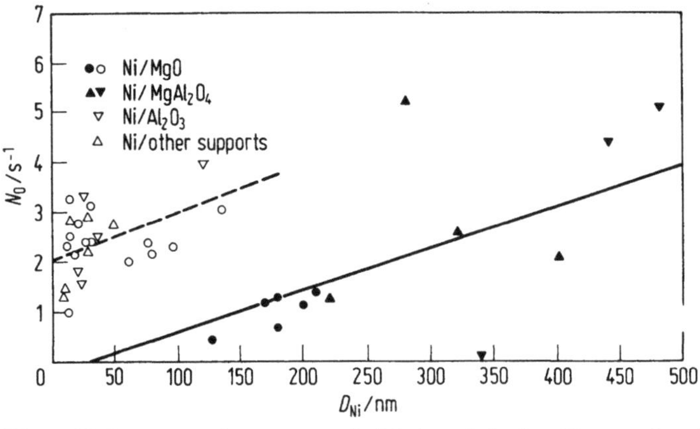

Figure 25. Turn-over frequency and nickel particle size. Steam reforming of C_2H_6. H_2O/C = 4, 0.1 MPa, 773 K. Open symbols: non-sintered catalysts. Closed symbols: catalysts after long time exposure to process conditions at p = *ca.* 3 MPa and T = *ca.* 1073 K. (Ref. [7, 184, 245])

correlate linearily [184] with activities for steam reforming of naphtha at 3 MPa and 773 K.

Figure 25 shows that N_0 remains within the range 1–4 s^{-1} for a large range of nickel crystal size from *ca.* 8 to 150 nm. The data [184] include alkali-free catalysts based on different support materials. Results from a systematic series of Ni/MgO catalysts in which the nickel content was varied, are extracted from Figure 25 and shown in Figure 26. The activity per gram of catalyst shows an optimum as does the nickel area. Similar results have been reported in the literature [177, 178, 179, 180].

The results may show a slight tendency for increasing N_0 with increasing particle size, however with a low correlation coefficient of 0.7. This would

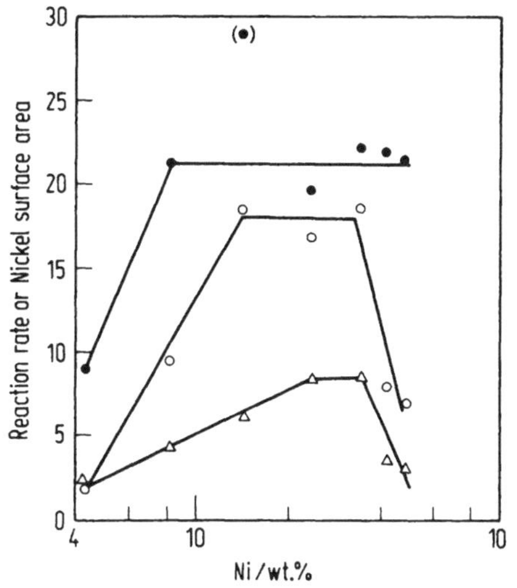

Figure 26. Influence of nickel content on nickel surface area and catalyst activity. Ni/MgO catalysts. Steam reforming of C_2H_6. $H_2O/C = 4$, $p_{H_2O}/p_{H_2} = 10$, 0.1 MPa, 773 K. ○; $r_i/10$ mol g^{-1} h^{-1}; ●; $r_s/100$ mol m^{-2} h^{-1}; ×; Ni area/m^2 g^{-1}. (Reproduced with permission from ref. [7, 184])

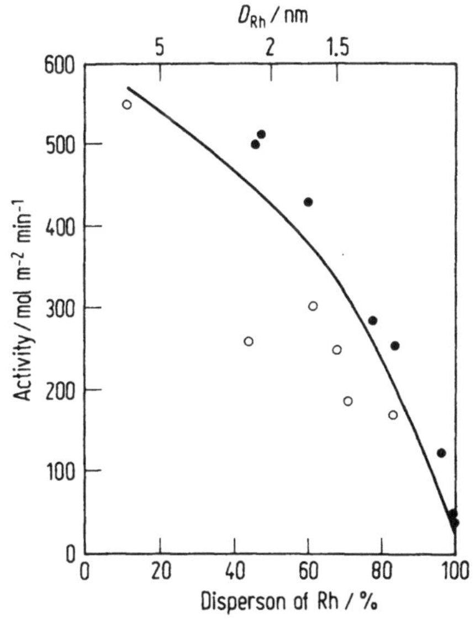

Figure 27. Specific activity and rhodium dispersion. Rh/MgO, $MgAl_2O_4$ catalysts. Steam reforming of *n*-heptane. $H_2O/C = 3$, 0.1 MPa, 823 K. (Reproduced with permission from ref. [247])

be in agreement with observations for ethane hydrogenolysis [201]. The tendency might explain the low N_0 observed by Brooks *et al.* [246] for 7 nm nickel particles ($N_0(C_2H_6) = ca.\ 1\ s^{-1}$), which compares with data for a corresponding crystal size in Figure 25.

The positive effect of crystal size could be explained by the resulting increase of large ensemble landing sites [195]. However, the nickel crystals in Figure 25 are nearly all above the range, where significant changes are normally observed and more data on crystals smaller than 7 nm are required. For rhodium crystals within this range (1.2–9 nm), Kikuchi *et al.* [247] observed a strong negative effect of dispersion on the reforming rate as shown in Figure 27.

Surface inhomogeneities may develop at large crystals as well [248], but attempts [184] in correlating the activity with the number of B_5-sites as determined by nitrogen adsorption are doubtful, because the method has been questioned [249].

2. *Supports and Alkali*

2.1. Kinetic Effects

As discussed above the support is probably involved in the reforming reaction by influencing the activation of steam. This is reflected by K_w or α_{H_2O} as illustrated in Table 12 for steam reforming of ethane [184]. Catalysts containing free magnesia show negative volues of α_{H_2O} down to —0.5, whereas catalysts based on magnesium aluminium spinel or alumina show zero or slightly positive values. This corresponds to the different values reported in Table 11.

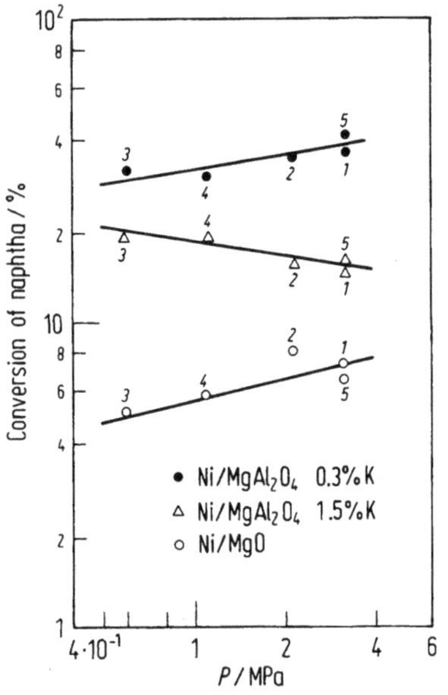

Figure 28. Impact of alkali on naphtha conversion. Steam reforming of naphtha (IBP/FBP = 313 K/ 393 K sp.gr. 0.674 g ml^{-1}), H_2O/C = 3.7, p_{H_2O}/p_{H_2} = 10, 773 K.
Different catalyst amounts as 1–2 mm particles
Ni/MgO: 0.5 g, SV = 8.9 g atom C g^{-1} h^{-1}
$Ni/MgAl_2O_4$ (1.5 wt% K): 10.7 g, SV = 0.05 g atom C g^{-1} h^{-1}
$Ni/Mg/Al_2O_4$ (0.3 wt% K): 11.7 g, SV = 0.05 g atom C g^{-1} h^{-1}
Conversions have not been corrected for different SV. The low activity in the presence of alkali is apparent as well as the different overall dependency of rate on pressure. (Reproduced with permission from ref. [7, 184])

Table 12. Kinetic parameters for steam reforming of ethane at atmospheric pressure[f]

Catalyst	Ni/ wt %	alkali/ wt %	Ni area $m^2\ g^{-1}$	N_0/ s^{-1}	Kinetic coefficients			Activation energy/ $kJ\ mol^{-1}$	
					$\alpha_{C_2H_6}$	α_{H_2O}	α_{H_2}		
Ni/MgO									
a1	25	0.07 Na	2.0	1.3	0.54[a]	—0.33[a]	0.2[a]	75.8	(0.6)[g]
a18	25	0.08 Na	3.5	1.3		—0.52[b]			
a19, a1 washed	25	<0.01	2.0	2.2		0.0[b]			
a21	25	<0.01	1.1	2.0		—0.3[c]			
a22	25	<0.01	4.2	3.1		—0.5[c]			
a23, a22 sintered	25	<0.01	1.1	3.1		—0.2[c]			
a34, a1 + K	25	0.14 K	3.0	0.3		—0.4[b]		79.2	(1.3)
a35, a1 + K	25	0.53 K	2.7	0.14		—1.8[e]		82.1	(2)
a37, a1 + Na	25	0.61 Na	2.4	0.14		—0.7[c]			
a13, Ni/Cu = 0.54	7.6	<0.01	(1.2)	0.02				102	(3)
$Ni/Mg\ Al_2O_4$									
b1		<0.01	0.44	4.0				76.6	(10)
b4		1.53 K	1.4	0.02	0.4[c]				
Ni/Al_2O_3									
c1	16.5	0.04 Na	7.1	0.9	0.6[c]	0.13[a]			
$Ni/MgO/Al_2O_3$ (MgO/Al_2O_3 = 1/1)									
d3	19.1	<0.01	14.5	1.3		—0.26[b]			
Ni/ZrO_2									
d4		<0.01	2.6	0.03				80.4	(4)

[a] accuracy ≦0.05; [b] accuracy 0.05–0.1; [c] accuracy 0.1–0.2; [d] accuracy 0.2–0.3; [e] accuracy 0.3–0.5
[f] measurements made at 773 K (*c.f.* ref. [184]); N_0 determined for H_2O/C_2H_6 = 8, H_2O/H_2 = 10
[g] figures in brackets indicate accuracy of activation energy

The addition of potassium causes α_{H_2O} to decrease significantly, whereas sodium has a less pronounced effect. The effect of alkali is reflected by the over-all pressure dependency (β in equation (67)) of rate for steam reforming of naphtha [184] as shown in Figure 28. A similar effect of alkali was reported by Bhatta *et al.* [233] (Table 11).

The data in Table 12 demonstrate that the activation energy is not affected by the addition of potassium in contrast to the effect of adding copper.

2.2. Turn Over Frequency

Catalysts based on a great number of different alkali-free supports show activities, N_0, within the same range (1 to 4 s^{-1}) [184] (Figure 25) such as Ni/MgO in various sintered forms and contents of alumina, $Ni/MgAl_2O_4$, Ni/γ-Al_2O_3, Ni/α-Al_2O_3, Ni/Cr_2O_3 and Ni/SiO_2. Some alkali-free nickel catalysts [184] show small values of N_0 such as Ni/SiO_2—Al_2O_3, Ni/SiO_2—MgO, Ni/TiO_2 ($N_0 = 0.2$–$0.4\ s^{-1}$) and Ni/ZrO_2 ($N_0 = 0.03\ s^{-1}$) and Ni/C ($N_0 = 0.004\ s^{-1}$).

The addition of potassium to a nickel catalyst may result in a decrease of N_0 with more than one order of magnitude [184] as demonstrated in Figure 29. The effect is stronger on Ni/MgO and $Ni/MgAl_2O_4$ than on Ni/Al_2O_3 and Ni/MgO—SiO_2 probably because the alkali is less reactive on the more acidic supports. The effect of sodium is significant but less pronounced than that of potassium.

In practice, this means that even trace amounts (~1000 wt.ppm) of alkali may bias activity tests. It might explain results reported by Ross *et al.* [250] (*c.f.* p. 53)

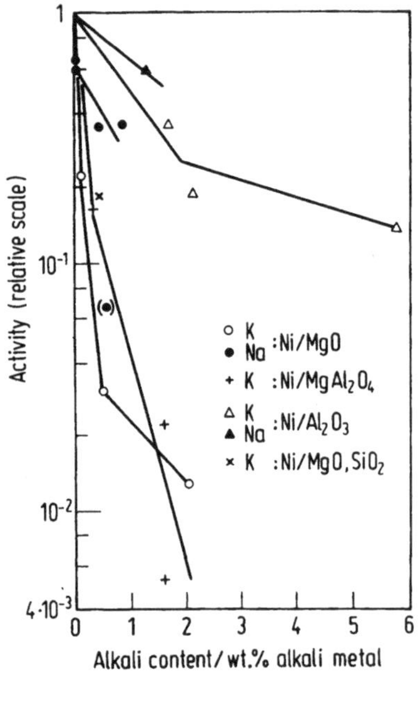

Figure 29. Turn-over frequency and alkali content. Steam reforming of C_2H_6. $H_2O/C = 4$, 0.1 MPa, 773 K. (Reproduced with permission from ref. [7, 184])

Table 13. Relative turnover frequencies for various reaction at atmospheric pressure[a]

	Catalyst[b]	Ni/wt. %	Ni area $m^2\,g^{-1}$	Reforming		Hydro-genolysis	Methanation of	Decomposition of
				C_2H_6 (773 K)	CH_4 (773.K)	C_2H_6 (573 K)	CO (523 K)	NH_3 (773 K)
a1	MgO, 0.07 wt.% Na	25	2.0	*1.0*	*1.0*	*1.0*	*1.0*	*1.0*
a22	MgO, <0.01 wt.% Na, K	25	4.2	2.4	2.0	4.2	9.1	0.9
a35	MgO, a1 + 0.53 wt.% K	25	2.7	0.03	0.09	0.08	0.002	0.9
b1	$MgAl_2O_4$, <0.01 wt.% Na, K	8	0.4	3.0	1.4			3.2
b5	$MgAl_2O_4$, 1.57 wt.% K	11	1.4	0.02	0.02			1.5
c1	η-Al_2O_3, 10 wt.% Na	17	7.0	0.7			2.7	0.4
c4	γ-Al_2O_3, <0.01 wt.% Na, K	44	11	2.5		8.7	20.6	0.5
c7	c1 + 5.8 wt.% K	17	3.4	0.09			0.001	0.5
d4	ZrO_2, <0.01 wt.% Na, K	16	2.6	0,02	0.01	0.08	0.8	0.4
d8	SiO_2, <0.01 wt.% Na, K	21	13	1.1			4.9	
d9	SiO_2/Al_2O_3, <0.01 wt.% Na, K	21	4.4	0.2			0.04	
d14	Carbon, <0.01 wt.% Na, K	14	(0.8)	0.04	0.04	0.01	0.4	0.5

[a] *c.f.* [7, 184]
[b] in all cases, nickel with the indicated support

The activity pattern for steam reforming of a number of nickel catalysts is compared with that for other reactions [184] in Table 13. The relative turnover frequencies show parallel behavior for steam reforming of ethane and methane. Hydrogenolysis of ethane shows a trend similar to the reforming reactions, whereas the effect of alkali is much more pronounced for the methanation reaction. Methanation shows large differences for different aluminas, whereas decomposition of ammonia appears unaffected by the addition of alkali and change of support.

The parallelism of ethane and methane reforming is more pronounced than with hydrogenolysis, and this may indicate that the reaction sequence for steam reforming is not simply related to that of hydrogenolysis. This may be ascribed to the different temperatures applied, which may also explain that methanation (at 523 K) appears more structure sensitive than reforming (at 773 K).

2.3. Origin of Effects

Balashova *et al.* [240] explained the low activity of Ni/C for steam reforming of cyclohexane compared to Ni/SiO_2 by a failing ability of coke for activation of steam, since both catalysts showed similar activities for dehydrogenation of cyclohexane. However, the data in Table 13 shows that Ni/C displays a relatively low activity also for hydrogenolysis, where steam is not involved.

A significant electronic interaction between the support and the surface atoms of nickel is very unlikely in view of the large nickel particles. Hence, the use of zirconia is not accompanied by any detectable change of the activation energy (*c.f.* Table 12). A strong metal support interaction as observed by Vannice *et al.* [251] for methanation on nickel-on-titania appears unlikely in view of the large quantities of steam present at steam reforming.

It appears more likely that the different turn over frequencies should be related to changes in surface structure of the nickel imposed by the support [184]. The number of optimum ensembles on the nickel surface may vary with crystal orientation or be related to epitaxial relations or abnormal shapes of the crystals. Shephard [136] considered the importance of various crystallite forms for the activity for hydrogenolysis. More studies are required to evaluate these effects.

The negative effect of alkali cannot be explained by the presence of steam and the large negative values of α_{H_2O} (Table 12), because the decrease in activity is also observed in absence of steam. The hydrogenolysis data in Table 13 are confirmed by kinetic studies of the effect of alkali on iron [252]. Moreover, Shephard [136] observed that removal of alkali from a Ni/Al_2O_3 catalyst resulted in higher activity for hydrogenolysis. For methanation, Schoubye [253] likewise reported low activities of alkali-promoted catalysts.

There is little evidence for an electronic interaction of alkali on nickel metallic state. Alkali caused no detectable change of the activation energy as observed for a copper nickel catalyst. Campbell *et al.* [254] reported similar observations for methanation on alkali doped nickel films. The decrease in activity was not accompanied by a change of the activation energy, thus indicating that the reaction mechanism remained the same.

The effect was explained by an electron ligand effect by which the adsorbed alkali atom (ion) donates extra electron density to the neighbouring nickel atoms. This results in increased bonding of adsorbed molecules (CO) and an increased dissociation rate. This is analogous to observations for dissociative adsorption of nitrogen on alkali promoted iron catalysts [255], but more studies are required to conclude whether this explanation applies for steam reforming. The possible role of alkali in stabilizing certain reaction intermediates is further discussed in relation to carbon formation (*c.f.* p. 92).

The high steam partial pressures make it unlikely that alkali can be present in a metallic state. However, the data in Figure 29 support the idea that the alkali partial pressure over the catalyst (and hence the surface coverage of alkali) is important. The less acidic supports have a looser bonding of alkali resulting in easier adsorption on the metal.

3. *Non-Nickel Catalysts*

3.1. Other Metals

The effect of the nature of the metal on the steam reforming of ethane [184] is illustrated in Table 14. Rhodium and ruthenium show N_0 about ten times higher than nickel, platinum and palladium. The low activity of cobalt is probably due to the process conditions with p_{H_2O}/p_{H_2} close to the equilibrium constant for oxidation of cobalt (Figure 15). However, when alloying cobalt and nickel, cobalt is stabilized and the resulting activity is close to that obtained on pure nickel.

The differences among group VIII metals appear less than observed at lower temperatures for hydrogenolysis [201] and methanation [251]. An almost parallel sequence of reactivity was reported for steam reforming of methane [256] and of toluene (steam dealkylation) [12] (*c.f.* Table 14).

Iron was studied by Münster *et al.* [257] as a catalyst for steam reforming of methane at strongly reducing conditions ($p_{H_2O}/p_{H_2} < 0.1$). The rate at 1023 K ($N_0 \sim$ *ca.* 1 s^{-1}) was comparable to that on nickel at 773 K. The activation energy was *ca.* 180 kJ mol^{-1} and the kinetics were represented by an expression close to equation (54).

Alloying nickel with group IB metals may cause the activity to dwindle. As shown in Table 14, copper has a larger effect on N_0 than variations among group VIII metals. A similar effect was observed for silver [258]. In contrast, small amounts of copper (1 %) were claimed [259] to improve the activity of nickel at high temperatures.

Rhodium, which appears to be the most active reforming metal (Table 14), shows a significant activity for aromatization in parallel to steam reforming [260]. The selectivity is influenced by crystal size [260] with corners and edges favoring aromatization, and by alloying with platinum [261]. Takami *et al.* [262] found $MgWO_4$ to be an advantageous support for large crystals of rhodium.

Table 14. Effect of metal on activity for steam reforming on relative turnover frequencies

		Pure metals									Alloys	
		Rh	Ru	Pd	Pt	Ir	Ni	Re	Os	Co	Ni, Co	Ni, Cu
ethane [184][a]	γ-Al_2O_3	13	9.5	1.0	0.9	—	*1.0*	0.2	—	1.4×10^{-3}	0.2[d]	1.0×10^{-2}[d]
toluene [12][b]	γ-Al_2O_3	11	5.8	3.8	2.3	1.6	*1.0*	—	0.3	—	—	—
methane [256][c]	SiO_2	1.6	1.4	0.6	0.5	0.7	*1.0*	—	—	nil	—	—

[a] $H_2O/C_2H_6 = 8$, $H_2O/H_2 = 10$, 773 K, 0.1 MPa, ($N_0(Ni) = 3.3\ s^{-1}$)
[b] $H_2O/C_7H_8 = 3.3$, 713 K, 0.034 MPa, ($N_0(Ni) = 5.5 \times 10^{-3}\ s^{-1}$: steam dealkylation, see equation (70)
[c] $H_2O/CH_4 = 3.0$, 773 K, 0.1 MPa
[d] MgO support

The precious metals have found special interest for the steam dealkylation of toluene

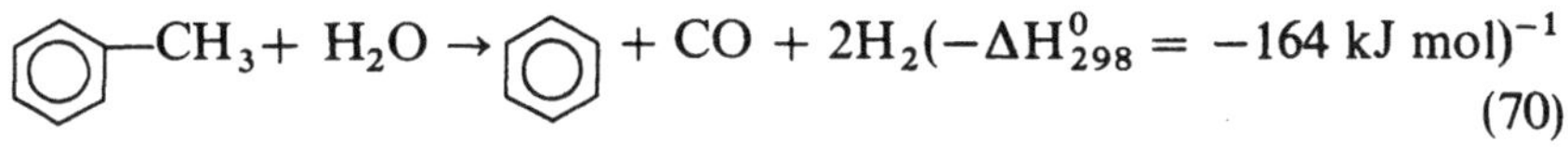

(70)

Rhodium has been used in most studies [12, 247, 263, 264]. Grenoble [12] found rhodium to be the most active, and that nickel has a very low selectivity. Palladium and platinum have a slightly better selectivity than rhodium, but the activity is inferior. Kikuchi *et al.* [247] observed that the selectivity depended on the dispersion of the rhodium crystals.

3.2. Non-Metal Catalysts

Difficulties in desulfurizing heavy feedstocks have lead to attemps to use non-metallic catalysts for steam reforming [13, 216, 265, 266]. The use of a calcium oxide catalyst was reported already in 1868 [23] (*c.f.* p. 11). Due to low activity the reactions are accompanied by steam cracking of the feedstock and a complete conversion into C_1 components may hardly be achieved.

Tomita *et al.* [266] studied the kinetics of methane conversion over a calcium aluminate catalyst. The activation energy was *ca.* 40 kJ mol^{-1} and the rate proportional to p_{CH_4}/p_{H_2O}. Carbon dioxide was claimed to be the primary product. At 1123 K the rate [267] ($N_0 \sim 10^{-4}$ s^{-1} for $H_2O/C = 4$, 0.1 MPa) was still inferior to the specific rate on nickel at 773 K.

4. *Poisons*

Sulfur is the most severe poison for steam reforming catalysts. The deactivation for steam reforming of ethane [184] appears to be linear with sulfur uptake (coverage), after an initial drop in activity as shown in Table 15.

Table 15. Poisoning of steam reforming on nickel catalyst[a]

Exp. No.	Poison		Sulfur capacity	Poison coverage[c]	Rate/mol g^{-1} h^{-1}	N_0/s^{-1}
	element	content/ wt. ppm	s_0/wt. ppm			
unused a1[b]	S	80	883	<0.1	0.241	1.3
4201	S	239	805	0.30	0.066	0.67
4202	S	360	805	0.45	0.053	0.75
4203	S	398	805	0.49	0.059	0.70
329	S	615	805	0.76	0.038	0.61
55	S	805	805	1.00	<0.001	—
133	Cl	1350	885	?	0.243	—
141	As	4200	885	?	0.055	—

[a] steam reforming of ethane at 773 K, 0.1 MPa: catalyst Ni/MgO (catalyst a1, Table 12, *c.f.* ref. [184])
[b] see Table 12 for details
[c] θ_9 for sulfur

At complete coverage, $s/s_0 = 1$, the catalyst is completely deactivated. At nigh temperatures (> 1.000 K) the rate was found to decrease with $(1 - \theta_9)^3$, (*c.f.* Fig. 64 [223]). Sulfur poisoning is discussed in detail in section 6.

The data in Table 15 indicate the effect of arsenic to be much less than that of sulfur. In contrast, Brigder *et al.* [269] observed in pilot tests that 50–100 ppm As_2O_3 on the catalyst affected the activity of an alkali-promoted catalyst. The reformer tube also picked up arsenic, which was transferred to subsequent charges of new catalysts and poisoned them too. Nielsen [245] found evidence that the poisoning effect of arsenic is due to alloying with nickels. The arsenic originates typically from the solution used in the carbon dioxide wash or from inpurities in some zinc oxide absorption masses.

Chlorine has no impact on the activity on a Ni/MgO catalyst according to the results in Table 15. However, for very large contents of chlorine (*i.e.* 1 wt% on the catalyst or *ca.* 1000 ppm in feed), Brigder *et al.* [268] were able to detect deactivation of an alkali-promoted catalyst. The difference between the data in Table 15 and the observations by Bridger *et al.* [268] might be ascribed to the presence of alkali affecting the adsorption of chlorine.

Other components (lead from the hydrocarbon feed, silica and other solids from steam, etc.) should be absent in the feed stream because they may eventually result in blockage of the catalyst pore system.

Phosphorous was reported to be a poison for reforming catalysts [5], whereas Cadmium [5] and zinc, which may escape from the sulfur removal system, have no dectable effect on the catalyst activity.

D. Catalyst Activity and Industrial Performance

1. Intrinsic and Effective Rate

The large catalyst particle sizes used in tubular reforming result in low effectiveness of the catalyst. The situation was analyzed by means of a computer program [269, 270] based on the intrinsic kinetics and the pore volume distribution. The program contained a model for the calculation of the restrictions to mass and heat transfer inside the catalyst particles and in the gas film surrounding the particles.

Figure 30 shows results from a typical steam naphtha reformer [234]. The mass transport restrictions are related mainly to pore diffusion, whereas heat transfer restrictions are located in the gas film. The naphtha is converted in the outer shell of the pellet, whereas the remaining part contains a gas equilibrated at the catalyst temperature. The strong endothermic reaction results in a temperature drop of *ca.* 10 K over the gas film, in spite of the high mass velocity.

As indicated in Figure 11 the bulk gas arrives rapidly a composition close to equilibrium when passing down the tube. This results in a decreasing reaction rate in the lower part of the tube according to equation (22), in spite of the increasing temperature. The diffusion rate and reaction rate have the same driving force ($C_{CH_4} - C_{CH_4,eq}$ or ΔT_R). This decreases more than k_{eff} increases. This is illustrated in Figure 31. It means that the concentration and

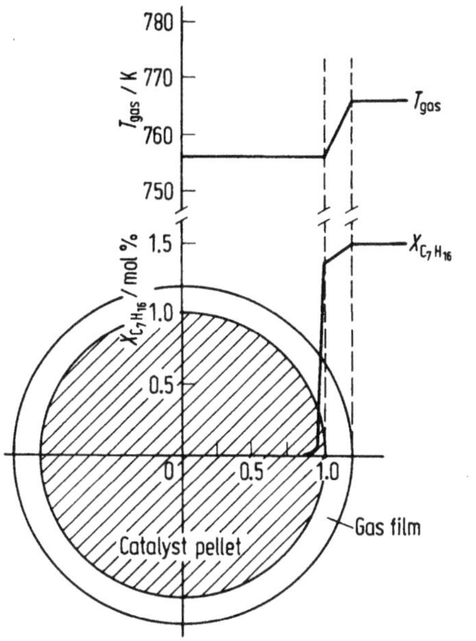

Figure 30. Concentration and temperature profiles inside and around catalyst pellet. Naphtha based hydrogen plant. $H_2O/C = 5.7$, $p_{exit} = 2.4$ MPa, $T_{exit} = 1093$ K, $Z = 1$ m from inlet. $d_{p,s} = 11$ mm. (Reproduced with permission from ref. [234])

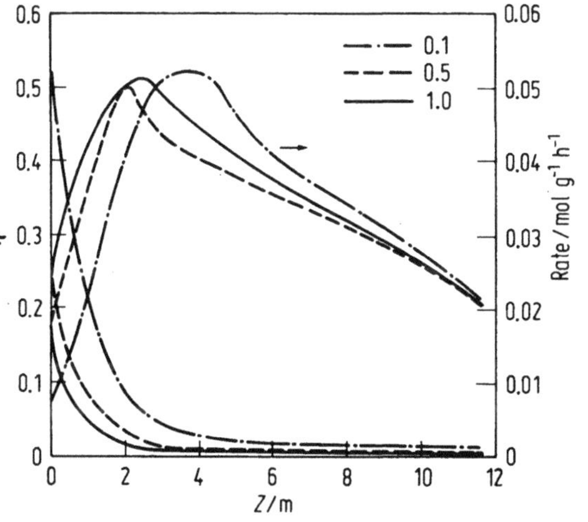

Figure 31. Reaction rate and effectiveness factor in industrial reformer. $p_{H_2O}/p_{CH_4} = 3.0$, $p_{exit} = 3.5$ MPa. $T_{exit} = 1083$ K. Calculation based on one-dimensional heterogeneous model accounting for interfacial and intraparticle gradients. Three levels of intrinsic activity of catalyst. Fixed temperature profile (corresponds to Figure 11)

temperature gradients in and around the pellets decrease with the distance from inlet. The calculated effectiveness factor η decreases simultaneously from 0.15 to 0.25 at the inlet to less than 0.01 at the exit. This may appear surprising in view of the decreasing rate, but for $\eta < 0.1$, η becomes propertional (*c.f.* p. 37) [133] to $\sqrt{(D_{eff}/k_i)\,(K_c'/(1 + K_c'))}$ which decreases with temperature independent of rate. The impact of the equilibrium form, $(K_c'/(1 + K_c'))$, becomes insignificant for methane reforming above 870 K (*c.f.* Table 2).

The low effectiveness factor means that rates can be represented by effective rate expressions similar to equations (30) and (68) with parameters to be determined for each catalyst and hydrocarbon in question. Therefore, a heterogeneous model for turbular reforming taking interfacial and intraparticle gradients into account may hardly be justified in view of the numerical problem compared to those associated with the pseudo-homogeneous models described in section 2.C.

Conversion profiles were shown in Figure 11 for a natural gas based reformer. Figure 32 shows axial conversion profiles in a naphtha based reformer for two different naphthas [216]. The naphtha is completely converted in the upper part of tube and the methane content approaches the equilibrium concentration.

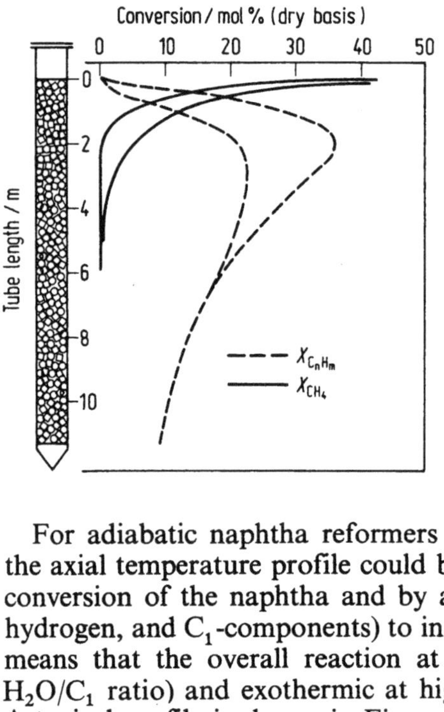

Figure 32. Conversion profiles in steam naphtha reformer. $H_2O/C = 3.5$, $p_{exit} = 3.2$ MPa, $T_{exit} = 1073$ K, $q_{av} = 80$ kw m^{-2}. Two different naphthas. (Reproduced with permission from ref. [216])

For adiabatic naphtha reformers Phillips *et al.* [271] demonstrated that the axial temperature profile could be calculated simply from the fractional conversion of the naphtha and by assuming all other components (steam, hydrogen, and C_1-components) to instantaneously react to equilibrium. This means that the overall reaction at low conversion is endothermic (large H_2O/C_1 ratio) and exothermic at high conversion (formation of methane). A typical profile is shown in Figure 42 (see later). This approach may also explain the axial methane profiles in Figure 32.

2. *Activity and Overall Conversion*

As discussed in section 2.D, many catalysts have sufficient activity to ensure a very close approach to equilibrium. Exception to this may be operation at special firing profiles (*c.f.* Figure 8), operation with strongly alkalized catalysts and low exit temperature (<970 K), and operation in the presence of substantial amounts of sulfur. In these situations it may be advantageous to increase the effectiveness factor (the activity) of the exit layer by decreasing the particle size (*c.f.* equation (30)).

The main impact of catalyst activity on reformer performance is associated with its influence on local phenomena as outlined in the following discussion.

3. *Activity and Tube Wall Temperature*

The balance between heat input through the reformer tubes and the heat consumption by the reaction as expressed in equation (15) is the central problem in steam reforming. The importance of the catalyst activity may be illustrated by a simplified example for an ammonia plant reformer (data as in Figures 11 and 31).

A differential segment dZ of the reformer tube is considered for fixed inlet gas composition at a distance from the inlet close to the position for maximum heat flux. The throughput (G_M) and heat flux (q) are increased proportionally and the resulting conversion and temperature increase is calculated for two catalyst activities. The results are shown in Figure 33. At low G_M(or q_{av}), the catalysts can easily establish the equilibrium ($\Delta T_R = 0$) at nearly adiabatic conditions ($T_{exit} < T_{inlet}$). When the load is increased, ΔT_R and T_{exit} increases reflecting how the transferred heat is utilized for the chemical reaction and heating-up of the gas. The differences between the two catalysts become significant, being most pronounced at q_{av} = 10–50 kW m^{-2} as also reflected by the different tube wall temperatures. At high loads the catalysts cannot absorp sufficient heat for reaction and the heating-up of the gas dominates at $q_{av} > 200$ kW m^{-2}.

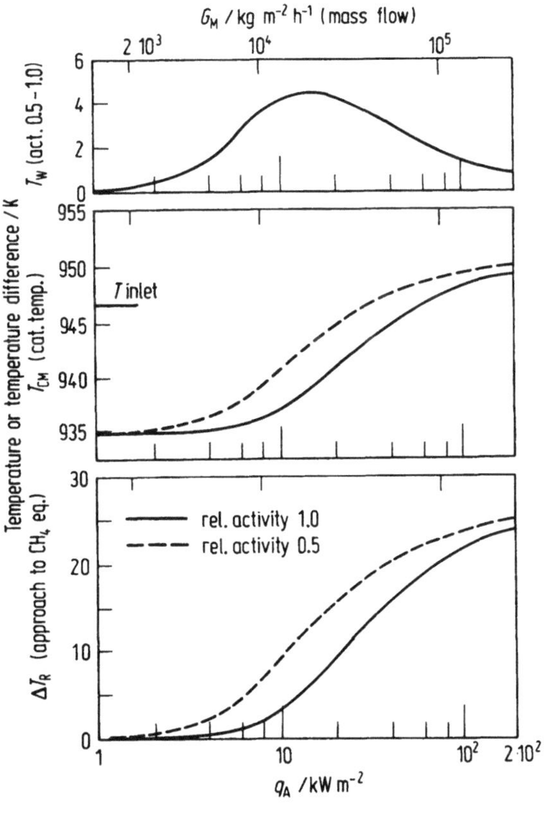

Figure 33. Catalyst performance and load. Ammonia plant reformer. p_{H_2O}/p_{CH_4} = 3.0, p_{exit} = 3.5 MPa, T_{exit} = 1083 K. Conditions as in Figures 11 and 31. Conversion and temperature increase over tube segment of 100 mm, 1 m from inlet. Calculation performed with two-dimensional pseudo-homogeneous model for fixed gas composition at segment inlet at different loads, G_M, and heat input, q_{av}, to tube segment. Two levels of effective catalyst activity

Figure 34 illustrates the overall effect of catalyst activity for tube wall temperature profile simulating a top fired furnace. An increase of the effective catalyst activity by a factor two causes the maximum tube wall temperature ($T_{w, max}$) to decrease by *ca.* 30 K. This could mean a factor of about three on tube life (*c.f.* Figure 10). At the same time high catalyst activity results in a lower catalyst temperature (T_c) in most of the tube. These effects are less pronounced at lower capacities (*c.f.* Figure 33).

The same is true when using reformer firing with no maximum in the T_w-profile (types II and III in Figure 8). In these types of reformers a fixed catalyst exit temperature dictates a minimum T_w to ensure the heat transfer near tube exit. The effect of a factor two in effective catalyst activity is however still significant in a side wall fired reformer (*i.e.*, nearly constant T_w, type II in Figure 8). The examples in Table 16 show

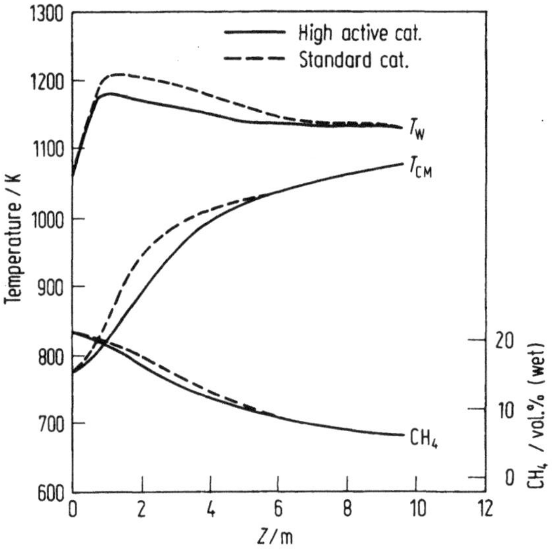

Figure 34. Axial Profiles in ammonia plant reformer at boosted capacity. $p_{H_2O}/p_{CH_4} = 3.5$, $p_{exit} = 3.3$ MPa, $q_{av} = 92$ kW m^{-2}, $Re_{inlet} = 9900$. High active catalyst filling has double effective activity at inlet and same activity at exit, as standard catalyst. (corresponds to Figure 50)

Table 16. Influence of catalyst activity on tube wall temperature and capacity[a]

1. Fixed values: 8.5% CH_4 at reformer exit: 100% capacity		
Catalyst activity (relative)	1.0	2.2
Maximum tube wall temperature, T_w/K	1158	1148
Catalyst outlet temperature, T_{exit}/K	1073	1068
2. Fixed values: 8.5% CH_4 at reformer exit: maximum tube wall temperature, 1158 K		
Catalyst activity (relative):	1.0	2.2
Capacity/%	100	112
Catalyst outlet temperature, T_{exit}/K	1073	1070

[a] data from primary reformer in a natural gas based ammonia plant (ref. [83]): $H_2O/C = 4.0$, $p_{exit} = 3.4$ MPa, $q_{av} = 58$ kW m^{-2}

that the higher activity may be utilized to either diminish the tube wall temperature by *ca.* 10 K at fixed ammonia production or alternatively to increase capacity by 12% at fixed tube wall temperature.

4. Activity and Radial Dispersion

Different activities result in different radial concentration gradients in the reformer tube (*c.f.* equation (16)). As the gas is heated up when approaching the tube wall $C_{CH_4,eq}$ decreases and consequently ΔT_R increases. The increased reaction potential may result in a decreasing hydrocarbon concentration when approaching the tube wall.

At the high mass velocities in modern high pressure reformers (G_M = 40,000 to 70,000 kg m^{-2} h^{-1}) the radial diffusivity, D_{er}, (equation (16)) will be too high to allow the development of significant radial concentration gradients and the hydrocarbon concentration remains almost constant as shown in Figure 50 (p. 89). However, in low pressure reforming as used for the manufacture of reducing gas, mass velocities are lower ($G_M \sim$ 7,000 to 15,000 kg m^{-2} h^{-1}) and significant radial gradients may develop depending on the catalyst activity as shown in Figure 51 (p. 89).

The effect of catalyst activity in decreasing the temperature level and, for some cases, in decreasing the hydrocarbon concentration close to the tube wall may be important for elimination of the risk for carbon formation. This will be discussed in the next section.

5. Carbon Formation on Steam Reforming Catalysts

A. Morphology and Mechanism

1. Different Routes to Carbon Formation

Steam reforming involves the risk of carbon formation, which may cause serious operational problems as discussed in section 2.D. Carbon may be formed *via* different routes, each influencing the morphology of the carbon. The most common types are:

— whisker-like carbon
— encapsulating carbon
— pyrolytic carbon

The main characteristics of the three routes are summarized in Table 17. The formation of bulk phase nickel carbide appears unlikely [272] at temperatures above 670 K as applied in steam reforming.

2. Whisker Carbon

It is well known that the reaction of hydrocarbons as well as carbon monoxide over transition metals can lead to the formation of filamentous carbons [9, 273, 274, 275]. The carbon grows typically in a whiskerlike structure with a nickel particle at the top as shown in Figure 35 and the strong whiskers may result in break-down of the catalyst particle.

Table 17. Different routes for carbon formation in steam reforming of hydrocarbons[a]

	Whisker carbon	Encapsulating polymers	Pyrolytic carbon
Formation	Diffusion of C through Ni-crystal: Nucleation and whisker growth with Ni-crystal at top	Slow polymerization of C_nH_m radicals on Ni-surface, into encapsulating film	Thermal cracking of hydrocarbon: Deposition of C-precursors on catalysts
Effects	No deactivation of Ni-surface: Break-down of catalyst and increasing Δp	Progressive deactivation	Encapsulation of catalyst particle: Deactivation and increasing Δp
Temperature range/K	>720	<770	>870
Critical parameters	High temperature: Low H_2O/C_nH_m: No enhanced H_2O adsorption: Low activity Aromatic feed	Low temperature Low H_2O/C_nH_m: Low H_2/C_nH_m: Aromatic feed	High temperature: High void fraction: Low H_2O/C_nH_m: High pressure: Acidity of catalyst

[a] ref. [216]

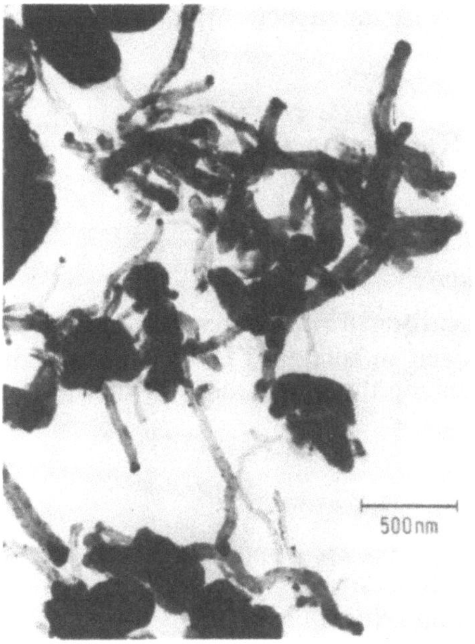

Figure 35. Electron micrograph of whisker-like carbon formed by decomposition of methane on nickel catalyst at 773 K. Catalyst A, Figure 16. The whiskers have a tubular appearance and most of the whiskers have a nickel particle at the end. It is remarkable that the diameter of the whisker is close to that of the nickel crystal. The nickel crystals show a preferred orientation and a nearly five fold symmetry (*c.f.* Figure 18). The dark spots in the whisker may be clusters of nickel. Reproduced with permission from ref. [277]

Studies in the electron microscope show the presence of tubular filaments with a co-axial channel. The whisker diameter is very close to that of the nickel crystal and the nickel crystal is very often pear shaped [276, 277] (*c.f.* Figure 35), which may indicate a reconstruction during carbon formation of the nearly ideally shaped nickel crystals (*c.f.* Figure 18). The reconstruction may be related to the strong interaction (wetting) of nickel with graphite surfaces observed by Baker *et al.* [278] for gasification studies.

The carbon structure is graphitic [279, 280] with the basal planes parallel to the long axis of the whisker. The whisker may contain small fragments of nickel. The formation of whisker carbon may be accompanied by the formation of flake-like carbon [218, 279].

The growth rate of the whisker is independent of time [273, 274] meaning that large amounts of carbon can accumulate [281]. This is in contrast to the carbon formation in catalytic cracking [282] where carbon deactivates the active site forming the carbon. In some situations, the whisker growth may cease because the nickel crystal becomes encapsulated in carbonaceous deposits [273, 274, 283, 284]. The whisker growth rate in the absence of steam is not significantly influenced by the support nor the presence of alkali [281, 285].

Typical plots of weight of carbon versus process time are shown in Figure 36 for TGA-studies [281] at steam reforming conditions with various hydrocarbons. After a certain induction period, t_0, the coking rate arrives at a constant value reflected by a straight line

$$c = k_c(t - t_0) \tag{71}$$

The extent of carbon formation depends strongly on the unsaturated character of the hydrocarbon.

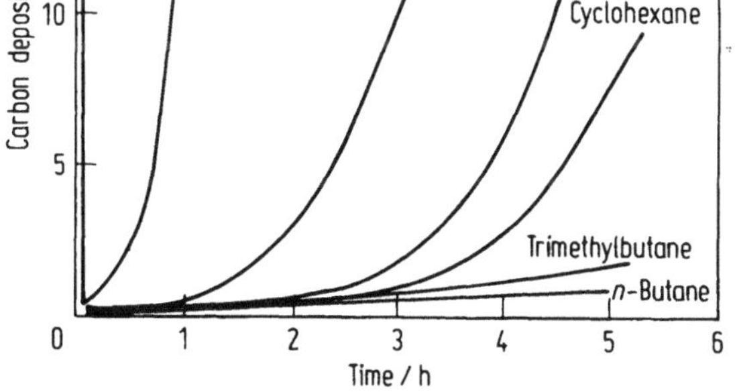

Figure 36. Carbon formation from different hydrocarbons. Thermogravimetric studies, 0.7 g Ni/MgO/catalyst. H_2O/C = 2, 0.1 MPa, 773 K. (Reproduced with permission from ref. [7, 281])

The rate of carbon formation shows a complex dependency on temperature as described by Trimm *et al.* [9, 286]. A typical Arrhenius plot is shown in Figure 37 for decomposition of butylene [286]. Similar plots were obtained for decomposition of other olefins and acetylene [283], as well as for conditions for tubular reforming [281]. Trimm *et al.* [9, 286] explained the appearance of Figure 37 as the formation of whisker carbon at low temperatures being replaced by carbon formation from thermal reaction at high temperatures. Studies carried out on carbon monoxide and methane in a smaller temperature range [288, 289] did not include the region for thermal reaction, but the same trend was observed as shown in Figure 38 for the decomposition of methane [289].

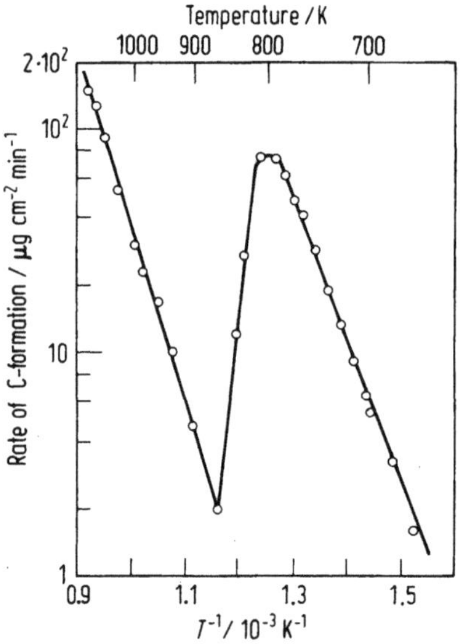

Figure 37. Temperature dependence of carbon deposition on nickel foil. *n*-butene, $p_{C_4H_8}$ = 13 kPa, p_{H_2} = 3.2 kPa. (Reproduced with permission from ref. [9])

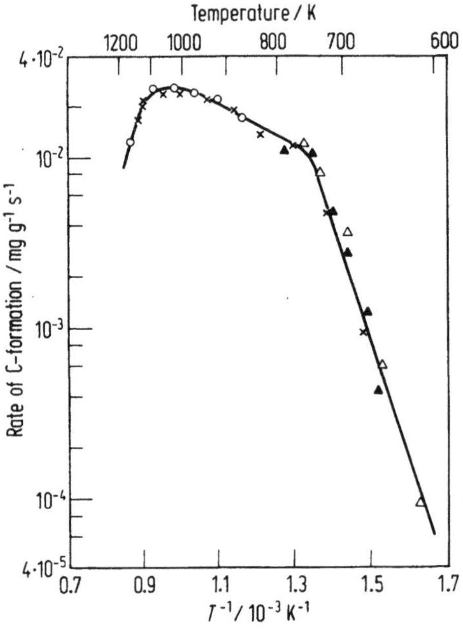

Figure 38. Temperature dependence of methane decomposition. Catalyst 0.7 g Ni/MgO, p_{CH_4} = 2.9 kPa. p_{H_2} = 0 [289]

The negative activation energy is explained [274, 287] by adsorption effects in a rate determining surface reaction. The position of the rate maximum depends on type of reaction, whereas the activation energy is found to be in the range 125 to 140 kJ mol^{-1} for a number of hydrocarbons [9] and carbon monoxide [288] as well. This activation energy is in good agreement with the activation energy for diffusion of carbon through nickel (84 kJ mol^{-1}) [290] plus the enthalpy of solution of carbon in nickel (*ca.* 42 kJ mol^{-1}) [291].

Baker *et al.* [284] were able to observe the growth directly by controlled atmosphere electron microscopy and the observed rate was close to the estimated diffusion rate of carbon through the nickel crystals [284]. Therefore, there is general acceptance [273, 274] for a growth model in which diffusion through the nickel particle is involved, as illustrated in Figure 39, and that the diffusion is rate determining at low temperatures (*i.e.*, below the maximum).

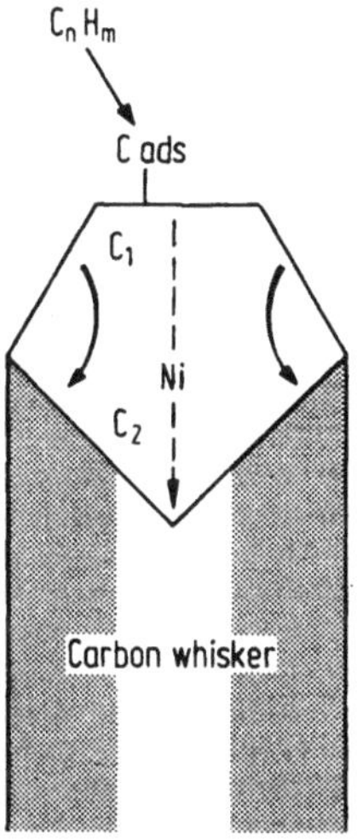

Figure 39. Whisker growth by diffusion through nickel particle.
$r_c \sim D_{C,\,Ni} \times \frac{C_2 - C_1}{d_{Ni}}$

Baker *et al.* [273, 284] suggested that the exothermic surface reaction resulted in a temperature gradient across the nickel particle working as driving force for the whisker growth. Carbon segrates at the cold rear of the particle, where the solubility is smaller.

This mechanism has been questioned [274] because the decomposition of some of the carbon-forming gases (butylene, methane) is endothermic.

Moreover, calculations by Holsten [292] indicated the absence of any significant temperature difference (less than 0.1 K) between a metal particle and its support as a result of catalytic reaction on the metal.

Alternative explanations [274] were based on different solubilities of carbon at the metal-gas interface and on the nickel-graphite interface [293] or based on assumptions of special nucleation sites [294].

The diffusion model (Figure 38) may explain the induction period, t_0, as the time required to saturate the nickel crystal with carbon prior to nucleation. This is in accordance with the observation [276, 281] that small crystals react faster than the larger ones. Surface structure (exposed planes, steps, grain boundaries etc.) may influence the nucleation of carbon [295, 296] or segration of carbon from the nickel crystal [297] and thereby the induction period.

Kinetic studies have emphasized on the rate of carbon formation, and only a few have dealt with the nucleation [298] although the understanding of the nucleation process is more important in attempts to eliminate carbon formation. In kinetic terms this is equivalent to prolonging t_0 to infinity.

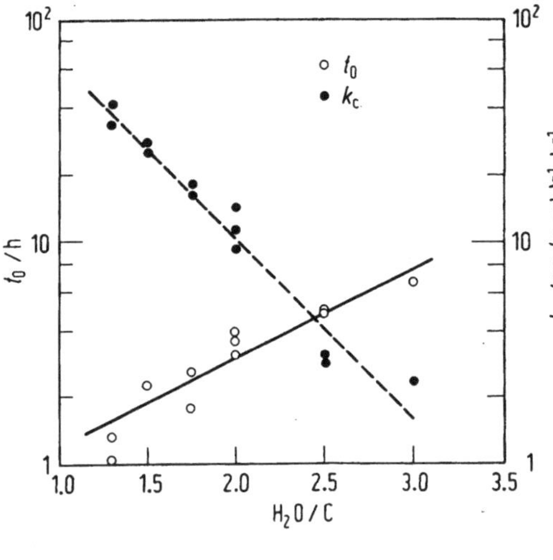

Figure 40. Carbon formation and steam to carbon ratio. Steam reforming of *n*-heptane. Thermogravimetric studies. 0.7 g Ni/MgO catalyst 6.2 g of C_7H_{16} h^{-1}, $p_{C_7H_{16}} = 6$ kPa, $p_{H_2} = 1.7$ kPa, 773 K. (Reproduced with permission from ref. [7, 281])

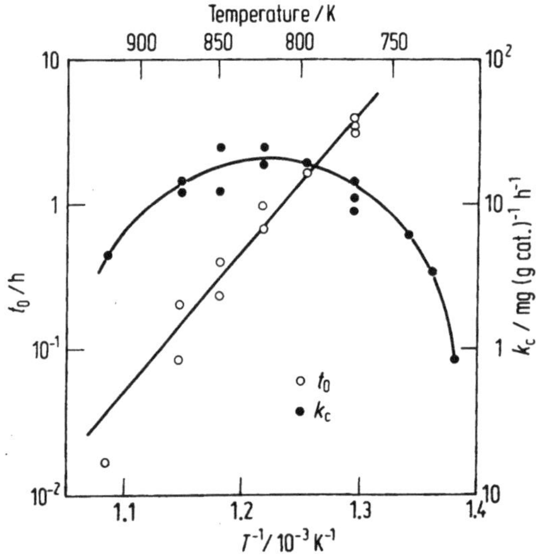

Figure 41. Carbon formation and temperature. Steam reforming of *n*-heptane. Thermogravimetric studies. Conditions as in Figure 40. $H_2O/C = 2.0$, 773 K. (Reproduced with permission from ref. [7, 281])

At conditions for steam reforming [281], t_0 was found to decrease rapidly with temperature and to increase with the steam-to-carbon ratio as shown in Figures 40 and 41, respectively. Moreover, t_0 depends strongly on the type of hydrocarbon as apparent from Figure 36.

3. *Encapsulating Deposits*

The adsorbed hydrocarbon may react into a film of nonreactive deposits, which may encapsulate and deactivate the nickel surface. The phenomenon was observed in studies [9, 283, 284, 286] of the decomposition of pure hydrocarbons to form graphitic encapsulating carbon, and at steam reforming conditions at low temperatures to form gum-like material [40, 218, 238, 299].

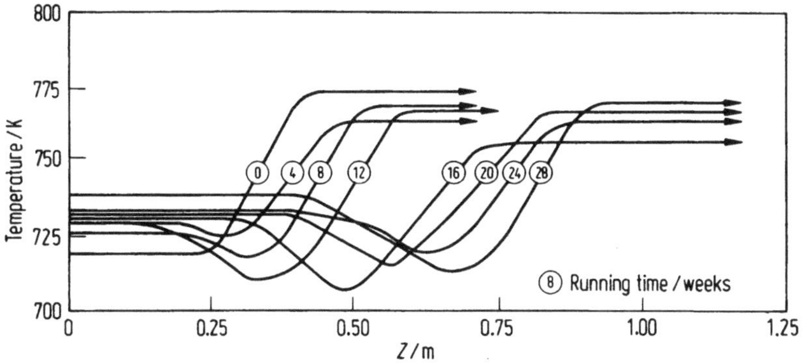

Figure 42. Progression of temperature profile in adiabatic steam reforming. Data from industrial operation with CRG-process. Feed: light naphtha, $H_2O/C = 1.6$, $p = 1.7$ MPa. (Reproduced with permission from ref. [40])

In adiabatic reformers the resulting deactivation causes a continuous movement of the temperature profile [40] in the flow direction as illustrated in Figure 42. Jackson *et al.* [218] investigated the deposits, which could be extracted from the spent catalyst with tetrachloromethane. The chemical structure of the polymer was apparently independent of the reacting hydrocarbon. IR- and mass spectra were consistent with a linear polymer of $-CH_2$-groups. This is in contrast to the analysis of extracts reported by Bhatta *et al.* [238], showing anthracene and other aromatics in accordance with results on extracts of industrially used catalysts. Apparently, the polymer observed by Jackson *et al.* [218] is slowly converted into an aromatic structure. This may also explain why the reforming reaction may occur on the carbonaceous overlayer on the nickel surface in pulsed-flow experiments by Jackson *et al.* [218] involving little ageing of deposits, whereas deactivation is observed at a slow rate in industrial operation. It is consistent with work by Frennet *et al.* [300] demonstrating that the adsorbed hydrocarbon species may be gradually dehydrogenated into non-reactive residues at a rate being slow compared to that of the reaction with the gas phase. The deactivation rate may be enhanced by presence of surface steps [301, 302].

Mosely *et al.* [299] studied process parameters influencing the deactivation rate, which can be expressed by a resistance number, R' defined as kg hydrocarbon feed processed per g catalyst deactivated. R' does not vary with pressure and it is favored by a high steam-to-carbon ratio, the presence of hydrogen in feed, and by high inlet temperatures, whereas high boiling point of the naphtha and in particular the content of aromatics result in smaller values of R'. Industrial operation at given conditions results in a nearly constant value of R' after an initial period with faster deactivation. Typical values are within the range 5–20 kg g^{-1}

4. Pyrolytic Carbon

The steam reforming reactions on the nickel surface may be accompanied by thermal cracking reactions (steams cracking), which may start at temperatures above *ca.* 920 K. In fact, a steam/naphtha reformer with a completely deactivated nickel catalyst will work as a steam cracker producing olefins [21, 192]. Therefore, the risk of carbon formation is to be analysed in the same way as for a steam cracker.

It is generally agreed that the gas film at the tube wall is overheated and acts as a source of radicals and coke precursors [9, 303, 304]. In steam crackers the tube skin temperature is the most important parameter determining the rate of coke formation. Moreover, the coking reactions are related to the so-called kinetic severity function (*KSF*) [192, 305, 306]

$$KSF = \int k(T)\,dt \tag{72}$$

which describes the residence time — temperature history of the reactants in a way that is consistent with kinetics. This means that for a given temperature profile the risk of carbon formation is increased with higher residence time. The catalyst filling has an influence on these parameters, first by

Figure 43. Coke deposit formed by pyrolysis of light paraffins. The coke was deposited in tubular reformer operating with nearly inactive catalyst at the inlet. The coke was deposited at the tube wall, the structure of which is visible on the coke

changing the film volume and secondly by influencing the residence time distribution *via* the void fraction [305]. Moreover, the nickel and the surface adicity of the catalyst will promote the formation of coke deposits from the tar-like intermediates [7, 20, 307].

The pyrolytic coke is normally found as dense shales on the tube wall as shown in Figure 43, or as deposits encapsulating the catalyst particles and eventually filling out the void between the particles.

5. *Model for Carbon Formation*

It can be useful to separate some effects of carbon formation in steam reforming by means of a simplified version of the sequence of reactions (55) to (59) of irreversible steps and assuming that thermal cracking can be disregarded

$$C_nH_m + * \xrightarrow{r_A} C_nH_x - *$$

$$C_nH_x - * \xrightarrow{r_H} CH_x - * \xrightarrow{r_r} \text{gas}$$

$$C_nH_x - * \xrightarrow{r_C} C - *_n \rightarrow [\text{Ni, C}] \rightarrow \text{C-whisker}$$

$$C_nH_x - * \xrightarrow{r_p} CH_z - * \rightarrow \text{"}C - *\text{" unreactive}$$

The rate of self-poisoning by "polymer" or "gum" formation is then

$$r_p = r_A - r_H - r_c \qquad (77)$$

The relative size of r_A, r_H, and r_c may result in different behavior. At normal operation r_c is zero and self-poisoning, resulting in progressive deactivation, characteristic for low temperature adiabatic reforming processes [40, 41] will be observed when r_A is larger than r_H. The activation energy for hydrocarbon adsorption [308] (r_A) on nickel (around 40 kJ mol^{-1}) is smaller than the activation energy for hydrocracking (r_H) of hydrocarbons on nickel (160–260 kJ mol^{-1}, depending on the hydrocarbon [309]). As a result r_H should become larger than r_A above a certain temperature T_p and self-poisoning should not occur. This reflects the situation typical for tubular reformers where stable performance is possible even under conditions for slow deposition of carbon (step (75)) [7, 281].

Pyrolytic carbon may be formed when unconverted hydrocarbons may pass to the hotter part of the reformer tube (*i.e.*, $T > 920$ K). This is unlikely with an active catalyst (*c.f.* Figure 32), but it can be provoked by sulfur poisoning of the catalyst.

Therefore, when carbon formation problems occur in a tubular reformer, the carbon is normally of the whisker type. It is evident that (apart from short upsets) conditions where whisker carbon is formed cannot be tolerated because of the consequent break-down of the catalyst pellets, increasing pressure drop and development of "hot tubes" (*c.f.* p. 29). The important problem is whether or not carbon is formed and not the rate. Therefore, the following chapter will discuss criteria for carbon-free operation.

B. Criteria for Carbon-free Operation

1. Carbon Formation by Reversible Reactions

1.1. Principle of Equilibrated Gas

Carbon formation can take place by the following reversible reactions

$$2\,CO \rightleftharpoons C + CO_2 \qquad (-\Delta H^0_{298} = 173\ \text{kJ mol}^{-1}) \tag{78}$$

$$CH_4 \rightleftharpoons C + 2\,H_2 \qquad (-\Delta H^0_{298} = -75\ \text{kJ mol}^{-1}) \tag{79}$$

This means that for a *fixed* gas composition of H_2, H_2O, CO, CO_2, and CH_4 there is a temperature, T_B, below which there is a thermodynamic potential (affinity) for the exothermic Boudouard reaction (78), and a temperature, T_M, above which there is an affinity for carbon formation by the endothermic decomposition of methane, reaction (79).

When a catalyst is present it is necessary to consider also the reforming and water gas shift equilibria (reaction (2) and (3)). The risk of carbon is then normally evaluated by means of the so-called *principle of equilibrated gas* [7, 85] which states:
Carbon formation is to be expected if the gas shows affinity for carbon after the establishment of the methane reforming and the shift equilibria.

Since the gas is at equilibrium it is sufficient to consider one of the two carbon-forming reactions (78) and (79). The principle is no law of nature as illustrated below. It is merely a rule of thumb, indicating process conditions which are critical for carbon formation.

According to the principle, the potential for carbon formation, $-\Delta G_e$, for a given feed gas composition should be calculated for the equilibrated gas to each temperature in the reactor, thus

$$-\Delta G_e = RT \ln (K_p/Q_{R,e}) \tag{80}$$

These calculations result in carbon limits, which may be expressed as upper and lower carbon limit temperatures, T_U and T_L, above or below which there is affinity for carbon. T_L and T_U are functions only of the atomic ratios O/C, H/C, and inert/C in the process stream and of the total pressure. The thermodynamic calculations are complex and are normally carried out by computer [18, 310].

The thermodynamic data to be used for calculations involving the carbon-forming equilibria (78) and (79) are influenced by the carbon modification involved [311, 277] as shown in Figure 44 for the Boudouard reaction (78). The equilibrium constant observed on the catalyst is smaller than that based on graphite. A similar result [277] was obtained for methane decomposition (79). This means that higher contents of carbon monoxide or methane are allowed before carbon formation. This effect can be ascribed to the whisker-

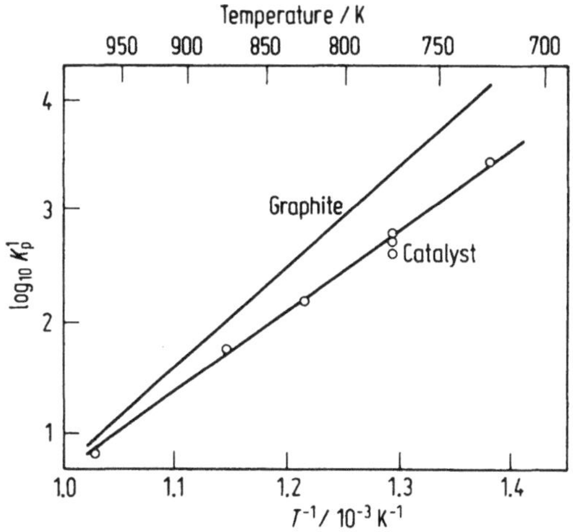

Figure 44. Decomposition of carbon monoxide on nickel catalyst. Temperature dependency of equilibrium constant. Thermogravimetric studies on Ni/$MgAl_2O_4$ catalyst K_p^1 calculated from p_{CO} and p_{CO_2} at equilibrium and for $a_c = 1$. K_p^1 for partial pressures in Mpa ($K_p \mp 0.1033 K_p^1$) (Reproduced with permission from ref. [7, 277])

like structure of the carbon formed on the catalyst [277]. The contribution from the surface energy compares with the observed deviation. The effect is favored by small nickel crystals [277] as the whisker diameter is close to that of the nickel crystal (*c.f.* Figure 35). The surface energy increases with decreasing whisker diameter and hence nickel particle size.

In a simplified model [277] assuming the whisker to be an infinite cylinder, the Kelvin equation becomes

$$\mu - \mu_0 = \frac{8M}{R_W \varrho} \tag{81}$$

The deviation from graphite data ΔG_C may be expressed by

$$\Delta G_C = \mu - \mu_0 + \mu^* \tag{82}$$

where μ^* is a contribution from structure defects compared to graphite. With $R_W = d_{Ni}/2$, equations (81) and (82) yield

$$\Delta G_C = \frac{a}{d_{Ni}} + b \tag{83}$$

Figure 45 shows a plot of equation (83) on data for CH_4 decomposition on various catalysts [277]. Thus, the largest deviations from graphite data are observed on catalysts with small nickel crystallites. The chemical composition of the catalyst appeared to have no influence on the observed equilibrium data [277]. ΔG_C depends on temperature [277] and at high temperatures the deviation from graphite data becomes insignificant.

Figure 46 shows a typical example of carbon limit temperatures as functions of O/C and H/C. The risk of carbon formation is reduced by increasing the steam-to-carbon ratio, and no carbon is expected at p_{H_2O}/p_{CH_4} higher than 1.2. This means that the principle will not predict carbon formation for normal steam reforming operations for ammonia, methanol or hydrogen. However, there may be a problem at conditions of O/C close

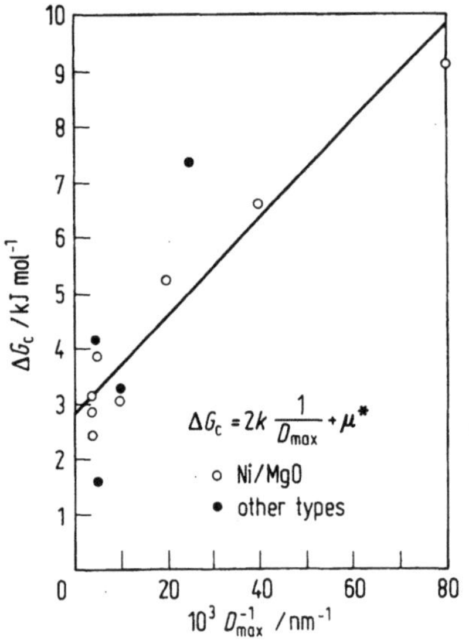

Figure 45. Deviation from graphite data and nickel crystal size. CH_4 decomposition on various catalysts in thermogravimetric studies. (Reproduced with permission from ref. [7, 277])

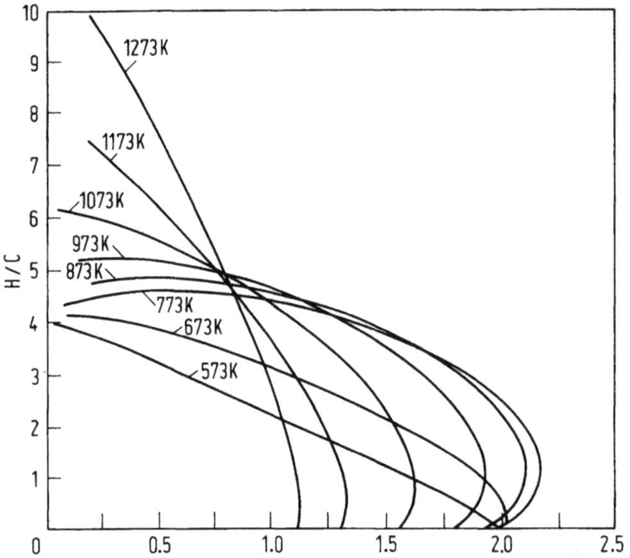

Figure 46. Carbon limits on nickel catalyst from principle of equilibrated gas. Example 1: O/C = 0.3, H/C = 0.6: Carbon formation >1073 K = T_U. Example 2: O/C = 1.53, H/C = 2.06: Carbon formation for T_U = 673 K < T < 1073 K = T_L.
The carbon limit temperatures T_U and T should not be confused with T_B and T_M. T_B and T_M were calculated for a given gas composition which was heated up or cooled down *without* reaction, whereas T_U and T_L imply equilibration of gas at all temperatures (see text for explanation of symbols)

to 1, this being optimum for the manufacture of reducing gas, or at low H/C as used for oxosynthesis (*c.f.* Table 3).

The principle of equilibrated gas is justified for tubular steam reforming. First, the feed gas arrives rapidly very close to equilibrium (*c.f.* Figures 11 and 32). Second, the effectiveness factor of the methane reforming reaction is less than 0.1 (*c.f.* Figure 30 and 31). This means that the gas in most of the catalyst pellet is at equilibrium. It has been indicated [312] that different diffusion rates of the individual gas components may results in O/C and H/C ratios in the interior of the pellet being different from those in the gas phase and hence in other carbon potentials than predicted from the equilibrated bulk gas phase. However, this effect remains insignificant for normal steam reforming conditions.

A direct support for the principle of equilibrated gas was obtained from thermogravimetric studies [85, 223], determining the critical p_{H_2O}/p_{CH_4} (or p_{CO_2}/p_{CH_4}) for onset of carbon formation. Results are shown in Table 18. The calculated affinities for carbon formation on the basis of the equilibrated gas and graphite data result in a value of $-\Delta G_e$ less than 4 kJ mol^{-1} (except for one measurement) which is comparable with the deviation from graphite data, ΔG_C, to be expected from the effect of the whisker structure mentioned above. The results were similar when carbon dioxide replaced steam. The value of $-\Delta G_a$, calculated from the exit, *i.e.*, the non-equilibrated bulk gas, was 20–40 kJ mol^{-1}.

Table 18. Carbon formation and equilibrated gas[a]

Catalyst temperature/K	886	977	1085	935
H_2O/CH_4 (Critical)	1.2	1.1	0.88	1.75[b]
$p_{H_2}^2/p_{CH_4}$ actual gas/MPa	0.0034	0.0055	0,0161	0.0438
$p_{H_2}^2/p_{CH_4}$ equil. gas ($= Q_{R,e}$)/MPa	0.221	0.722	1.099	0.306
K_p^1 graphite data	0.2660	0.819	2.468	0.4939
$-\Delta G_e^0 = RT \ln (K_p/Q_{R,e})$/kJ mol^{-1}	1.6	0.9	(7.3)	2.3

[a] thermogravimetric studies at 0.1 MPa, ref. [85, 223]
[b] CO_2/CH_4
The CH_4 flow was measured stepwise until on-set of carbon formation. Then the CH_4 flow was decreased for removal of carbon. $(H_2O/C)_{crit}$ exit was determined by interpolation

1.2. Carbon Activity at Steady State

The principle of equilibrated gas is no law of nature. Rates of carbon formation may be too slow. On the other hand, carbon formation may occur in spite of the principle if the *actual* gas in the bulk phase, and thus in the exterior of the catalyst pellet, shows affinity for carbon formation ($\Delta G_a < 0$). If so, methane may decompose to carbon instead of reacting with steam

$$CH_4 \begin{cases} \xrightleftharpoons{H_2O} \text{Gas} \\ \rightleftharpoons \text{Carbon} \end{cases} \qquad (84)$$

Of course, this is not possible in a closed thermodynamic system, but in an open system carbon may be stable in a steady state and the accumulation of carbon may continue. Carbon formation is then a question of kinetics and the local approach to the reforming equilibrium.

The surface intermediate $CH_x - *$ may react either to gas *via* step (74) or to whisker carbon *via* step (75). Carbon will nucleate if the steady state activity (concentration) of carbon dissolved in the nickel crystal, a_c^s, exceeds the activity at saturation a_c^L. At equilibrium with a given composition at the gas phase (not necessarily an equilibrated gas), the carbon activity a_c^{eq} is expressed by

$$a_c^{eq} = K_{79} a_c^L \frac{p_{CH_4}}{p_{H_2}^2} \quad (85)$$

in which K_{79} is the equilibrium constant for reaction (79) (based on whisker carbon). It may be questionated whether a_c^2 represents the activity determined from solubility data. Schouten *et al.* [206] found that the nickel crystal could dissolve larger amounts of carbon when the crystal was not annealed. Although, in principle, a_c^{eq} should be referred to carbon dissolved in nickel, for simplicity it will be referred to graphite in the following.

Following the procedure by Williams *et al.* [313], the steady state activity, a_c^s, can be expressed by balancing the rate of carbon formation without the presence of steam with the rate of the gasification of adsorbed carbon atoms

$$r_c = k_c p_{CH_4} p_{H_2}^{\alpha'} - k_{-c} a_c p_{H_2}^{(2+\alpha')} = k_{-c} p_{H_2}^{(2+\alpha')} (a_c^{eq} - a_c) \quad (86)$$

$$r_g = k_g a_c [\text{mari}] = k_g a_c \frac{K_w (p_{H_2O}/p_{H_2})}{1 + K_w (p_{H_2O}/p_{H_2})} = k_g a_c K_w' \left(\frac{p_{H_2O}}{p_{H_2}}\right)^{\alpha} \quad (87)$$

in which [mari] is the surface concentration of the mari, *v.i.z.* $O - *$.

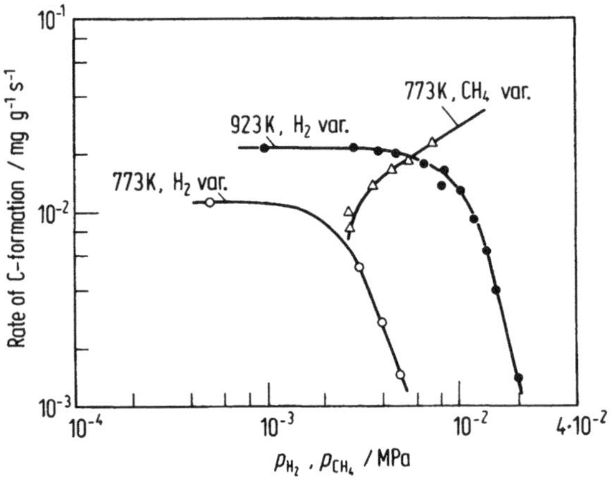

Figure 47. Influence of hydrogen on rate of methane decomposition. Thermogravimetric studies. Rates corrected for approach to equilibrium: $r_c = r_{c,obs}/(1 - Q_R/K_p)$ 0.7 g $Ni/MgAl_2O_4$ catalyst, $p = 0.1$ MPa, $p_{CH_4} = 2.9$ kPa. $p_{H_2} = 0$ when p_{CH_4} varied: balance N_2. [289])

Williams *et al.* [313] used data for r_c obtained on iron films. The data shown in Figure 47 were obtained on a Ni/MgO catalyst [289]. Hydrogen strongly retards the coking rate with α_{H_2} approaching —3.5 to —3.7. This can be understood in terms of the sequence (50) to (53) by assuming that the complete dissociation of the methane molecule into adsorbed carbon atoms followed by incorporation into the bulk phase requires a large ensemble $*_m$

$$CH_4 \rightarrow CH_x - *_n \rightarrow C - *_m \rightarrow [Ni, C] \rightarrow C \text{ whisker} \quad (88)$$

If so, r_c without steam present becomes

$$r_{c+} = \frac{k p_{CH_4}}{(1 + K \sqrt{p_{H_2}})^m} \qquad (89)$$

For $m = 7$ as suggested by Frennet *et al.* [195], for complete dissociation of methane, α_{H_2} approaches -3.5, which is close to the trend in Figure 47.

A simplified expression for a_c^s can be derived for a_c^s, *i.e.* when $r_g = r_c$. With $a_c = a_c^s$ and $\alpha' = -3.5$ equations (86) and (87) yield

$$a_c^s = \frac{a_c^{eq}}{1 + \frac{k_g}{k_{-c}}\left(K_w' \frac{p_{H_2O}}{p_{H_2}}\right)^{\alpha} p_{H_2}^{1.5}} \qquad (90)$$

carbon may be formed when $a_c^s > 1$. This means that carbon may be eliminated even when the actual gas shows affinity ($a_c^{eq} > 1$) for decomposition of methane (equation (79)).

The strong retarding effect [289] on coking rate from methane of small amounts of steam and carbon dioxide is demonstrated in Figure 48.

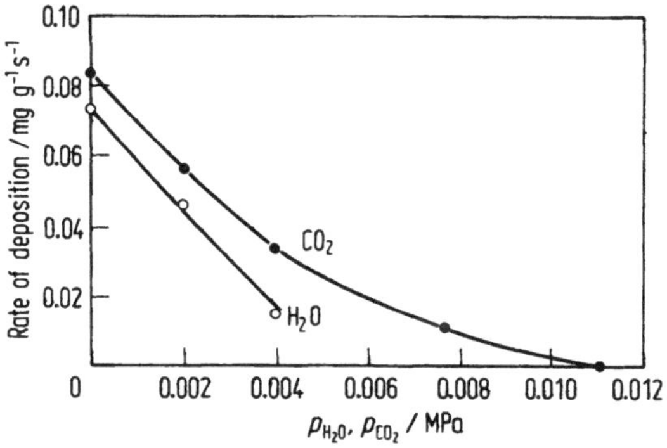

Figure 48. Influence of steam and carbon dioxide on methane composition. Thermogravimetric studies. Catalyst 0.7 Ni/MgO, $p = 0.1$ MPa. Composition of H_2O- and CO_2-free gas (vol%: H_2: 35, CH_4: 12, N_2: 53. [289]

A large value of K_w decreases the steady state activity a_c^s. This can be achieved by means of alkali, magnesia etc. (*c.f.* Table 12); however, the effect may vanish at high temperatures where K_w is smaller [85].

Another kinetic approach [20, 179] to the problem has been to consider bulk carbon as an intermediate

$$CH_4 \begin{cases} \rightarrow \text{carbon} \\ \quad \downarrow H_2O \\ \rightarrow \text{gas} \end{cases} \qquad (91)$$

Carbon is not formed when the rate of gasification of bulk carbon is greater than the rate of carbon formation. However, since bulk carbon is not present at coke-free operation, this approach appears awkward. If the argument refers to the adsorbed intermediates (carbon atoms), the two approaches are identical.

1.3. "Hot Tubes": Radial Gradients

One criterion for operation without carbon formation will be that the actual gas composition at any position in the reformer tube shows no affinity for methane decomposition. This is equivalent to disregarding the denominator in equation (90).

As mentioned above, the gas is close to reforming equilibrium after the upper 1–2 meters of the catalyst bed. However, this one-dimensional description does not consider the significant radial temperature gradients in the catalyst bed resulting from the high heat transfer rate. The radial temperature gradient may cause a significant deviation from the reforming equilibrium close to the tube wall (*c.f.* Figure 50).

In a simplified approach the situation corresponds to heating the average gas composition from the average gas temperature to the temperature at the wall without reaction. Thereby, the gas may be heated to a temperature above T_M (*c.f.* p. 82) above which methane may crack into carbon.

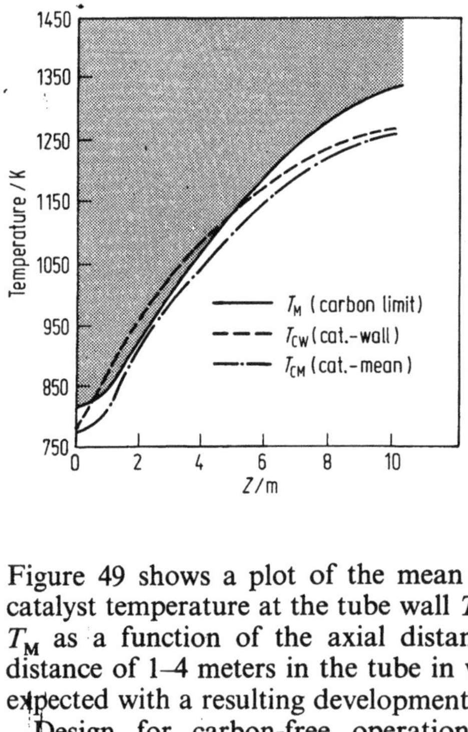

Figure 49. Carbon limit and catalyst temperature. Steam reforming for reducing gas. p_{H_2O}/p_{CH_4} = 1.3, p_{exit} = 0.4 MPa, q_{av} 72 kw m^{-2}. Potential for carbon for $T_{cat} > T_M$. (Reproduced with permission from ref. [85])

Figure 49 shows a plot of the mean temperature of the catalyst T_{CM}, the catalyst temperature at the tube wall T_{CW}, and the carbon limit temperature T_M as a function of the axial distance. The plot shows $T_{CW} > T_M$ at a distance of 1–4 meters in the tube in which zone carbon formation is to be expected with a resulting development of a "hot band" (*c.f.* p. 29).

Design for carbon-free operation should therefore aim at keeping $T_{CW} < T_M$. This can be done by decreasing $T_{CW} - T_{CM}$ in the critical zone, which is equivalent to decreasing the heat flux. This illustrates the importance of the reformer design for coke-free operation.

A more quantitative approach requires the use of a two-dimensional pseudo-homogenous model (*c.f.* p. 26), describing the mass and heat transport in the reformer tube.

Typical radial profiles for an ammonia plant reformer are shown in Figure 50, which corresponds to the data in Figure 34 for the less active catalyst. The catalyst temperature exceeds the carbon limit temperature T_M close to the tube wall ($T_{CW} > T_M$) which means a potential for carbon formation by methane decomposition. With the highly active catalyst the temperature level in the reformer is lowered and calculations similar to those in Figure 50 show no potential for carbon formation.

Figure 50 shows that T_M is nearly constant with radial distance, which reflects nearly constant gas composition over the tube cross section. This is due to the high Reynold's number. In reducing gas plants, operating at low pressure (*c.f.* Table 3) and low Reynolds number, significant radial concentration gradients may develop. This is illustrated in Figure 51. The

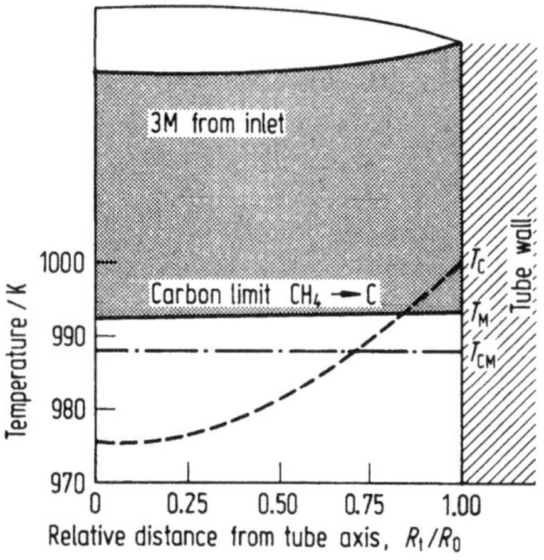

Figure 50. Radial gradients and carbon formation. Ammonia plant reformer at brosted capacity. Conditions as Figure 34. Standard catalyst 3 m from inlet. Potential for carbon formation for $T_{cat} > T_M$

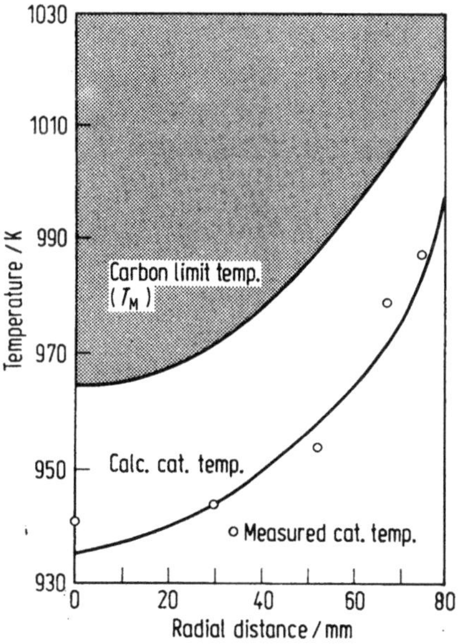

Figure 51. Radial gradients and carbon formation reducing gas conditions. $p_{H_2O}/p_{CH_4} = 1.35$, $p_{exit} = 0.44$ MPa, $T_{exit} = 1073$ K, $Re_{inlet} = 1000$, 4 m from inlet. The points show measurements from monotube pilot plant, whereas curves result from calculations with a two-dimensional pseudo-homogeneous model. The catalyst temperature stays below the carbon limit temperature, T_M, and there is no risk of carbon formation. [86]

catalyst is able to decrease the méthane content close to the tube wall, and hence T_M increases towards the tube wall and the potential for carbon formation is eliminated.

In conclusion, the radial concentration and temperature gradients in a tubular reformer may create local areas with potential for carbon formation, although not predicted by the principle of equilibrated gas. Therefore, design of reformers for high heat flux requires sophisticated calculations and availability of highly active catalysts.

2. Carbon Formation by Irreversible Reactions

2.1. Catalyst Selectivity

If the feed gas contains higher hydrocarbons the reaction scheme (84) is replaced by the irreversible reactions

$$C_nH_m \begin{array}{l} \xrightarrow{H_2O,\,H_2} \text{gas} \\ \longrightarrow \text{carbon} \end{array} \tag{92}$$

The hydrocarbon may decompose into carbon instead of reacting with steam, although thermodynamics in terms of the principle of equilibrated gas predicts no carbon formation. The actual gas has a potential for carbon formation as long as unconverted higher hydrocarbons are present, *i.e.* a_c^{eq} remains larger than one. Therefore, the conservative criterion applied above for the steam reforming of methane (p. 88) would predict carbon formation. The elimination of carbon can be obtained by depressing the steady state activity a_c^s as expressed by equation (90) for steam reforming of methane.

For higher hydrocarbons equation (90) is modified. Step (88) is replaced by

$$\begin{aligned} &C_nH_m + 2* \xrightarrow{k_A} [C_nH_z - *_2] + \frac{m-z}{2} H_2 \\ &[C_nH_z - *_2] + (nm-2)* \xrightarrow{k_c} n[C - *_m] + \frac{z}{2} H_2 \\ &\qquad \rightarrow [Ni, C] \rightarrow C, \text{ whisker} \\ &[C_nH_z - *_2] + n* \xrightarrow{k_H} [CH_x - *_n] + [C_{n-1}H_{z'} - *_2] + \frac{z - z' - x}{2} H_2 \end{aligned} \tag{93}$$

Hydrogen may be a reactant in the latter step. In a simplified creatment, considering the adsorbed hydrocarbon radical and empty sites only, and assuming $k_H \gg k_c$, r_c (without steam present) can be approached by $r_c = k_c \frac{k_A}{k_H} p_{C_nH_m} p_{H_2}^{\alpha}$, whereas r_g can be obtained from equation (87). On this basis a_c^s or the risk for carbon formation can be expressed by

$$a_c^s \approx \frac{k_A}{k_H} \frac{k_c}{k_g K_w^{\alpha''}} \frac{p_{C_nH_m}}{p_{H_2O}^{\alpha''}} p_{H_2}^{\alpha} \tag{94}$$

Carbon may be formed for $a_c^s > 1$, and carbon formation is then a question of kinetics and hence of the catalyst selectivity for the gasification step.

The risk of carbon formation depends on the type of hydrocarbon, the catalyst and of course the process conditions. The data in Figure 36 illustrated that the risk of carbon formation strongly depends on the unsaturated character of the hydrocarbon. In terms of equation (93) ethylene and benzene chemisorb more easily (high values of k_A) and hydrocrack less easily (low values of k_H) than do paraffins. In conclusion, the content of aromatics is a most critical parameter to evaluate for carbon-free operation in steam reforming of naphtha.

The influence of the catalyst is related to the value of k_g and K_w, expressing the catalyst activity and the steam adsorption [7], respectively. It has been speculated whether the crystal size may play a role as well. For given catalyst composition (*i.e.* K_w), Borowiecki [314] obtained results which indicated that the carbon formation was depressed when using smaller nickel crystals. The spill-over of water from the support was assumed to be less effective on large nickel crystals.

Normal alumina-based catalysts for steam reforming of natural gas have low values of K_w and cannot operate with naphtha.

One solution has been to promote catalysts with alkali [20, 179]. Steam adsorption can be significantly improved by the addition of alkali, which results in a high value of K_w and consequently negative reaction order with respect to steam (*c.f.* Table 12 and Figure 28). However, this effect is partly lost by a drastic decrease of the catalyst activity k_g by the addition of alkali (*c.f.* Figure 29). Therefore, the naphtha conversion over alkali-promoted catalysts in tubular reformers will start later and be completed at temperatures of about 920 K at which the reactions on the surface are accompanied by the cracking reactions in the gas phase.

Another solution has been the use of active magnesia for enhanced steam adsorption [7, 234]. A moderate value of K_w is obtained at the same time as a high value of k_g is maintained due to the absence of alkali.

It has been argued [315] that steam reforming of naphtha proceeds *via* the formation of carbon

$$C_nH_m \xrightarrow{1} \text{carbon} \xrightarrow[H_2O]{2} \text{gas} \tag{95}$$

meaning that the rate of carbon formation is $r_1 - r_2$. A good catalyst should have high activity for gasification (r_2) and apparently low activity for the break-down of the hydrocarbon (r_1). The effect of alkali should be to speed up step 2 in reaction (95). This mechanism has difficulties in explaining the function of an alkali-free catalyst [7, 20] and it finds little support in experiments on carbon formation on nickel. Bernardo [316] and Figueiredo *et al.* [237] studied the rates of carbon formation and gasification at 823 K on the same catalyst and found that the rate of carbon formation calculated as $r_1 - r_2$ exceeded significantly the observed rate of carbon formation at steam reforming conditions.

Moreover, Trimm *et al.* [9, 317] suggested that nickel-catalysed steam gasification of carbon proceeded at the same rate as diffusion of carbon through nickel, which implies a mechanism being the reverse of the whisker growth mechanism (*c.f.* section 5.A.2.). This was confirmed by *in situ* studies by electron microscopy by Baker *et al.* [273, 284, 318]. It appears unlikely that promoters (like alkali) act directly as catalysts for the gasification since the rate determining step is diffusion of carbon through nickel. It is more likely that the promoters interfere with the reaction chain leading to surface carbon [7, 9] as described in sequence (73) to (75) rather than reacting with coke already produced.

The promoting role of alkali and magnesia in steam reforming on nickel catalysts should probably be ascribed to their ability to increase the steam adsorption. The well known effect of alkali on the gasification of carbon [319, 320] is hardly involved in normal steam reforming, although alkali may work at high temperatures as a reforming catalyst by the regasification model *via* carbon formed by thermal cracking [179].

Alkali has been claimed [321] to stabilize CH_x—* and to retard its conversion to C—*. This assumption may be supported by studies of Cimino *et al.* [252] of hydrogenolysis on alkali-promoted iron. Moreover, steam gasification in the presence of alkali results in high methane yields [320]. At steam reforming conditions, exchange studies by Ross *et al.* [232] on a coprecipitated catalyst (containing alkali) showed deuterium exchange with the adsorbed hydrocarbon in contrast to results on an impregnated (alkali free) catalyst (*c.f.* p. 53 and p. 62). This might indicate the presence of a weaker adsorption of the hydrocarbon in the presence of alkali. The stabilization of reaction intermediates may be a result of an electron ligand effect by alkali [254, 255] (*c.f.* p. 65). However, further studies are required to explore whether this effect has any substantial influence on the mechanism of carbon formation in steam reforming. Thus, alkali has no significant influence on coking rates in the absence of steam [281, 285] (*c.f.* p. 75).

2.2. Critical Steam to Hydrocarbon Ratio

In an empirical approach, carbon may be expected below a certain *critical* steam-to-hydrocarbon ratio. At this ratio the induction period, t_0, (*c.f.* equation (71)) becomes so small that carbon formation can be expected. Assuming semilogarithmic dependency of t_0 with reciprocal temperature and with steam-to-carbon ratio as indicated in Figures 40 and 41, the critical ratio can be expressed by

$$(p_{H_2O}/p_{C_nH_m})_{crit} = \frac{a}{T} + b \qquad (96)$$

in which a and b are positive constants depending on catalyst type and hydrocarbon.

Hence, the critical ratio was found to increase rapidly with the temperature and to be influenced by the type of hydrocarbon and by the catalyst. This means that for given process conditions, feedstock and catalyst, there is a temperature T_p, below which self-poisoning will occur (*c.f.* p. 81) and a temperature T_C, above which whisker carbon is formed. The hydrocarbons (naphtha) should be converted in the gap between these two temperatures to ensure trouble-free operation.

In tubular reformers, the *actual* steam-to-hydrocarbon ratio increases with the distance from the inlet as the hydrocarbon is being converted [216, 234]. The actual profile of this ratio depends on the partial pressures of reactants, space velocity, temperature profile, reactivity of hydrocarbon, catalyst activity and poisoning level of the catalyst.

If the actual ratio is lower than the critical ratio, carbon formation is expected as shown in Figure 52. The critical and actual steam-to-hydrocarbon ratios have been plotted as a function of the average catalyst temperature. However, as described above (section V.B.1.3.) this one-dimensional description of a reformer is too simplified. Again, the radial temperature gradients result in higher potential for carbon formation at the tube wall [234]. The critical ratio increases significantly when approaching the tube wall because of the increasing temperature, whereas the actual ratio remains nearly constant because of the small concentration gradients.

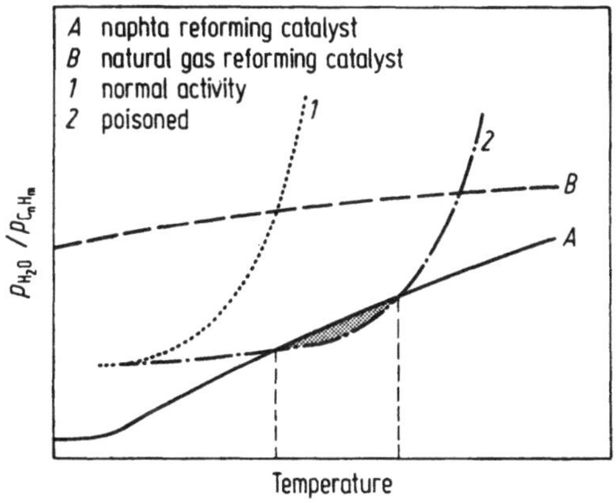

Figure 52. Critical and actual steam-to-hydrocarbon ratio. Tentative plot of temperature dependence. Potential for carbon formation when the actual ratio is lower than the critical ratio. (Reproduced with permission from ref. [234])

With a smaller heat flux the overheating of the average gas at the tube wall is less, which results in smaller potential for carbon formation. This situation changes with the actual position and is most critical for steam/naphtha reforming in the upper part of the tube. Therefore, in order to operate with a high average flux it is important to optimize the heat flux distribution.

C. Regeneration of Coked Catalyst

The discussion has shown that formation of bulk carbon can have different effects on the operation of a tubular reformer.

Rapid formation of whisker carbon can result in spalling of the catalyst and accumulation of carbon. The spalling can result in break-down of the outer part of the particle into powder. The result is an increasing pressure drop and the development of hot tubes.

However, if whisker carbon is formed at a very slow rate, significant amounts may build up in the catalyst without harming the performance of the reformer. This is illustrated in Figure 53 showing analyses of the radial distribution of coke in a ring-shaped catalyst pellet, which was exposed to coking in a pilot test [7, 281]. The data show a strong accumulation of coke close to the external surface, which is not surprising in view of the low effectiveness factor for the reforming reaction at industrial conditions (*c.f.* Figure 30). The catalyst in Figure 53 was able to operate with no sign of deactivation nor increased pressure drop [281].

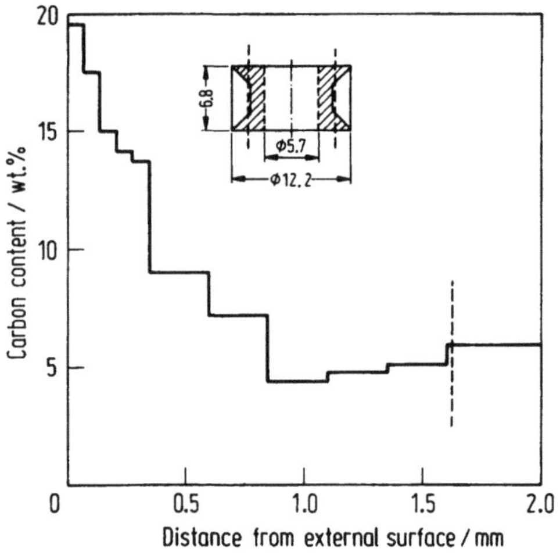

Figure 53. Radial carbon profile in a ring-shaped steam reforming catalyst. Samples for analysis were taken with a lathe as indicated. (Reproduced with permission from ref. [7, 281])

The formation of pyrolytic carbon can result in coke deposits at the tube wall and consequently a reduced heat transfer coefficient, which may cause the development of a "hot band". The catalyst particles can be encapsulated, which results in increased pressure drop. However, the encapsulation does not harm the catalyst particles.

Therefore, it is possible in many situations to regenerate the catalyst and to re-establish satisfactory performance of the reformer. The whisker carbon is highly reactive and may be gasified by means of hydrogen, carbon dioxide, and steam with nickel as catalyst [317, 318]. Steam is the most effective gasifying agent [322]

$$C + H_2O \rightleftharpoons CO + 2\,H_2 \quad (-\Delta H^0_{298} = -132\ \mathrm{kJ\ mol^{-1}}) \qquad (97)$$

Therefore, non-aged whisker carbon may be removed simply by increasing the steam-to-carbon ratio [318, 7], at the same time maintaining the catalyst in a reduced state.

However, the whisker structure will collapse with time and be converted into a more dense layer of carbon, which may be difficult to remove under reducing conditions. Regeneration in steam can be carried out at temperatures around 870–970 K or higher depending on the ageing of the deposits [7, 9].

However, with the addition of a small percentage of air the burn off of carbon is easily performed at a temperature above *ca.* 720 K [7]. The addition of air should be well controlled to minimize the local overheating of the catalyst caused by the heat produced from oxidation of the nickel and the carbon.

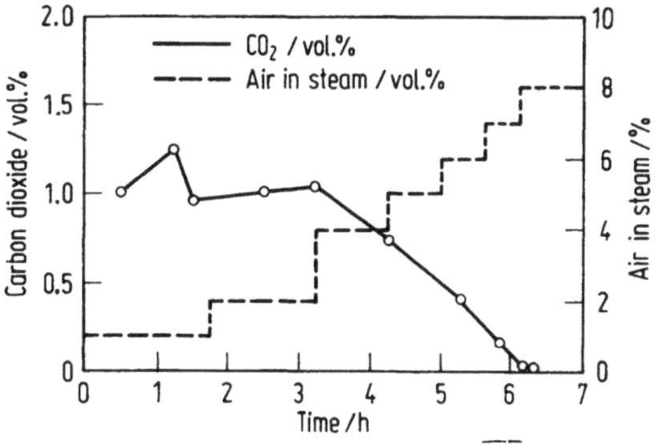

Figure 54. Regeneration for carbon formation in industrial reformer. Analyses of dry exit gas. *Ca.* 0.4 ton of steam per ton of catalyst per h. T_{cat} = 723–873 K. p = *ca.* 0.6 MPa. (Reproduced with permission from ref. [323])

In an industrial plant the progress of the regeneration can easily be followed by analysis of the exit gas from the reformer. An example [323] is shown in Figure 54. The addition of air is increased as the production of CO_2 decreases. In this way overheating can be controlled.

6. Impact of Sulfur on Steam Reforming

A. Sulfur Uptake

1. Adsorption Isotherms

Sulfur is a severe poison for steam reforming catalysts because sulfur compounds are strongly chemisorbed on the metal surface. The catalysts may deactivate (*c.f.* Table 15) as a result of small impurities of sulfur compounds present in the reactants or incorporated in the catalyst during its preparation. Normally, sulfur will be present in the feedstock which can be purified effectively over activated zinc oxide with or without preceeding hydrogeneration over a sulfided Co—Mo catalyst.

Poisoning effects are often correlated with the poison concentration in the feed steam, which, of course, is the important parameter in practical operation. However, in a more detailed analysis this approach can hardly be justified for other than isothermal tests in gradientless reactors. The adsorption equilibrium depends on temperature and composition of the gas phase, which varies through the reactor as well as within the single catalyst pellet. Therefore, it appears more rational to correlate the deactivation with the amount of poison present on the catalyst rather than with the poison concentration in the feed stream. However, the correlation between sulfur in the feed and in the catalyst may be complex as illustrated below.

Hydrogen sulfide chemisorbs dissociatively on nickel (*c.f.* p. 43)

$$H_2S + * \leftrightarrows S - * + H_2 \tag{98}$$

Stable saturation uptakes of sulfur are observed at values of p_{H_2S}/p_{H_2} from $1\text{–}10 \times 10^{-6}$ up to $100\text{–}1000 \times 10^{-6}$ above which bulk sulfide is formed. The saturation layer was described earlier (p. 43). Below a certain p_{H_2S}/p_{H_2} the saturation layers becomes unstable and the equilibrium coverage is dependent on the p_{H_2S}/p_{H_2} and temperature. This is normally described in terms of an adsorption isotherm and the isosteric heat of chemisorption.

A number of recent investigations [174, 145, 324] have found the isosteric heat of adsorption to be approximately 155 kJ mol^{-1}, decreasing slowly at high coverages. This value should be compared to the heat of formation for bulk sulfide (Ni_3S_2) being *ca.* 75 kJ mol^{-1} of S. This demonstrates that the adsorbed sulfur is strongly bound to the surface of nickel and that the "two-dimensional sulfide" can be stable at conditions where bulk sulfide does not exist.

However, attempts [158, 162, 174, 145] have not been successful in correlating data with a Langmuir-like isotherm

$$\frac{\theta}{1-\theta} = B \exp(-\Delta H^0_{ads}/RT)\, f(Y) \tag{99}$$

where for Langmuir adsorption, $f(Y) = Y = p_{H_2S}/p_{H_2}$.

This failure is not surprising in view of the mechanism of the adsorption, violating the assumptions for the Langmuir isotherm (*c.f.* p. 43).

Alstrup *et al.* [324] suggested the following Temkin-like isotherm

$$Y = \exp(\Delta H^0_0(1 - a\theta_s)/RT - \Delta S^0/R) \tag{100}$$

which was used to correlate data obtained at very different conditions as shown in Figure 55. Equation (100) implies an entropy of adsorption being independent of θ, which is in accordance with a "bulk"-like behavior of the chemisorbed layer (*i.e.* a "two-dimensional" sulfide).

The isobars in Figure 55 show straight lines in the Y versus T plot. The deviations at low temperature may be explained by adsorption on the support [324, 178]. The plotted lines correspond to the constants $-\Delta H^0_0 = 280$ kJ mol^{-1}, $\Delta S^0 = -19$ J mol^{-1} K^{-1} and $a = 0.69$, which result in the equation [325]

$$\theta_s = 1.45 - 9.53 \times 10^{-5} T + 4.17 \times 10^{-5} T \ln Y \tag{101}$$

This expression is not valid for θ_s close to zero and close to 1.0. However, it fits reasonably with other published data [145, 158, 174]. It is evident, however, that additional data, in particular for low coverages, are required for a more detailed description.

In principle, it should be expected that reaction (98) is influenced by the competing chemisorption reaction

$$* + H_2O = * - O + H_2 \tag{102}$$

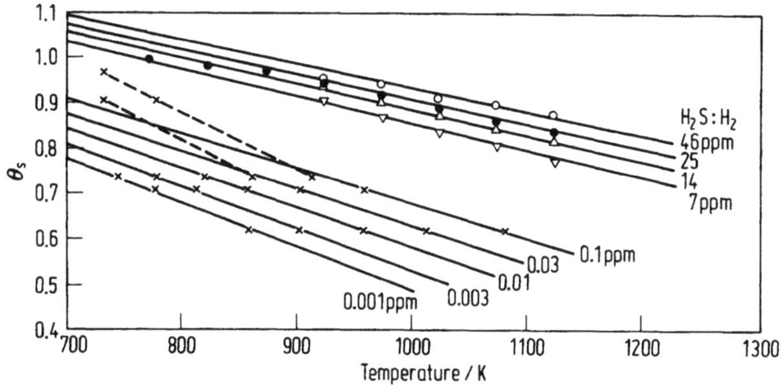

Figure 55. Isobars for chemisorption of hydrogen sulfide on nickel catalysts. The curves were calculated by equation 101. (Reproduced with permission from ref. [324])

However, experiments [185] in the range 823–1123 K, in which p_{H_2O}/p_{H_2} was varied from 0 to 10 and p_{H_2S}/p_{H_2} from 0.2×10^{-6} to 930×10^6, showed no significant influence of the p_{H_2O}/p_{H_2} ratio. This result may appear surprising as the heat of chemisorption of oxygen [308] is high (*ca.* 430 kJ mol^{-1} of O_2). However, this value is less than the heat of chemisorption for sulfur which is about 480 kJ mol^{-1} of S_2 (as calculated from ΔH^0_{ads} for H_2S).

2. H_2S Chemisorption at Industrial Conditions

The distribution of sulfur on the catalyst in a tubular reformer is complex even when chemisorption equilibrium is established. This is due to significant axial and radial gradients of temperature and of hydrogen partial pressure as well as gradients inside the pellets. However, if radial gradients and pellet gradients are neglected, axial profiles of sulfur coverage at equilibrium may be estimated on basis of equation (101). As an example some calculated profiles [325] for a typical natural gas based ammonia plant reformer are shown in Figure 56. It was assumed for simplicity that the axial profiles of temperature and hydrogen partial pressure were constant for all sulfur contents in the feed. It is evident that the poisoning will influence the hydrogen and temperature profiles as well.

The data in Figure 56 demonstrate the influence of temperature and hydrogen partial pressure on the equilibrium coverage. It is clear that there is no non-zero fixed limit of sulfur content in the feed below which poisoning will not occur. The usual sulfur limit quoted in literature and elsewhere (less than 0.2 ppm S) stems from the mid-sixties where this was the analytical limit for the methods available at that time. With a good desulfurization system sulfur contents much lower can easily be obtained. Hence the reaction

$$ZnO + H_2S \rightleftharpoons ZnS + H_2O \ (-\Delta H^0_{298} = 75 \text{ kJ mol}^{-1}) \qquad (103)$$

shows an equilibrium constant at 573 K, $K_p = 5.9 \times 10^6$. For high-quality zinc oxide absorption masses it is possible also to establish the chemi-

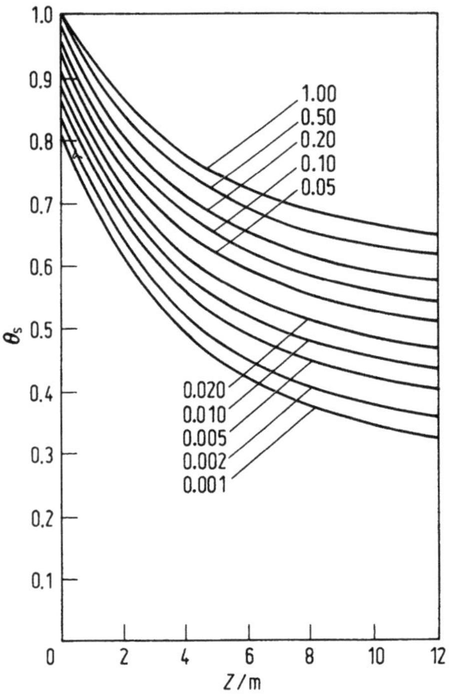

Figure 56. Sulfur coverage of nickel surface of catalyst in tubular reformer. Ammonia plant reformer $p_{H_2O}/p_{CH_4} = 3.3$, $p_{exit} = 3.4$ MPa, $T_{inlet}/T_{exit} = 778/1073$ K. Equilibrium profiles calculated from equation 101 are shown for different sulfur contents (vol ppm) in natural gas. (Reproduced with permission from ref. [325])

sorption equilibrium corresponding to p_{H_2O}/p_{H_2S} one to two orders of magnitude below data for the bulk phases. This means that with a light feedstock sulfur contents could be decreased to less than 10 ppb.

3. Dynamics of Poisoning

The chemisorption of hydrogen sulfide on nickel is very rapid with a sticking coefficient of close to 1.0 for less than 70% of a full monolayer [162]. This suggests no barrier to adsorption and dissociation until saturation of the surface is approached. However, in industrial reactors diffusion restrictions are present for the chemisorption process. Therefore, the rate of poisoning should be evaluated in terms of fixed bed adsorption, and poisoning models neglecting the pore diffusion [326, 327] are of less relevance for industrial conditions.

The pore diffusion restrictions in the sulfur adsorption in a single pellet has a complex influence on the transient sulfur profiles in the reactor and a mathematical model is required to evaluate more exactly the time for full saturation and the break-through curves of sulfur. A model based on a Langmuir-type adsorption was described by Christiansen *et al.* [328]. The transient profiles are calculated with fixed conversion and fixed axial temperature profile. For the tubular steam reforming reaction this represents an acceptable approximation since a decrease in catalyst activity caused by the sulfur poisoning will be compensated for by increasing the tube wall temperature in order to keep the outlet conditions fixed. The model also assumes that the temperature and the concentration of hydrogen do not vary in the catalyst particle.

The model has been improved by introducing the Temkin-like isotherm [325] (equation (101)), and Figure 57 shows calculated break-through curves in a steam reformer of a naphtha-based ammonia plant. The relative concentration of hydrogen sulfide at the tube exit is plotted as function of time. It is seen that the break-through occurs immediately for sulfur contents in the naphtha higher than 0.1 wt. ppm, indicating the serious effect of a sulfur peak in the reformer feed on downstream catalysts.

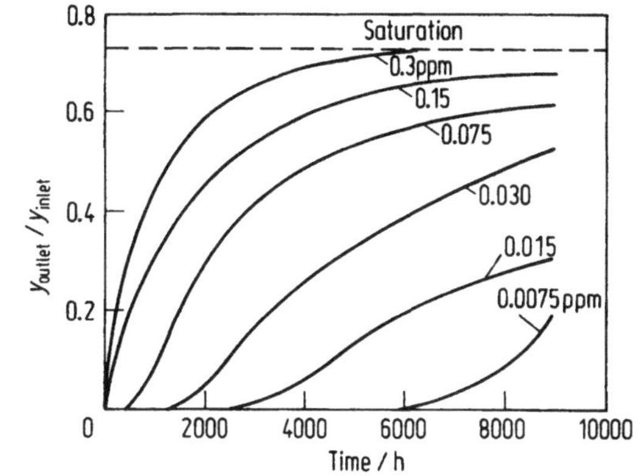

Figure 57. Break-through curves for hydrogen sulfide. Model for transient profiles in poisoning of tubular reformer. Naphtha based ammonia plant. $H_2O/C = 3.5$, $p_{exit} = 3.4$ MPa, $T_{exit} = 1073$ K, $G_M = 46{,}400$ kg m^{-2} h^{-1}. Catalyst: 16/6.5–16 mm rings, $s_0 = 1500$ wt. ppm. Curves for different sulfur contents in naphtha (wt. ppm). (Reproduced with permission from ref. [325])

Figure 58 shows catalyst poisoning profiles in the tubes. This is a plot of the average sulfur coverage as a function of the axial distance with time as a parameter. It is apparent that there is no front of sulfur moving through the bed and that the approach to saturation is slow, since the main part of the sulfur passes through the bed without adsorption.

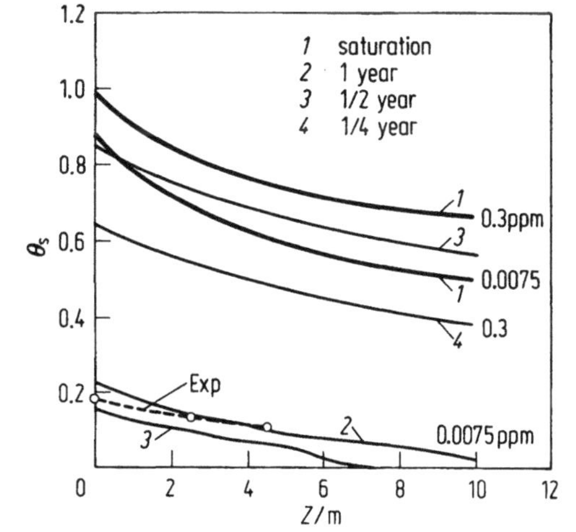

Figure 58. Axial profiles of sulfur uptake on catalyst. Model for transient profiles in poisoning of tubular reformer. Conditions as Figure 57. Curves for different sulfur content in naphtha feed (wt. ppm). Dotted line: measured data on used catalyst (Table 19). (Reproduced with permission from ref. [325])

Figure 59 shows calculated profiles in a catalyst pellet. A core/shell behavior of the poisoning is apparent.

The data on used catalyst shown in Table 19 were obtained from a plant which had operated steadily for one year [7] at conditions close to those used in the calculation example. The catalyst had been exposed to sintering during operation as reflected by the calculation of the sulfur coverages. The average sulfur contents are plotted in Figure 58 and the data fit well with the calculated data for 0.0075 wt. ppm sulfur in the feed.

Table 19. Sulfur contents of steam reforming catalyst from industrial ammonia plants[a]

Distance from inlet. of cat bed/m	Sulfur coverage, s/s_0 [b]			Sulfur capacity[b], s_0/wt. ppm	Nickel area/ m^2
	mean	external layer	core		
0	0.19	0.47	0.14	1868	4.25
2.5	0.14	1.37	0.10	1174	2.67
4.5	0.11	0.22	0.11	1032	2.35

[a] obtained after one year of stable operation, corresponding to the example in Figures 57–59 [7]

[b] One catalyst ring from each sample was divided into 3 parts for the following analyses:
1. mean sulfur content,
2. sulfur contents of external layer (about 10 wt. %) and of remaining core,
3. determination of sulfur capacity (Ni area)

Even after one year's operation a significant radial gradient is present in the catalyst particles at the inlet (Table 19). This may partly be ascribed to a higher hydrogen partial pressure inside the pellet because of the low effectiveness factor of the reforming reaction (*c.f.* p. 68). Results from a microprobe analysis are shown in Figure 59. Again, the data fit well with the calculated data for 0.0075 wt. ppm sulfur in the feed.

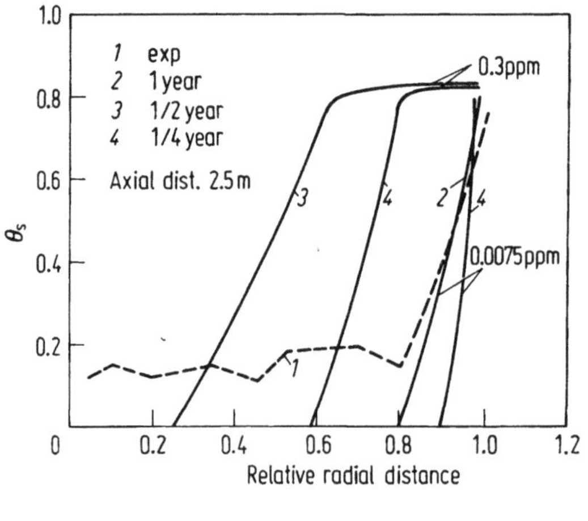

Figure 59. Sulfur gradients in catalyst pellet. Model for transient profiles in poisoning of tubular reformer. Conditions as Figure 57. Distance from inlet = 2.5 m. Curves for different sulfur contents in naphtha feed (wt ppm). Dotted line: measured data on used catalyst (Table 19). (Reproduced with permission from ref. [325])

B. Effects of Sulfur Poisoning

The sulfur poisoning is reversible as shown in Figure 60, representing data from a differential and gradientless reactor system [329]. The catalyst regains its initial activity after removal of the sulfur compounds in the feed.

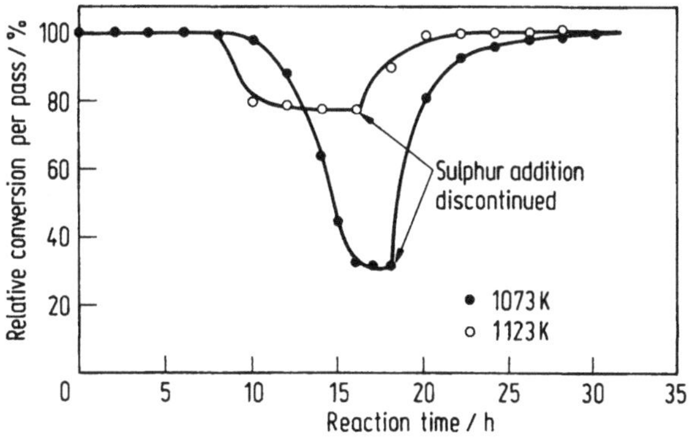

Figure 60. Sulfur poisoning of steam reforming catalyst. Time variations in relative activity. Differential recycle reactor. Catalyst: Ni/SiO_2, $p_{H_2O}/p_{CH_4} = 2.7$, sulfur addition 120 to 140 mg S m^{-3} CH_4. (Reproduced with permission from ref. [329])

For industrial reformers, the deactivation from sulfur poisoning is most dominating at the inlet of the reformer as illustrated in Figure 56 and 58. At high sulfur coverages this may result in increased tube wall temperatures and eventually in unsatisfactory approach to equilibrium in the exit gas. Sulfur poisoning can easily result in carbon formation indirectly since the catalyst activity as well as the tube wall temperature are important parameters in the development of hot tubes (*c.f.* p. 89 and p. 93). If higher hydrocarbons are present in the feed gas they may pass over a deactivated catalyst to hotter parts of the reformer tube where conditions for carbon formation are more critical on the nickel surface as well as in the gas phase *via* thermal cracking to olefins (*c.f.* p. 80).

C. Regeneration of Sulfur-poisoned Catalysts

1. Regeneration in Reducing Atmosphere

In principle it should be possible to remove sulfur from a poisoned catalyst simply by decreasing the sulfur content of the feed because the chemisorption of hydrogen sulfide on the nickel surface is reversible. This may well be achieved in experiments using a high flow rate and showing no diffusion restrictions (*c.f.* Figure 60). However, on an industrial scale this method will normally result in a slow regeneration because the rate of diffusion-controlled elution decreases exponentially with time [330, 7].

The exponential decay is even true at conditions with no diffusion restrictions and at which the gas is leaving the reactor saturated with hydrogen sulfide. Assuming a linear adsorption isotherm, it can easily be shown that the sulfur leaving the catalyst decreases exponentially with time [7].

Therefore, the rate of sulfur removal by means of a desorption process decreases with time, and only if bulk phase reactions are involved, would the sulfur removal rate be constant, assuming that equilibrium is established. The use of high temperatures during regeneration in a reducing atmosphere will have an effect because of the high heat of chemisorption. However, this may often be difficult at the inlet of the reactor (where sulfur is normally accumulated), because of temperature restrictions in preheating of the feed gas.

It was discussed previously (p. 96) that the p_{H_2O}/p_{H_2} ratio has no influence on chemisorption equilibrium [178]. Therefore, no regeneration effect is to be expected when increasing the ratio p_{H_2O}/p_{H_2} provided the catalyst is still in a reduced state.

2. Regeneration in Oxidizing Atmosphere

The data in Figure 61 summarize a series of regeneration tests [7, 185], using a sulfur-poisoned nickel catalyst. The degree of regeneration is expressed by the ratio s/s_1, *i.e.* the ratio of the sulfur content after and before the treatment.

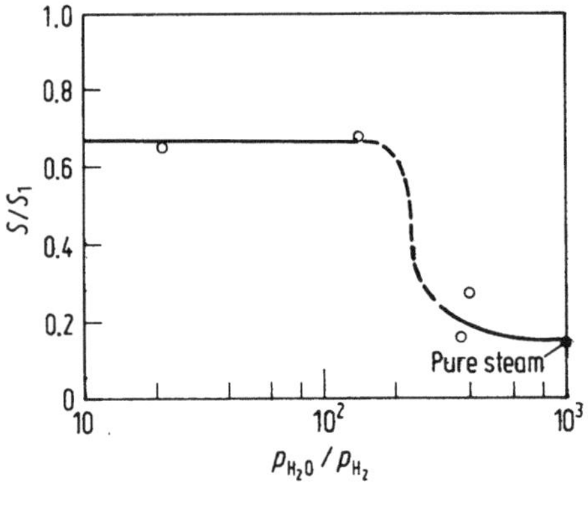

Figure 61. Influence of H_2O/H_2 on regeneration of nickel catalyst. Catalyst: 0.5 g $Ni/MgAl_2O_4$ (s_1 = 963 wt. ppm), 973 K, 3 mol H_2O h^{-1}, duration of each test: 4 hours. (Reproduced with permission from ref. [7, 185])

When the ratio p_{H_2O}/p_{H_2} was increased no improvement was observed as expected from the chemisorption experiments. However, at p_{H_2O}/p_{H_2} above 150–250 a significant change of the regeneration degree of the catalyst is obtained as shown in Figure 61. Increased sulfur removal is achieved at a p_{H_2O}/p_{H_2} which may be close to the equilibrium constant for the oxidation of the catalyst (*c.f.* Figure 15).

Analysis of the exit gas showed the presence of sulfur dioxide as well as hydrogen sulfide, which may indicate the following reaction pattern

$$Ni - S + H_2O \rightleftharpoons NiO + H_2S \quad (104)$$

$$H_2S + 2\,H_2O \rightleftharpoons SO_2 + 3\,H_2 \quad (-\Delta H^0_{298} = -207\ \text{kJ mol}^{-1}) \quad (105)$$

This sequence does not include the chemisorption equilibrium (98) and this may be the reason for the improved regeneration. The equilibrium constant,

K_p^1, of reaction (105) at 973 K is 3.5×10^{-7}, ($K_p = 3.5\times10^{-8}$), and consequently even a small amount of hydrogen will inhibit the conversion of hydrogen sulfide. Therefore, sulfur removal following this reaction pattern requires a total oxidation of the catalyst. If some part of the nickel surface is still exposed to the gas, hydrogen formed by equation (102) will cause hydrogen sulfide to be retained at the surface by the chemisorption reaction (98).

The regeneration in steam is very sensitive to the composition of the support because the sulfur dioxide may react with the support. The results in Figure 62 [7, 185] show that the sulfur content of catalysts promoted with sodium and potassium is nearly unaffected by the treatment with steam. Sulfur dioxide reacts with the alkali forming alkali sulfate.

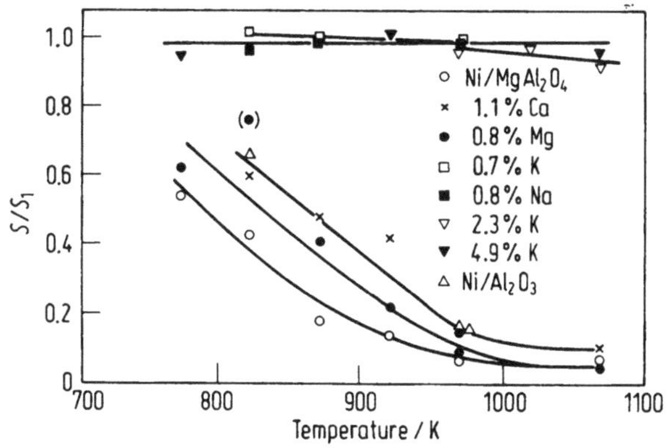

Figure 62. Regeneration of nickel catalysts with steam. 0.5 g catalyst. 3 mol H_2O h^{-1}, 3 hours. (Reproduced with permission from ref. [7, 185])

The strong temperature dependence of the results obtained with the catalyst containing no alkali may indicate a bonding of sulfur dioxide to the catalyst support at low temperatures. Therefore, high-temperature steaming is normally required to obtain sufficient removal of the sulfur on the catalyst.

When the catalysts were exposed to steam containing air (vol %), formation of sulfates [7, 185] took place on all catalysts resulting in only limited sulfur removal.

3. *Regeneration by Mild Reduction of Oxidized Catalysts*

In principle, it should be possible to remove sulfur retained by the support (or promoter) by means of a reduction. This presumes first that the support will not retain the hydrogen sulfide formed by the reduction, and secondly that the reduction is performed under conditions at which nickel oxide is reduced at a rate significantly lower than the rate of formation of hydrogen sulfide. If not so, hydrogen sulfide is chemisorbed at the nickel surface [7, 185].

D. Sulfur-passivated Steam Reforming

It is well known that the presence of more than 50 ppm of sulfur in the feed stream to steam crackers significantly reduces the coke formation on the tube walls [331, 332]. Moreover, the sulfur may inhibit the carburization of the construction materials [333]. For nickel catalysts a similar effect is observed. At complete coverage the sulfur blocks the nickel surface which means that adsorbed carbon atoms cannot be dissolved in the nickel crystal [334] and that the whisker growth mechanism is blocked [150, 281] (*c.f.* Figure 39).

For the steam reforming of methane a complete coverage of the catalyst with sulfur results in total deactivation (*c.f.* Table 15). However, it was observed [85] that carbon-free steam reforming operation could be obtained with partly poisoned catalysts at conditions which would otherwise result in carbon formation. A series of experiments were performed with different sulfur contents in the feed, and at conditions for which the principle of equilibrated gas (*c.f.* p. 82) predicts carbon formation at temperatures below about 1100 K. The results shown in Table 20 indicate a certain activity for methane reforming and the existence of a threshold content of sulfur below which rapid carbon formation occurs.

Table 20. Optimum sulfur content for sulfur passivated reforming[a]

Exp. No.	1	2	3	4
S in feed/vol ppm H_2S	28	14	5	1
Duration/h	47	42	95	5
T_{exit}/K	1218	1217	1221	1221
CH_4 (dry exit)/%	0.71	0.70	0.36	—
Carbon formation[c]	no	no	no	yes[b]

[a] Bench scale tests: d_t = 22.6 mm,
Catalyst: Ni/Al_2O_3 as 4×4 mm cylinders, vol = 405 cm^3, p_{exit} = 0.3 MPa, T_{in} = 793 K,
Feed flows (mol/h): H_2O = 0.9, H_2 = 4, $CO = CO_2 = CH_4$ = 2.7 (Ref. [85, 223])
[b] carbon formation was rapid, and no gas analyses were taken
[c] $-\Delta G_e > 0$ for $T < 1113$ K

Other studies [85, 223] showed that a supersaturation with $-\Delta G_e$ as high as about 30 kJ mol^{-1} was required for the formation of carbon. This should be compared with the low values of $-\Delta G_e$ for sulfur-free tests shown in Table 18. The supersaturation depends on sulfur coverage [223] as illustrated in Figure 63.

It was shown [223] that the retarding effect of sulfur on carbon formation is a dynamic phenomenon. Sulfur inhibits the rate of carbon formation more than the rate of the reforming reactions.

Apparently, the ensembles of free nickel atoms available at high coverages of sulfur are sufficient for the conversion of adsorbed methane with steam but they are too small to allow the normal dissolution-precipitation nucleation of

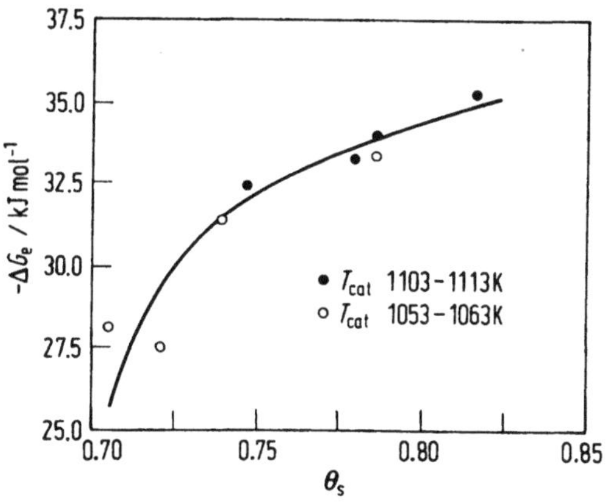

Figure 63. Supersaturation for start of carbon formation as function of sulfur coverage. Thermogravimetric studies. $-\Delta G_e$ was calculated for CH_4 decomposition on basis of equilibrated gas at $(p_{H_2O}/p_{CH_4})_{crit}$. $(p_{H_2O}/p_{CH_4})_{crit}$ was determined as the ratio below which carbon was formed and above which carbon was removed. [223]

the carbon whisker. It is evident that the concentration of the large ensemble will decrease drastically with sulfur coverage.

A sulfur-passivated process suffers from low catalyst activity which results in large deviations from the reforming equilibrium. Therefore, high reaction temperatures are required. The sulfur passivated catalyst operates at a coverage close to one and accordingly the available surface is determined by the chemisorption equilibrium for hydrogen sulfide (equation (101)). This is reflected by a high activation energy [223] of 266 kJ mol^{-1} which includes a contribution from the heat of adsorption. This is illustrated in Figure 64.

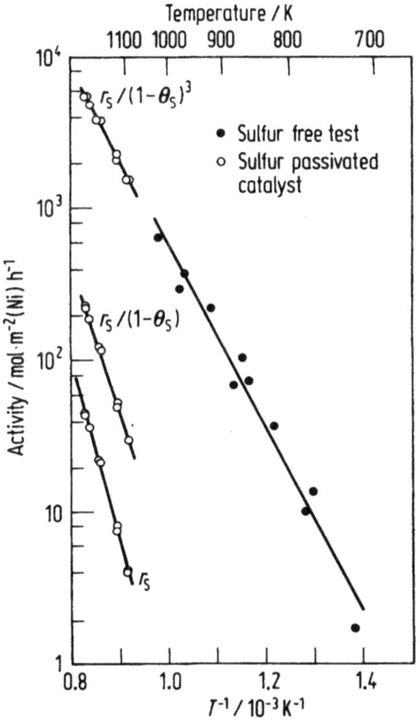

Figure 64. Sulfur passivation and specific activity. Ni/Al_2O_3 catalyst. r_s is reffered to total nickel area. θ_s is calculated by means of equation 101. Experiment with sulfur passivation: $p_{H_2O}/p_{CH_4} = 1.0$. $p_{H_2O}/p_{H_2} = 5.0$, $p_{H_2S}/p_{H_2} = 2.8 \times 10^{-5}$, 0.1 MPa. Experiment at sulfur-free conditions: $p_{H_2O}/p_{CH_4} = 0.94$, $p_{H_2O}/p_{H_2} = 2.5$, 0.1 MPa. (Reproduced with permission from ref. [223])

The estimated specific rates based on free nickel surface area are comparable to those for sulfur-free operation when it is assumed [223] that the rate decreases with $(1 - \theta_s)^3$. The agreement indicates that the same type of ensembles (probably 3 nickel atoms) are involved and that the sulfur poisoning is caused by a blockage of nickel sites. This result corresponds to results from methane exchange studies by Chahar *et al.* [198], who found no impact of sulfur on the deuterium exchange ratio.

The effect of eliminating carbon formation by a partial sulfur passivation is utilized in steam reforming for manufacture of reducing gas [60, 85] and in the production of carbon monoxide rich gases for chemical syntheses. Operation on feed gases containing substantial amounts of higher hydrocarbons may present a problem for a sulfur-passivated process, because these hydrocarbons may crack thermally into olefins, which may form pyrolitic carbon (*c.f.* p. 80).

7. Concluding Remarks

Steam reforming is an essential process in the manufacture of synthesis gas and hydrogen from hydrocarbons. Its applications are wide-spread in the petrochemical industry as well as in the energy related industry. There may be many future areas in which steam reforming may play a role.

Industrial operation involves a coupling of heat transfer and the strongly endothermic reaction. The balance between heat input through the reformer tubes and the heat consumption by the reaction is a central problem in steam reforming.

The catalyst properties are dictated by the severe operating conditions involving high temperatures, high steam partial pressure and high mass velocity. The catalyst is normally based on nickel with relatively little dispersion on a ceramic-like support of large pellet size. The effectiveness factor is very low, but the activity is normally sufficient for complete conversion.

High catalyst activity results in lower tube wall temperatures and hence longer tube life. The catalyst may also inhibit the formation of carbon, which cannot be allowed in industrial operation. This can be achieved by increased adsorption of steam on the catalyst.

The catalyst is very sensitive to sulfur poisoning which may also affect the formation of carbon.

Although steam reforming is a well established process there are still areas which require further clarification such as:

- heat transfer in packed tubes at the very high Reynolds number applied in steam reforming
- sintering of nickel crystals on supported catalysts and its impact on the specific activity
- the effect of alkali on decreasing the specific activity of nickel
- the mechanism of nucleation of carbon
- the role of the sublayer in chemisorption of sulfur and in the transformation of adsorbed carbon atoms into dissolved carbon atoms.

Acknowledgements

Thanks are given to many of my colleguas for suggestions and stimulating discussions, in particular to Dr. L. J. Christiansen, Mr. H. J. Hansen and Ms. B. Nielsen. Ms. U. Ebert Petersen, Ms. U. C. Jørgensen, and Ms. G. Werther kindly assisted in preparation of the manuscript.

Symbols

A	pre-exponential factor in rate equation
A_D	pre-exponential factor for diffusion rate $D_{eff} = A_D \exp(-E_D/RT)$
$A_{i,w}$	pre-exponential factor for intrinsic rate on reactor weight basis
B	adsorption coefficient
C	molar concentration
C_{LM}	constant depending on tube material, equation (9)
D	dispersion of nickel, equation (34)
D_B	bulk diffusion coefficient
D_{eff}	effective diffusion coefficient
D_{er}	effective diffusivity in catalyst bed in radial direction, equation (18)
D_K	Knudsen diffusion coefficient
$D_{C,Ni}$	diffusion coefficient of carbon in nickel
E_a	apparent activation energy
E_D	activation energy for diffusion
G	free energy
$-\Delta G_a$	potential for carbon formation reaction of actual gas
ΔG_c	deviation of free energy from graphite data, defined by equation (82)
$-\Delta G_e$	potential for carbon formation reaction (78) or (79) after equilibration of reactions (2) and (3)
G_M	mass velocity
H	enthalpy
K_c	equilibrium constant, concentration basis
K_c'	modified equilibrium constant for methane reforming, $K_c' = K_c \frac{C_{H_2O}}{C_{H_2}^2}$ (*c.f.* equation (30))
K_H	equilibrium constant for hydrogen adsorption on catalyst
K_p	equilibrium constant (standard state: 1 atm, 298 K)
K_p'	equilibrium constant (standard state: 0.1 MPa, 298 K)
K_w, K_w'	equilibrium constant for steam adsorption on catalyst
M	molecular weight
N	turn-over frequency
N_0	turn-over frequency, $p_{C_nH_m} = 0.01$ MPa, 773 K (molecular basis)
N_0'	turn-over frequency, $p_{C_nH_m} = 0.01$ MPa, 773 K (C-atom basis)
Nu	Nusselt number, equation (21)
$Pe_{h,r}$	Peclet number for heat transport in radial direction, equation (20)
$Pe_{m,r}$	Peclet number for mass transport in radial direction, equation (18)
P_{LM}	Larson-Miller parameter, equation (9)
Pr	Prandtl number ($c_p \cdot \mu/\lambda_g$), equation (21)

Q	transferred heat, duty
Q_R	reaction quotient
$Q_{R,e}$	reaction quotient for reaction (79) after equilibration of reactions (2) and (3)
R	radius
R'	resistance number characterizing rate of gum formation
Re	Reynolds number for packed bed ($Re = G_M d_P/\mu$)
R_p	pore radius
R_t	radius of tube
R_w	radius of coke whisker
S_{BET}	BET surface area of catalyst particle
S_{Ni}	nickel surface area
S_P	external surface area of catalyst particle
SV	space velocity
T	temperature
T_B	carbon limit for Boudouard reaction (78), fixed gas composition, C, for $T < T_B$
T_c	carbon limit for higher hydrocarbon, C, for $T > T_c$
T_L	lower carbon limit of equilibrated gas, C, for $T < T_L$
T_u	upper carbon limit of equilibrated gas, C, for $T > T_u$
T_m	melting point
T_M	carbon limit for methane decomposition (79), fixed gas composition, C, for $T > T_M$
T_{CM}	mean catalyst temperature
T_{CW}	catalyst temperature at tube wall
T_p	temperature below which self poisoning occurs
T_W	temperature of outer tube wall
ΔT_R	temperature approach to reforming equilibrium (2)
ΔT_S	temperature approach to shift reaction (3)
U	overall heat transfer coefficient
V	volume
V_p	volume of catalyst particle
X	conversion
X_{eq}	conversion at equilibrium
Y	p_{H_2S}/p_{H_2}
Z	axial distance
a, b	constants
a_c^{eq}	carbon activity at equilibrium
a_c^L	carbon activity at saturation of nickel crystal
a_c^s	carbon activity at steady state
c	carbon content on catalyst
c_p	heat capacity
d_{ex}	external diameter of catalyst ring
d_{in}	internal diameter of catalyst ring
d_t	internal tube diameter
d_{Ni}	diameter of nickel crystal

d_p	particle diameter
$d_{p,s}$	$(= 6V_p/S_p)$ equivalent particle diameter (surface basis)
$d_{p,v}$	equivalent particle diameter (volume basis)
f	friction factor, equation (14)
f'	conversion factors, Table 2
$f(p_{H_2O}, p_{H_2})$	function of partial pressures of steam and hydrogen
g	accelleration of gravity
h	height of catalyst ring
k	rate constant
k_A	rate constant for adsorption of hydrocarbons
$k_{i,v}$	intrinsic rate constant on catalyst bed volume basis
$k_{i,w}$	intrinsic rate constant on catalyst weight basis
k_c	rate constant for carbon formation
k_{eff}	effective rate constant
k_g	rate of gasification
k_H	rate constant for hydrocracking of adsorbed radicals
k_r	rate constant for gasification step
k_v	rate constant on reactor volume basis
m, n	numbers
p	pressure
Δp	pressure drop
q	heat flux
q_{av}	average heat flux
r	reaction rate
$r_{eff,v}$	effective reaction rate per reactor volume
$r_{eff,w}$	effective reaction rate on catalyst weight basis
r_A	adsorption rate of hydrocarbons
r_c	rate of carbon formation
r_H	rate of hydrocracking (hydrogenolysis)
r_i	initial rate (mol g^{-1} h^{-1})
r_p	rate of self-poisoning
r_s	specific rate (mol m^{-2} (Ni) h^{-1})
r_v	rate of reaction per unit reactor volume
r_w	rate per unit catalyst weight
s	sulfur content (uptake) of catalyst (wt/wt)
s_0	sulfur capacity (wt/wt)
s_1	sulfur content before regeneration (wt/wt)
t	time
t_0	induction period, equation (71)
t_R	time for tube rupture, equation (10)
t_t	thickness of tube wall
u_s	superficial velocity
y	mole fraction
$\alpha, \alpha', \alpha''$	kinetic orders
α_w	wall coefficient for heat transfer, equation (19)
$\alpha_{w,t}$	turbulent part of α_w, equation (21)
β	reaction order with respect to total pressure

γ	surface tension
ε	void fraction of catalyst bed
ε_p	porosity of catalyst particle
η	effectiveness factor
θ	coverage
λ_{er}	effective thermal conductivity in catalyst bed in radial direction, equation (17)
$\lambda_{r,t}$	turbulent part of λ_{er}
λ_g	thermal conductivity of gas
μ	viscosity
ϱ_B	bulk density of catalyst bed
ϱ_g	gas density
ϱ_p	density of catalyst particle
σ	stress
τ	tortuosity factor
ψ	shape factor, $d_{p,s} = \psi d_{p,v}$

References

1. Sinfelt, J. H.: In: Catalysis, Science and Technology (Anderson, J. R.; Boudart, M., eds.). Springer, Berlin Heidelberg New York, 1981. Vol. 1 p. 257
2. Kirk, R. E.; Othmer, D. F.; Scott, J. D.; Standen, A.: Encyclopedia of Chemical Technonology. Interscience Inc., Vol. 8, 792 (1952)
3. Padovani, C.: Inst. Gas Ing. J. **4**, 760 (1964)
4. Jockel, H.: In: Ullmann Encyklopädie der Technischen Chemie. Weinheim, 1977, Verlag Chemie, Vol. **14**, p. 403
5. Bridger, G. W.; Chinchen, G. C.: In: Catalyst Handbook, Wolfe Scientific Books, London 1970, p. 64
6. Bridger, G. W.: In: Catalysis Vol. **3**, (Kemball, C., and Dowden, D. A., eds.) (Specialist Periodical Reports), London 1980, The Chemical Society, p. 39.
7. Rostrup-Nielsen, J. R.: Steam Reforming Catalysts, Danish Technical Press Copenhagen 1975
8. Ross, J. R. H.: In: Surface and Defect Properties of Solids, (Roberts, M. W., and Thomas, J. M., eds.) (Specialists Periodical Reports), London 1974, The Chemical Society, Vol. **4**, p. 34
9. Trimm, D. L.: Catal. Rev.-Sci. Eng. **16**, 155 (1977)
10. van Hook, J. P.: Catal. Rev.-Sci. Eng. **21**, 1 (1981)
11. Bartholomew, C. H.: Catal. Rev.-Sci. Eng‘ **24**, 67 (1982)
12. Grenoble, D. C.: J. Catal. **51**, 203 (1978)
13. Tomita, T.; Kitagawa, M.: Chem. Ing. Tech. **49**, 469 (1977)
14. Dry, M. E.: In: Catalysis, Science and Technology, (Anderson, J. R. and Boudart, M., eds.), Springer, Berlin Heidelberg New York 1981, Vol. 1, p. 159
15. Lihou, D. A.: Chem. Process Eng. **46**, 487 (1965)
16. Morse, P. L.: Hydrocarbon Process **52**, (1) 113 (1973)
17. Hampson, G. M.: Chem. Eng. (Lomdon) p. 523 (1979)
18. Kjær, J.: Computer Methods in Gas Phase Thermodynamics, Haldor Topsøe, Vedbæk 1972
19. Rostrup-Nielsen, J. R.: Ammonia Plant Saf. **15**, 82 (1973)
20. Andrew, S. P. S.: Ind. Eng. Chem. Prod. Res. Develop. **8**, 321 (1969)
21. Froment, G. F.: Chem. Eng. Sci. **36**, 1271 (1981)
22. Davy, H.: Phil. Trans. Roy. Soc. London, **107**, 77 (1817)
23. Tessie Du Motay, M., Maréchal, M.: Bull. Soc. Chim. France **9**, 334 (1868)

24. DRP 51572, 1889
(Mond, L. and Langer, C.), Chem. Zentralbl. II, 32 (1890)
25. Lang, J.: Z. Phys. Chem. **2**, 161 (1888)
26. DRP 229406, 1909
(Dieffenbach, O., and Moldenhauer, W.)
27. DRP 296866, 1912
(BASF (Mittasch, A., and Schneider, C.))
28. Sabatier, P.: La Catalyse en Chimie Organique, Librarie Polytéchnique, Paris 1920
29. Neumann, B.; Jacob, K.: Z. Elektrochem. **30**, 557 (1924)
30. Fischer, F.; Tropsch, H.: Brennst. Chem. **3**, 39 (1928)
31. Liander, H.: Trans. Faraday Soc. **25**, 462 (1929)
32. Fraser, O. B. J.: Trans. Electrochem. Soc. **71**, 425 (1937)
33. U.S. Pat. 1 934 836, 1927
(I.G. Farben, (Wietzel, G., Haller, W., and Hennicke, W.))
34. U.S. Pat. 2 056 911, 1932
(I.G. Farben (Schiller, G., and Wietzel, G.))
35. Byrne, Jr. P. J.; Gohr, R. J.; Haslam, R. T.: Ind. Eng. Chem. **24**, 1129 (1932)
36. Gard, N. R.: Nitrogen **39**, 25 (1966)
37. Topsøe, H. F. A.; Fossum Poulsen, H.; Nielsen, A.: Chem. Eng. Prog. **63**, 67 (1967)
38. Cockerham, R. G.; Percival, G.: Trans. Inst. Gas Eng. **107**, 390 (1956/57)
39. BRIT. Pat. 820 257, 1958
(Gas Council, (Dent, F. J., Cockerham, R. G., and Percival, G.))
40. Davies, H. S.; Humphries, K. J.; Hebden, D.; Percy, D. A.: Inst. Gas Eng. J. **7**, 708 (1967)
41. Jockel, H.; Triebskorn, B. E.: Hydrocarbon Process. **52**, (1) 93 (1973)
42. Ishiguro, T.: Hydrocarbon Process. **47**, (2) 87 (1968)
43. Czuppon, T. A.; Buvidas, L. J.: Hydrocarbon Process. **58**, (9) 197 (1979)
44. Lastigzon, J.: Technical Bull. T-22, International Fertilizer Development Center (UNIDO) 1981
45. Corneil, H. G.; Heinzelmann, F. J.: In: Hydrogen, Production and Manufacture, (Smith, W. N., and Santangelo, J. G., eds.). ACS-symposium Series 116, Amer. Chem. Soc. Washington 1980, p. 147
46. Rose, F.: Stahl. Eisen **95**, 1012 (1975)
47. Manley, J.: Metals Bull., Dec. 1981, p. 49
48. Congram, G. E.: Oil Gas J. **74**, 74 (1976)
49. Dybkjær, I.: Chem. Econ. Eng. Review **13**, 17 (1981)
50. Chang, C. D.: Catal. Rev.-Sci. Eng. **25**, 1 (1983)
51. Gilles, E. A.: Chem. Eng. Prog. **76** (10) 88 (1980)
52. Minet, R. G.; Olesen, O.: In: Hydrogen Production and Marketing, (Smith, W. N., and Santangelo, J. G., eds.)). ACS-symposium Series 116, Amer. Chem. Soc. Washington 1980, p. 67
53. Fedders, H.; Harth, R.; Höhlein, B.: Nucl. Eng. Des. **34**, 119 (1975)
54. Jørn, E.; Skov, A.; Harms, H.; Höhlein, B.: Oil and Gas J. **78**, 120 (1980)
55. Miyasugi, T.; Kosaka, S.; Mukai, K.; Suziki, A.: J. Japan Petrol. Inst. **25**, 260 (1982)
56. Chem. Eng. News. July 30, 26 (1979)
57. Richardson, J. T.: (to be pusblished)
58. Sjöström, K.: Ind. Eng. Chem. Process Des. Dev. **19**, 148 (1980)
59. Nielsen, A.; Hansen, J. B.; Houken, J.; Gam, E. A.: Plant/Oper. Progr. **1**, (3) 186 (1982)
60. Topsøe, H.; Rostrup-Nielsen, J. R.: Scan. J. Metallurgy **8**, 168 (1979)
61. Töpfer, H. J.: Gas Wasserfach **117**, 412 (1976)
62. Suga, Y.; Mori, T.: In: Reduccion directa '80, Proc. 4th ILAFA-Direct Reduction Congress ILAFA Buenos Aires 1980
63. Zardi, U.; Antonini, A.: Nitrogen **122**, 33 (1979)
64. Raghuraman, K. S.; Johansen, T.: In: Proc. Symp. on "Science and Catalysis and its Application in Industry", FPDIL, Sindri 22–24 Feb. 1979, p. 418
65. Nutting, J.: In: Materials Technology in Steam Reforming Processes, (Edelnau, C., ed.). Pergamon Press, London 1966, p. 11
66. Estruch, B.; Lyth, C.: ibid. p. 29

67. Kawai, T.; Takemura, K.; Shibasaki, T.; Mohri, T.: Ammonia Plant Saf. **22**, 119 (1980)
68. Zeis, L. A.; Heinz, E.: Ammonia Plant Saf. **12**, 55 (1970)
69. Salot, W. J.: Ammonia Plant Saf. **13**, 119 (1971)
70. Appl, M.; Gössling, H.: Chem. Z. **96**, 135 (1972)
71. Blythe, B. M.; Sampson, R. W.: Oil and Gas J. **72**, 91 (1974)
72. Singh, C. P. P.; Saraf, D. N.: Ind. Eng. Chem. Process Des. Dev. **18**, 1 (1979)
73. Roesler, F. C.: Chem. Eng. Sci. **22**, 1325 (1967)
74. Hottel, H. C.; Cohen, E. S.: AIChE Journal **4**, 3 (1958)
75. Vercammen, H. A. J.; Froment, G. F.: In: Proc. 5th Int. Symp. Chemical Reaction Engineering — Houston, (Weekmann Jr., V. W., and Luss, D., eds.) ACS Symposium series 65, Amer. Chem. Soc. Washington 1978, p. 271
76. Lenoir, J. M.: Hydrocarbon Process. **48** (10) 87 (1969)
77. Blythe, B. M.; Sampson, R. W.: Oil and Gas J. **72**, 60 (1974)
78. Topsøe, H.: Inst. Gas Eng. J. **6**, 401 (1966)
79. Hyman, M. H.: Hydrocarbon Process. **47**, (7) 131 (1968)
80. Grover, S. S.: Hydrocarbon Process. **49**, (4) 109 (1970)
81. Golebiowski, A.; Wasala, T.: Int. Chem. Eng. **13** 133 (1973)
82. Froment, G. F.; Bischoff, K. B.: Chemical Reactor Analysis and Design, John Wiley & Sons. New York 1979
83. Rostrup-Nielsen, J. R.; Wrisberg, J.: Proc. Int. Conf. on Natural Gas Processing and Utilization, Dublin 1976, p. 5–53
84. Froment, G. F.: Ind. Eng. Chem. **59**, 19 (1967)
85. Rostrup-Nielsen, J. R.; Christiansen. L. J.: In: Proc. 6th Simposio Ibero-Americano de Catalise, 1978. Rio de Janeiro 1978, p. 1615
86. Christiansen, L. J.; Tøttrup, P. B.: (to be pusblished)
87. Villadsen, J.; Michelsen, M. L.: Solution of Differential Equation Model by Polynomial Approximation. Prentice-Hall, New York 1977
88. Sundaram, K. M.; Froment, G. F.: Chem. Eng. Sci. **35**, 364 (1980)
89. Schlünder, E. U.: In: Chemical Reaction Engineering Reviews-Houston ACS Symposium Series 72, (Luss, D., and Weekman, Jr., V. W., eds.), Amer. Chem. Soc. Washington 1978, p. 110
90. Bridger, G. W.: Chem. Process Eng. **53**, 38 (1972)
91. Jones, F. W. S.: Ammonia Plant Saf. **12**, 61 (1970)
92. Nicklin, T.; Whitaker, R. J.: Inst. Gas Eng. J. **8**, 15 (1968)
93. Brit. Pat. 1 182 829, 1967
(Haldor Topsøe (Rostrup-Nielsen J. R.))
94. Dowden, D. A.; Schnell, C. R.; Walker, G. T.: 4th Int. Congr. Catal., Moscow 1968, preprint 62
95. Trimm, D. L.: Design of Industrial Catalysts, Elsevier, Amsterdam 1980, p. 169
96. Williams, A.; Butler, G. A.; Hammonds, J.: J. Catal. **24**, 352 (1972)
97. Pedersen, K.; Skov, A.; Rostrup-Nielsen, J. R.: Prepr. ACS Div. Fuel Chem. **25** (2), 89, (1980)
98. Erekson. E. J.; Sughrue, E. L.; Bartholomew, C. H.: Fuel Process. Technol. **5**, 91 (1981)
99. Tiwari, R. N.; Sengupta, P. K.; Chakravorty, B. K.; Hazra, P. K.: Fertilizer Technol. **12**, 352 (1975)
100. Cooke, P. W.; Hareshape, J. N.: Trans. Faraday Soc., **43**, 43 (1947)
101. Gaugin, R.; Graulier, M.; Papee, D.: Adv. Chem. Ser. **143**, 147 (1975)
102. Fleming, H. W.; Cromeans, J. S.: Ammonia Plant Saf. **13**, 85 (1971)
103. Aleksić, B. R.; Stefanović, M. M.; Aleksić, B. D.; Vujović-Djordjević, B. D.; Arandjelović, P. S.; Jovanović, N. N.; Mihajlović, L. M.: Chem. Tech. **28**, 89 (1976)
104. Barin, I.; Knacke, O.: Thermochemical properties of inorganic substances, Springer, Berlin Heidelberg New York 1973, C. Supplement 1977
105. Chown, J.; Deacon, R. F.: Trans. Brit. Ceram. Soc. **63**, 91 (1964)
106. Raymond, R. J.; Brindley, G. W.: Trans. Faraday Soc. **61**, 1017 (1965)
107. Takemura, Y.; Yamamoto, K.; Morita, Y.: Int. Chem. Eng. **7**, 737 (1967)
108. Alzarmora, L. E.; Ross, J. R. H.; Kruissink, E. C.; van Reijnen, L. L.: J. Chem. Soc. Faraday Trans. I **77**, 665 (1981)

109. Blume, H.; Becker, K.; Engels, S.; Hille, J.; Maye, H.; Richter, K.: Z. anorg. allg. Chem. **405**, 211 (1974)
110. Simosova, L. G.; Dzis'ko, V. A.; Borisova, M. S.; Karakchiev, L. G.; Olen'kova, I. P.: Kinet. Katal. **14**, 1566 (1973)
111. Lo Jacono, M.; Schiavello, M.; Cimino, A.: J. Phys. Chem. **75**, 1044 (1971)
112. Nikolajenko, V.; Danes, Vl.: Kinet. Katal. **7**, 1025 (1966)
113. Hahn, Jr., W. C.; Muan, A.: J. Phys. Chem. Solids **19**, 338 (1961)
114. Ross, J. R. H.; Steel, M. C. F.: J. Chem. Soc. Faraday Trans. I **69**, 10 (1973)
115. Vignes, S.; Jeannot, F.: Rev. Inst. Fr. Petrole. Ann. Combust. Liquides, **21**, 875 (1966)
116. Fricke, R.; Weitbrecht, G.: Z. Elektrochem. **48**, 87 (1942)
117. Rankin, J. D.; Livingstone, J. G.: Ammonia Plant Saf. **23**, 203 (1981)
118. Bandrowski, J.; Bickling, C. R.; Yang, K. H.; Hougen, O. A.: Chem. Eng. Sci. **17**, 379 (1962)
119. Boldyrev, V. V.; Bulens, M.; Delmon, B.: The Control of the Reactivity of Solids, Elsevier, Amsterdam 1979
120. Noskova, S. P.; Bovisova, M. S.; Dzis'ko, V. A.; Alabucher, Y. A.: Kinet. Katal. **15**, 592 (1974)
121. Arora, B. R.; Banerjee, A. K.; Mandal, N. K.; Ganguli, N. C.; Sen, S. P.: Technol. **10**, 193 (1973)
122. Banerjee, A. K.; Nardu, S. R.; Ganguli, N. C.; Sen, S. P.: Technol. **11**, 162 (1974)
123. Sen, S. P.; Majumdar, D. S.; Hazra, P. K.; Prasad, R.; Gosh, S. K.; Bhattacharyya, N. B.: Technol. **11**, 149 (1974)
124. Eischens, R. P.: In The Physical Basis for Heterogeneous Catalysis, (Drauglis, E., and Jaffee, R. E., eds.), Plenum, New York 1975, p. 485
125. Nowak, E. J.; Koros, R. M.: J. Catal. **7**, 50 (1967)
126. Takemura, Y.; Morita, Y.; Yamamoto, K.: Int. Chem. Eng. **6**, 725 (1966)
127. Nikolajenko, V.; Danes, Vl.; Krivanek, M.: Kinet. Katal. **7**, 816 (1966)
128. Tikkanen, M. H.; Rosell, B.-O.; Wiberg, Ö.: Acta Chem. Scan. **17**, 513 (1963)
129. Blank, S. L.; Pask, J. A.: J. Amer. Ceram. Soc. **52**, 669, (1969)
130. Pettit, F. S.; Randklev, E. H.; Felten, E. J.: J. Amer. Cer. Soc. **49**, 199 (1966)
131. Tournayan, L.; Frety, R.; Charcosset, H.; Trambouze, Y.: Ind. Chim. Belg. **38**, 496 (1973)
132. Leva, M.: Ind. Eng. Chem. **40**, 747 (1948)
133. Satterfield, C. N.: Mass Transfer in Heterogeneous Catalysis, MIT Press, Cambridge 1970, p. 138
134. Schoubye, P.: Chem. Eng. World, **5**, 40 (1970)
135. Coenen, J. W. E.; Linsen, B. G.: In Physical and Chemical Aspects of Adsorbents and Catalysts, (Linsen, B. G., ed.), Academic Press, London, New York 1970, p. 471
136. Shephard, F. E.: J. Catal. **14**, 148 (1969)
137. Sørensen, O.; Topsøe, He.; Villadsen, Jø.: (unpublished results)
138. Yacaman, M. J.; Fuentes, S.; Dominquez, J. M.: Surface Sci. **106**, 472 (1981) p. 290
139. Smith, D. J.; Marks, L. D.: Phil. Mag. A **44**, 735 (1981)
140. Anderson, J. R.: Structure of Metallic Catalysts, Academic Press, New York 1975,
141. Beeck, O.: Advan. Catal. **2**, 151 (1950)
142. Selwood, P. W.: J. Catal. **42**, 148 (1976)
143. Smith, J. S.; Thrower, P. A.; Vannice, M. A.: J. Catal. **68**, 270 (1981)
144. Pannell, R. B.; Chung, K. S.; Bartholomew, C. H.: J. Catal. **46**, 340 (1977)
145. McCarthy, J. G.; Wise, H.: J. Chem. Phys. **72**, 6332 (1980)
146. Erley, W.; Wagner, H.: J. Chem. Phys. **72**, 2207 (1980)
147. Gundry, P. M.; Tompkins, F. C.: Trans. Faraday Soc. **53**, 218 (1957)
148. Della Gatta, G.; Fubini, B.; Ghiotti, G.; Morterra, C.: J. Catal. **43**, 90 (1976)
149. Beeck, O.; Ritchie, A. W.: Discuss. Faraday Soc. **8**, 159 (1950)
150. Klemperer, D. F.; Stone, F. S.: Proc. Roy. Soc. **A243**, 375 (1957)
151. Lapujoulade, J.; Neil, K. S.: Surface Sci. **35**, 288 (1973)
152. Martin, G. A.; Ceaphalan, N.; de Montgolfier, P.; Imelik, B.: J. Chim. Phys. **70**, 122 (1973)
153. Batholomew, C. H.; Pannell, R. B.: J. Catal. **65**, 390 (1980)

154. Eberhardt, W.; Greuter, F.; Plummer, E. W.: Phys. Rev. Lett. **46**, 1085 (1981)
155. Nørskov, J. K.: Phys. Rev. Lett. **48**, 1620 (1982)
156. Vaska, L.; Selwood, P. W.: J. Amer. Chem. Soc. **80**, 1331 (1958)
157. Coenen, J. W. E.; Wells, P. B.: In Proc. 3rd Int. Symp. Scientific Bases for the Preparation of Heterogeneous Catalysts, (Delmon, B. ed.) (in press)
158. Rostrup-Nielsen, J. R.: J. Catal. **11**, 220 (1968)
159. Den Besten, I. E.; Selwood, P. W.: J. Catal. **1**, 93 (1962)
160. Saleh, J. M.; Kemball, C.; Roberts, M. W.: Trans. Faraday Soc. **57**, 1771 (1961)
161. Perdereau, M.: C. R. Acad. Sci. Ser. C **267**, 1107 (1968)
162. Bartholomew, C. H.; Agrawal, P. K.; Katzer, J. R.: Advan. Catal. **31**, 135 (1983)
163. Rosenqvist, T.: J. Iron Steel Inst. London **176**, 37 (1954)
164. Edmonds, T.; McCarroll, J. J.; Pitkethly, R. C.: J. Vac. Sci. Technol. **8**, 68 (1971)
165. Perdereau, M.; Oudar, J.: Surface Sci. **20**, 80 (1970)
166. Erley, W.; Wagner, H.: J. Catal. **51**, 229 (1978)
167. Demuth, J. E.; Jepsen, D. W.; Marcus, P. M.: Phys. Rev. Let. **32**, 1182 (1974)
168. Delescluse, P.; Masson, A.: Surface Sci. **100**, 423 (1980)
169. Demuth, J. E.; Rhodin, T. N.: Surface Sci. **45**, 249 (1974)
170. Weeks, S. P.; Plummer, E. W.: Chem. Phys. Lett. **48**, 601 (1977)
171. Rhead, G. E.: Surface Sci. **47**, 207 (1975)
172. Oudar, J.: Catal. Rev.-Sci. Eng. **22**, 171 (1980)
173. McCarthy, J. G.; Wise, H.: J. Phys. Chem. **76**, 1162 (1982)
174. Perdereau, M.: Surface Sci. **24**, 239 (1971)
175. Oliphant, J. L.; Fowler, R. W.; Pannell, R. N.; Bartholomew, C. H.: J. Catal. **51**, 229 (1978)
176. de Rosset, A. J.; Finstrom, C. G.; Adams, C. J.: J. Catal. **1**, 235 (1962)
177. Puri, V. K.; Mahapatra, H.; Ghorai, D. K.; Ganguli, N. C.; Sen, S. P.: Proc. Indian Nat. Sci. Acad. **41A**, 30 (1975)
178. Rostrup-Nielsen, J. R.: Chem. Eng. World **5**, 50 (1970)
179. Bridger, G. W.: Ammonia Plant Saf. **18**, 24 (1976)
180. Bhattacharyya, N. B.; Majumdar, D. S.; Hazra, P. K.; Prasad, R.: Proc. Indian Nat. Sci. Acad. **41A**, 90 (1975)
181. Martin, G. A.; Mirodatos, C.; Praliaud, H.: Appl. Catal. **1**, 367 (1981)
182. Bartholomew, C. H.; Farrauto, R. J.: J. Catal. **45**, 41 (1976)
183. Hermans, L. A. M.; Geus, J. W.: In: Proc. 2nd Int. Symp. on Scientific Bases for the Preparation of Heterogeneous Catalysts (Louvain-la-Neuve, 1978). (Delmon, B., Grange, P., Jacobs, P., and Poncelet, G., eds.), Elsevier, Amsterdam 1979, p. 113
184. Rostrup-Nielsen, J. R.: J. Catal. **31**, 173 (1973)
185. Rostrup-Nielsen, J. R.: J. Catal. **21**, 171 (1971)
186. Moayeri, M.; Trimm, D. L.: J. Chem. Soc. Faraday Trans. I **73**, 1245 (1977)
187. Ruckenstein, E.; Pulvermacher, B.: AIChE Journal **19**, 356 (1976)
188. Wynblatt, P.; Tal-Mocu. A.: In Sintering and Catalysis. Materials Science Res. **10**, 83 (1975)
189. Ruckenstein, E.; Pulvermacher, B.: J. Catal. **37**, 416 (1975)
190. Borisova, M. S.; Fenelonov, V. B.; Dzisko, V. A.; Simonova, L. G.: Kinet. Katal. **17**, 653 (1976)
191. Sememov, N. N.: Einige Probleme der chemischen Kinetik und Reaktionsfähigkeit, Akademie Verlag, Berlin 1961
192. Albright, L. F., Crynes, B. L.; Corcoran, W. H., eds.: Pyrolysis, Theory and Practice, Academic Press New York 1983
193. Karim, G. A.; Metwally, M. M.: Adv. Hydrogen Energy Syst. **2**, 937 (1979)
194. Kemball, C.: Catal. Rev. **5**, 33 (1971)
195. Frennet, A.; Lienard, G.; Crucq, A.; Degols, L.: J. Catal. **53**, 150 (1978)
196. Leach, H. F.; Mirodatos, C.; Whan, D. A.: J. Catal. **63**, 138 (1980)
197. Martin, G. A.; Imelik, B.: Surface Sci. **42**, 157 (1974)
198. Chahar, B. S.; Hightower, J. W.: J. Catal. (in press)
199. Martin, G. A.: J. Catal. **60**, 345 (1979)
200. Babernics, L.; Guczi, L.; Matusek, K.; Sarkany, A.; Tétényi, P.: In: Proc., 6th Int.

Congr. Catal. London 1976. (Bond, G. C., Wells, P. B., and Tompkins, F. C., eds.), The Chemical Society, London 1976, p. 456
201. Sinfelt, J. H.: Catal. Rev. **3**, 175 (1969)
202. Boudart, M.: AIChE J. **18**, 465 (1972)
203. Machiels, C. J.; Anderson, R. B.: J. Catal. **58**, 268 (1979)
204. Maire, G.; Anderson, J. R.; Johnson, B. B.: Proc. Roy. Soc. **A 320**, 227 (1970)
205. Legare, P.; Maire, G.: J. Chim. Phys. Physicochim, Biol. **68**, 1206 (1971)
206. Schouten, F. C.; Gijzeman, O. L. J.; Bootsma, G. A.: Surface Sci. **87**, 1 (1979)
207. Martin, G. A.; Dalmon, D. A.: C. R. Acad. Sci. Ser. C **286**, 127 (1978)
208. Lehwald, S.; Ibach, H.: Surface Sci. **89**, 425 (1979)
209. McNaught, W. G.; Kemball, C.; Leach, H. F.: J. Catal. **34**, 98 (1974)
210. Matsumoto, H.: Shokubai **18**, 71 (1976)
211. Balashova, S. A.; Slovokhotova, T. A.; Balandin, A. A.: Izv. Akad. Nauk. SSSR. Ser. Khim. **2**, 275 (1965)
212. Phillips, T. R.; Mulhall, J.: Turner, G. E.: J. Catal. **15**, 233 (1969)
213. Traply, G.; Parlagh, G.; Rácz, G.; Steingaszner, P.; Székely, G.: Acta Chim. Acad. Sci. Hung. **88**, 235 (1976)
214. Yoshitomi, S.; Morita, Y.; Yamamoto, K.: Bull. Jap. Petrol. Inst. **4**, 15 (1962)
215. Treiger, L. M.; Rabinovitch, G. L.; Moslanski, G. N.: Dokl. Akad. Nauk. SSSR **209**, 130 (1973)
216. Rostrup-Nielsen, J. R.; Tøttrup, P. B.: In: Symposium on Science of Catalysis and its Application in Industry FPDIL, Sindri 1979, p. 380
217. Traply, G.; Parlagh, G.; Rácz, G.; Steingaszner, P.; Székely, G.: Acta. Chim. Acad. Sci. Hung. **88**, 223 (1976)
218. Jackson, S. D.; Thomson, S. J.; Webb, G.: J. Catal. **70**, 249 (1981)
219. Schnell, C. R.: J. Chem. Soc. (B) 158 (1970)
220. Akers, W. W.; Camp, D. P.: AIChE J. **1**, 471 (1955)
221. Bodrov, I. M.; Apel'baum, L. O.; Temkin, M. I.: Kinet. Katal. **8**, 821 (1967)
222. Allen, D. W.; Gerhard, E. R.; Linkins Jr., M. R.: Brit. Chem. Eng. Proc. Tech. **17**, 605 (1972)
223. Rostrup-Nielsen, J. R.: J. Catal. **85**, XXX (1984)
224. Bodrov, I. M.; Apel'baum, L. O.; Temkin, M. I.: Kinet. Katal. **5**, 696 (1964)
225. Bodrov, I. M.; Apel'baum, L. O.: Kinet. Katal. **8**, 379 (1967)
226. Bodrov, I. M.; Apel'baum, L. O.; Temkin, M. I.: Kinet. Katal. **9**, 1065 (1968)
227. Khomenko, A. A.; Apel'baum, L. O.; Shub, F. S.; Temkin, M. I.: Kinet. Katal. **11**, 1480 (1970)
228. Khomenko, A. A.; Apel'baum, L. O.; Shub, F. S.; Snagovskii, Y. S.; Temkin, M. I.: Kinet. Katal. **12**, 423 (1971)
229. Boudart, M.; Djega-Mariadasson, G.: Kinetics of Heterogeneous Catalytic Reactions, Princeton University Press (in press), Princeton 1983
230. Temkin, M. I.: Int. Chem. Eng. **11**, 709 (1971)
231. Agranat, B. D.; Leibush, A. G.; Semenov, V. P.: Kinet. Katal. **17**, 1241 (1976)
232. Ross, J. R. H.; Steel, M. C. F.; Zeini-Isfahani, A.: In Mechanisms of Hydrocarbon Reactions. (Marta, F.; Kalló, D.; eds.). Budapest 1975
233. Bhatta, K. S. M.; Dixon, G. M.: Ind. Eng. Chem. Prod. Res. Develop. **8**, 324 (1969)
234. Rostrup-Nielsen, J. R.: Chem. Eng. Prog. **73**, (9) 87 (1977)
235. Licka, S.: Thesis, Vyscká Skola Chemico-Technologická Prag 1971
236. Meshenko, N. T.; Veselov, V. V.; Shub, F. S.; Temkin, M. I.: Kinet. Katal. **18**, 962 (1977)
237. Figueiredo, J. L.; Trimm, D. L.: Rev. Port. Quim. **19**, 363 (1977)
238. Bhatta, K. S. M.; Dixon, G. M.: Trans. Faraday Soc. **63**, 2217 (1967)
239. Saito, M.; Tokuno, M.; Ichiro, A.; Morita, Y.: Kogyo Kagaku Zasshi **73**, 2405 (1970)
240. Balashova, S. A.; Slovokhotova, T. A.; Balandin, A. A.: Kinet. Katal. **7**, 303 (1966)
241. Kato, A.; Ogino, Y.: Bull. Jap. Petrol. Inst. **15**, 115 (1973)
242. Tøttrup, P. B.: Appl. Catal. **4**, 377 (1982)
243. Allen, D. W.; Gerhard, E. R.; Linkins Jr., M. R.: Ind. Eng. Chem. Proc. Design Devel. **14**, 256 (1975)

244. Grenoble, D. C.; Estadt, M. M.; Ollis, D. F.: J. Catal. **67**, 90 (1981)
245. Nielsen, B.: (unpublished results)
246. Brooks, C. S.; Golden, G. S.; Lemkey, F. D.: Surface Technol. **11**, 333 (1980)
247. Kikuchi, E.; Ito, K.; Morita, Y.: J. Japan Petrol. Inst. **22** (2) 73 (1979)
248. Somorjai, G. A.: J. Elektrochem. Soc. **124**, 205 C (1977)
249. Nieuwenhuys, B. E.; Sachtler, W. M. H.: J. Colloid Interface Sci. **58**, 66 (1977)
250. Ross, J. R. H.; Steel, M. C. F.; Zeini-Isfahani, A.: J. Catal. **52**, 280 (1978)
251. Vannice, M. A.: Catalytic Activation of Carbon Monoxide on Metal Surfaces. In: Catalysis: Science and Technology vol. 3 (Anderson, J. R. and Boudart, M. eds.), Springer-Verlag, Berlin Heidelberg New York, 1982, p. 139
252. Cimino, A.; Boudart, M.; Taylor, H.: J. Phys. Chem. **58**, 796 (1954)
253. Schoubye, P.: J. Catal. **14**, 238 (1969)
254. Campbell, C. T.; Goodman, D. W.: Surface Sci. **123**, 413 (1982)
255. Ertl, G.; Weiss, M.; Lee, S. B.: Chem. Phys. Lett. **60**, 391 (1979)
256. Kikuchi, E.; Tanaka, S.; Yamazaki, Y.; Morita, Y.: Bull. Japan Petrol. Inst. **16**, 95 (1974)
257. Münster, P.; Grabke, H. J.: Bunsenges, Phys. Chem. **84**, 1068 (1980)
258. Veijola, V.; Kinnari, P.; Alaaho, P.; Harkonen, M.: Kem.-Kemi **3**, 173 (1976)
259. Barcicki, J.; Denis, A.; Grzegorizyk, W.; Nazimek, D.; Borowiecki, T.: React. Kinet. Catal. Lett. **5**, 471 (1976)
260. Kikuchi, E.; Ito, K.; Ino, Y.; Morita, Y.: J. Catal. **46**, 382 (1977)
261. Kikuchi, E.; Ito, K.; Morita, Y.: Wasada Daigaku Rikogaku Kenkyusho Hokoku **83**, 86 (1978)
262. Takami, K.; Igarashi, A.; Ogino, Y.: Bull. Japan Petrol. Inst. **19**, 37 (1977)
263. Rabinovitch, G. L.; Mozhaiko, V. N.: Neftekhimiya **15**, 373 (1975)
264. Koechloefl. K.: Proc. 6th Int. Congr. Catal. London 1976 (Bond, G. C.; Wells, P. B.; and Tompkins, F. C.; eds.). The Chemical Society, London 1976, p. 1122
265. Morita, Y.; Osawa, N.; Fukase, S.; Yamakazi, S.; Kikuchi, E.: J. Japan Petrol. Inst. **20**, 142 (1977)
266. Tomita, T.; Moriya, A.; Shinjo, T.; Kikuchi, K.; and Sakamoto, T.: J. Japan Petrol. Inst. **23**, 69 (1980)
267. Tomita, T.: (private communication)
268. Bridger, G. W.; Wyrwas, W.: Chem. Process Eng. **48**, 101 (1967)
269. Kjær, J.: Computer Methods in Catalytic Reactor Calculations, Haldor Topsøe, Vedbæk 1972
270. Jarvan, J. E.: Ber. Bunsenges. Phys. Chem. **74**, 142 (1970)
271. Phillips, T. R.; Yarwood, T. A.; Mulhall, J.; Turner, G. E.: J. Catal. **17**, 28 (1970)
272. Coad, J. P.; Riviere, J. C.: Surface Sci. **25**, 609 (1971)
273. Baker, R. T. K.; Harris, P. S.: In: Chemistry and Physics of Carbon Vol. 14 (Walker, P. L., Jr.; Thrower, P. A., eds.), Dekker, New York 1978, p. 83
274. Rostrup-Nielsen, J. R.; Trimm, D. L.: J. Catal. **48**, 155 (1977)
275. Hofer, L. J. E.; Sterling, E.; McCartney, J. T.: J. Phys. Chem. **59**, 1153 (1955)
276. Baker, R. T. K.; Harris, P. S.; Thomas, R.; Waite, R.: J. Catal. **30**, 86 (1973)
277. Rostrup-Nielsen, J. R.: J. Catal. **27**, 343 (1972)
278. Baker, R. T. K.; Sherwood, R. D.; Simoens, A. J.; Derouane, E. G.: In: Metal-Support and Metal-Additive Effects in Catalysis. (Imelik, B.; Naccache, C.; Courdier, G.; Praliaud, H.; Meriaudeau, P.; Gallezot, P.; Martin, G. A.; and Vedrine, J. C.; eds.), Elsevier, Amsterdam 1982, p. 149
279. Baird, T.; Fryer, J. R.; Grant, B.: Carbon (Oxford) **12**, 591 (1974)
280. Robertson, S. D.: Carbon (Oxford) **8**, 365 (1970)
281. Rostrup-Nielsen, J. R.: J. Catal. **33**, 184 (1974)
282. Butt, J. B.: Adv. Chem. Ser. **109**, 259 (1972)
283. Figueiredo, J. L.; Trimm, D. L.: Proc. 4th London Int. Carbon-Graphite Conf. 1974, p. 314
284. Baker, R. T. K.; Barber, M.; Harris, P.; Feates, F.; Waite, R.: J. Catal. **26**, 51 (1972)
285. Figueiredo, J. L.; Trimm, D. L.: J. Appl. Chem. Biotechnol. **28**, 611 (1978)
286. Lobo, L. S.; Trimm, D. L.: J. Catal. **29**, 15 (1973)

287. Lobo, L. S.; Trimm, D. L.; Figueiredo, J. L.: Proc. 4th Int. Congr. Catalysis Palm Beach 1972, Hightower, J. E., (ed) North-Holland Publ., Amsterdam 1973, p. 1125
288. Tøttrup, P. B.: J. Catal. **42**, 29 (1976)
289. Rostrup-Nielsen, J. R.: (unpublished results)
290. Massaro, T. A.; and Petersen, E. E.: J. Appl. Phys. **42**, 5534 (1971)
291. Lander, J. J.; Kern, H. E.; Beach, A. L.: J. Appl. Phys. **23**, 1305 (1952)
292. Holstein, W. L.; Boudart, M.: Lat. Am. J. Chem. Eng. Appl. Chem. (in press)
293. Wada, T.; Wada, H.; Elliott, J. F.; Chipman, J.: Metal. Trans. **4**, 2199 (1971) Lat. Am. J. Chem. Eng. Appl. Chem. (in press)
294. Nishiyama, Y.; Tamai, Y.: J. Catal. **33**, 98 (1974)
295. Bernardo, C.; Trimm, D. L.: Carbon (Oxford) **14**, 225 (1976)
296. Cunningham, R. E.; Gwathmey, A. T.: Advan. Catal. **9**, 25 (1957)
297. Eizenberg, M.; Blakely, J. M.: J. Chem. Phys. **71**, 3467 (1979)
298. Lobo, L. S.: Acta Cient, Venez., Supl. **24**, 219 (1973)
299. Moseley, F.; Stephens, R. W.; Stewart, K. D.; Wood, J.: J. Catal. **24**, 18 (1972)
300. Frennet, A.; Lienard, G.: Surface Sci. **18**, 80 (1969)
301. Presland, A. E. B.; Walker, Jr.: P. L.: Carbon (Oxford) **7**, 1 (1969)
302. Whalley, L.; Davis, B. J.; Moss, R. L.: Trans. Far. Soc. **66**, 3143 (1970)
303. Lahaye, J.; Badie, P.; Ducret, V.: Carbon (Oxford) **15**, 87 (1977)
304. Sundaram, K. M.; Froment, G. F.: Chem. Eng. Sci. **34**, 635 (1979)
305. de Blieck, J. L.; Grosens, A. G.: Hydrocarbon Process. **50**, (3) 76 (1971)
306. Zdonik, S. B.; Green, E. J.; Haller, L. P.: Oil Gas J. **65**, 96 (1967)
307. Mizuno, K.; Ikeda, M.; Imokawa, T.; Take, J.; Yoneda, Y.: Bull. Chem. Soc. Japan **49**, 1788 (1976)
308. Hayward, D. O.; Trapnell, B. M. W.: Chemisorption Butterworths, London 1964
309. Freel, J.; Galway, A. K.: J. Catal. **10**, 277 (1968)
310. Cohon, J. W.: Hydrocarbon Process. **60**, (1) 177 (1981)
311. Dent, F. J.; Cobb, J. W.: J. Chem. Soc. **2**, 1903 (1929)
312. Ovenston, A.; Wells, J. R.: Chem. Eng. Sci. **37**, 1 (1982)
313. Williams, D. S.; Möller, R.; Grabke, H. J.: High Temp. Sci. **14**, 33 (1981)
314. Borowiecki, T.: Appl. Catal. **4**, 223 (1982)
315. Misra, J.; Pandey, N. K.; Mukherjee, D. K.; Sen, S. P.: In: "Symposium on Science of Catalysis and its Application in Industry", FPDIL, Sindri 1979, p. 325
316. Bernardo, C.: Ph. D. Thesis (Univ. of London) 1977
317. Figueiredo, J. L.; Trimm, D. L.: J. Catal. **40**, 154 (1975)
318. Baker, R. T. K.; Sherwood, R. D.: J. Catal. **70**, 198 (1981)
319. Otto, K.; and Shelef, M.: Proc. 6th. Int. Congr. Catalysis, London, 1976, (Bond, G. C.; Wells, P. B.; and Tompkins, F. C., eds.), The Chemical Society, London 1977, p. 1082
320. Cabrera, A. L.; Heinemann, H.; Somerjai, G. A.: Chem. Phys. Letters **81**, 402 (1981)
321. Sen, S. P.: (private communication)
322. Bernardo, C.; Trimm, D. L.: Carbon (Oxford) **17**, 115 (1979)
323. Rostrup-Nielsen, J. R.: In: Progress in Catalyst Deactivation (Figueiredo, J. L., ed.) Martinus Nijhoff Publ. The Hague 1982, p. 127
324. Alstrup, I.; Rostrup-Nielsen, J. R.; Røen, S.: Appl. Catal. **1**, 303 (1981)
325. Rostrup-Nielsen, J. R.: In: "Progress in Catalyst Deactivation" (J. L. Figueiredo, ed.), Martinus Nijhoff Publ. The Hague 1982, p. 209
326. Wentrcek, P. W.; McCarthy, J. G.; Ablow, C. M.; Wise, H.: J. Catal. **60**, 257 (1980)
327. Bartholomew, C. H.; Wheaterbee, G. D.; Jarvi, G. A.: J. Catal. **60**, 257 (1979)
328. Christiansen, L. J.; Andersen, S. A.: Chem. Eng. Sci. **35**, 314 (1980)
329. Morita, S.; Inoue, T.: Int. Chem. Eng. **5**, 180 (1965)
330. Gorring, R. L.; De Rosset, A.: J. Catal. **3**, 341 (1964)
331. Mol, A.: Hydrocarbon Process. **53**, (2) 115 (1974)
332. Trimm, D. L.; Holmen, A.; Lindvang, O.: J. Chem. Tech. Biotechnol. **31**, 311 (1981)
333. Grabke, H. J.; Möller, R.; Schnaas, A.: Werkstoffe u. Korrosion **30**, 794 (1979)
334. Sickafus, E. N.: Surface Sci. **19**, 181 (1970)

Chapter 2

Automobile Catalytic Converters

Kathleen C. Taylor

Physical Chemistry Department General Motors Research Laboratories
Warren, Michigan 48090, USA

Contents

1. Introduction

Catalysts have been widely used to lower the emissions of carbon monoxide (CO) and hydrocarbons (HC) in the exhaust of automobiles in the United States since the introduction of 1975 models in the fall of 1974. These catalysts, contained in so-called catalytic converters in the exhaust system of automobiles, promote the oxidation of CO and HC to CO_2 and H_2O under net oxidizing conditions (*e.g.* A/F > 14.6[1]). Until 1978, emission control requirements for nitrogen oxide (NO_x) emissions were met through non-catalyst technology, primarily exhaust gas recirculation (EGR) [1, 2]. Starting with some vehicles sold in California in 1977, NO_x emissions from gasoline engines have been subject to catalytic control. The catalyst here has the additional function to promote the reduction of NO to N_2 via reaction of NO with hydrogen or CO. Catalyst systems designed to reduce NO_x are considerably more complex than the earlier control systems. For example, the control system introduced by General Motors on some 1978 model year cars has closed-loop air-fuel ratio control (closed-loop fuel metering system, exhaust gas oxygen sensor, and an electronic control unit) as well as a three-way catalyst which simultaneously promotes the conversion of HC, CO, and NO_x [3]. Stringent federally mandated emission control requirements of 1 gram per mile (g mi^{-1}) for NO_x have led to the further application of three-way catalysts. This review will emphasize the state-of-the-art of catalytic control of automobile exhaust emissions since 1978, specifically three-way catalysts. A recent review by J. Kummer covers part of this period and earlier years [4]. Other reviews of this subject are listed in the reference section [5–14].

2. Emission Regulations in the United States

Table 1 lists the passenger car emission control requirements (current as of September, 1982) for all passenger cars sold in the United States except where waivers have been granted by the Environmental Protection Agency for specific vehicles. The law is currently under review (1982) so these requirements may be changed for future years. The 1981 exhaust standards represent a reduction from uncontrolled 1960 levels of 96% for HC, 96% for CO, and 76% for NO_x.

Exhaust emissions of an automobile are compared with the standard according to a preproduction certification procedure. Certification requires an 80,450 km (50,000 mi) durability test on a prototype car for each engine

[1] A/F ≡ air-to-fuel ratio

family, drivetrain, and catalyst combination. The durability mileage is accumulated on an approved route according to a pre-established driving schedule, the AMA cycle. Prior to mileage accumulation and at set mileage intervals during the AMA, emission tests are conducted on the vehicle. The term 1975 Federal Test Procedure (FTP) is commonly used to refer to the emission test (also called CVS-CH test) by which the vehicle emissions are compared with the emission control requirements. The 1975 FTP is a standardized 23 cycle driving schedule for which vehicle preconditioning, driving speed, and test duration are specified. The vehicle is driven on a dynamometer, and the exhaust emissions are sampled and analyzed for HC, CO, and NO_x. The analysis techniques used are nondispersive infrared for CO, flame ionization for HC, and chemiluminescence for NO_x. The exhaust emission test results determined according to the 1975 FTP between 8045 and 80,450 km are plotted versus mileage; and the ratio of the least squares estimate of the

Table 1. Passenger car regulatory exhaust emission control requirements in the United States. Emissions are expressed in grams per mile (g mi^{-1})

Model Year	1975–76	1977–79	1980	1981–82	1983
HC	1.5	1.5	0.41	0.41	0.41
	CA 0.9[c]	CA .41	CA 0.39	CA 0.39	CA 0.39
CO	15	15	7.0	3.4[a]	3.4
	CA 9.0	CA 9.0	CA 9.0	CA Opt.[d]	CA 7.0
NO_x	3.1	2.0	2.0	1.0[b]	1.0[b]
	CA 2.0	CA 1.5	CA 1.0	CA Opt.[d]	CA 0.4

[a] Possible 2 year waiver to 7 g mi^{-1}.
[b] Possible waiver to 1.5 g mi^{-1} for diesel or innovative technology through 1984.
[c] CA designates California.
[d] CA option to 1981 Federal standards: If select to meet CO 7.0 and NO_x 0.7 in 1981, must also select these standards in 1982. If select to meet the 1981 Federal standards, must meet CO 7.0 and NO_x 0.4 in 1982.

Additional Car Requirements

1981 — No crankcase emissions allowed.
Tampering by service industry, dealers, etc. prohibited. Fuel filler must exclude leaded fuel nozzles (catalyst veh.). Exhaust standards apply to diesel with test modification.
Assembly line test requiremet-SEA.
Parameter adjustment requirements; idle mixture and choke.

1982 — Diesel particulate std. 0.6 g mi^{-1}.
High altitude stds.: Exh. HC 0.57, CO 7.8, NO_x 1.0 g mi^{-1}, evap. 2.6 g per test; altitude waiver stds.: CO 11, NO_x 1.5.

1984 — All cars meet stds. at all altitudes.

1985 — Diesel particulate std. 0.2 g mi^{-1}.

California — In Addition to Federal Car Requirements

1981 — End-of-line exhaust test.
Fuel filler specs. for vapor recovery-individual veh. delay till 1982 possible depending on extent of body changes.
Higher optional std. for 100,000 mi certification.
Restrictions on allowable maintenance.
"Fixed" idle mixture required.
HFET NO_x std. 1.33 × FTP std.
Must "meet" CO standard to 6000 ft.

mean emission at 80,450 km to the least squares estimate of the mean emission at 6436 km is called the deterioration factor (df) [15]. The deterioration factors so established are then used to predict the 80,450 km emissions of data vehicles which have been driven for just 6436 km (4000 mi) by multiplying the 6436 km emissions by the deterioration factor. For a vehicle to be certified by the EPA, the extrapolated emission numbers cannot exceed the requirements shown in Table 1.

Emission control requirements have been established in the U.S. for medium and light-duty trucks (Table 2) and for heavy-duty vehicles. Requirements for trucks sold in California are stricter and have different vehicle weight classifications.

Table 2. Light/Medium Duty Truck Regulatory Emission Control Requirements[a]

	1981–1983	1984
HC	1.7	0.8
CO	18	10
NO_x	2.3	2.3

[a] Units are g mi^{-1}. Federal definition of light duty trucks is ≦8500 gross vehicle weight.

The combined effect of new stricter emission control requirements for 1981, changes in the certification regulations over the years, different regulations for vehicles granted waivers, and changes in vehicles in order to simultaneously improve fuel economy and exhaust emissions is the use of many different catalysts by automobile manufacturers on 1981 and 1982 model year vehicles. For example, the total amount of noble metal and the Pt/Pd/Rh ratios differ among catalysts. Some vehicles have single three-way catalytic converters while others have dual-bed catalytic converters for which the first bed operates as a three-way converter, and air is added before the second bed which contains an oxidation catalyst. Catalyst supports are both alumina coated ceramic monoliths and alumina pellets.

3. Exhaust Emission Characteristics

The engine-out exhaust emissions of CO, HC, and NO_x vary as a function of air-fuel ratio as well as several other parameters such as ignition timing and EGR. Here we shall consider only the relationship to air-to-fuel ratio (A/F). A general relationship between engine out CO, HC, and NO_x emissions and A/F is shown in Figure 1. Engine operation at lean A/F (net oxidizing condition) results in lower HC and CO emissions and more O_2. These conditions favor subsequent catalytic oxidation reactions. At lean A/F the exhaust contains insufficient reducing agents to react with all the O_2 *and* all the NO. By operating closer to the stoichiometric A/F more NO can react but the volume of CO which must be oxidized increases. The stoichiometric A/F occurs at about 14.6. Here, the concentrations of oxidizing gases and reducing gases are matched, and equilibration of the exhaust mixture would yield only CO_2, H_2O, and N_2. The application of three-way catalytic converters has this objective. Three-way catalysts operate in a narrow A/F band between 14 and 15.

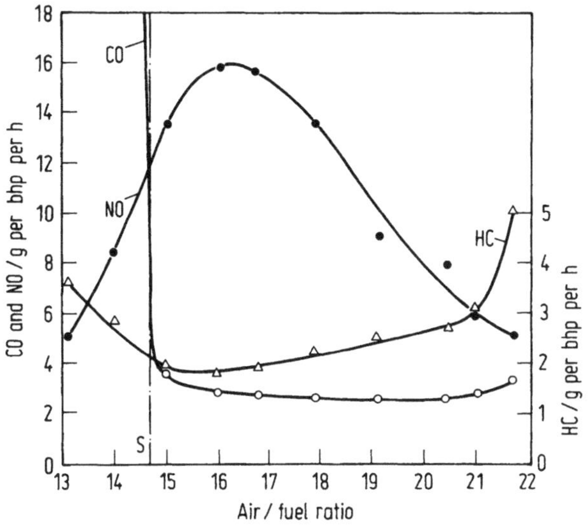

Figure 1. The effects of air-fuel ratio on hydrocarbon, carbon monoxide, and nitric oxide exhaust emissions. (Reproduced with permission from ref. [11])

The exhaust hydrocarbons are a mixture of olefins, paraffins, and aromatics ranging from C_1 to C_8. Table 3 lists the important exhaust hydrocarbons [16]. The total and relative concentrations will vary depending upon the A/F and spark timing [16]. Other exhaust constituents include hydrogen which is present at a concentration of about 1/3 that of CO. Water vapor and carbon dioxide concentrations are each about 10% by volume. Sulfur dioxide emissions are 20 ppm for a corresponding sulfur level in the fuel of 0.03 wt%. The use of alternative fuels such as ethanol and methanol leads to an increase in the fraction of exhaust hydrocarbons emitted as aldehydes.

Other exhaust constituents which are significant because they can lead to deterioration of catalyst performance are phosphorus and lead. Phosphorus is present in the fuel (typically 0.02–0.1 mg P l^{-1}) but can also be derived from oil consumption. Motor oil typically contains 1.2 g P l^{-1}. Lead (as tetraethyl lead) is present at low levels (1 mg Pb l^{-1}) in unleaded fuel.

Engine-out emissions of HC, CO, and NO_x measured during FTP tests have been observed to fall within the following ranges: 1.3 to 2.5 g mi^{-1} HC, 10 to 17 g mi^{-1} CO, and 0.8 to 2.0 g mi^{-1} NO_x [17]. In order for these engine-out emissions to fall below the 1981 Federal emission control requirements (0.41 HC, 3.4 CO, 1.0 NO_x), percentage conversions of 70 to 90 percent for HC, 66 to 80 percent for CO, and 0 to 50 percent for NO_x are required.

Table 3. Important Exhaust Hydrocarbons. Summary of Results by Model Year

	1970	1972	1973	1974	1975	1976	1977
No. of Calif. Cars	—	1	—	1	5	—	4
No. of Fed. Cars	1	—	3	8	9	2	—
Mileage	3600	42105	19802	9162	16075	3218	9859
Carbon Percent of Total Hydrocarbon							
Nonreactive Hydrocarbons							
Acetylene	7.4	10.7	7.8	7.7	2.5	3.6	0.3
Methane	6.2	6.3	4.9	4.7	11.3	9.8	28.0
Benzene	6.4	3.8	3.4	4.9	3.7	4.8	1.6
Ethane	1.2	0.9	1.0	1.0	2.6	3.0	4.5
Propane	0.1	0.0	0.1	0.1	0.3	0.3	0.3
Nonreactive Total	21.4	21.7	17.1	18.4	20.4	21.6	34.7
Reactive Hydrocarbons							
Ethylene	10.1	15.5	12.8	12.4	7.4	7.2	7.8
Toluene	11.7	11.4	9.1	7.5	6.5	7.2	7.7
Xylenes	9.7	2.3	4.1	7.7	5.9	7.3	2.0
Propylene	8.0	9.6	6.3	6.1	3.1	3.9	1.5
Trimethylpentanes	2.1	3.0	5.5	4.0	6.4	5.0	15.6
n-Butane	1.6	7.0	2.0	4.7	5.3	3.9	7.2
i-Pentane	4.4	5.2	4.5	3.6	5.4	4.5	5.9
Butenes	6.0	2.3	2.9	4.0	2.5	4.2	0.4
Methylpentanes	2.0	1.2	2.9	2.0	3.0	3.2	1.5
n-Pentane	0.6	2.1	1.4	1.8	2.1	2.1	0.8
Ethylbenzene	1.7	0.6	1.1	1.7	1.4	1.9	0.6
i-Butane	0.8	2.3	0.8	0.6	1.0	0.8	0.3
Other Hydrocarbons	19.8	15.6	29.7	25.6	29.6	27.2	14.0
Reactive Total	78.6	78.3	82.9	81.6	79.6	78.4	65.3
Hydrocarbon Classes							
Total Parafins	23.9	34.5	36.8	33.2	52.9	46.4	74.3
Total Olefins	32.0	31.1	25.7	25.8	16.4	17.0	9.9
Acetylene	7.4	10.7	7.8	7.7	2.5	3.6	0.3
Total Aromatics	36.7	23.7	29.7	33.3	28.2	33.0	15.5

Included in the tailpipe emissions will be gases emitted at the start of the test while the catalyst is still below its operating temperature. HC and CO emissions from the first part of the 1975 FTP test are a significant fraction of the total tailpipe emissions.

The temperature of the exhaust gases vary depending upon time from start up, engine tuning, engine load, distance from the engine, and heat losses in the exhaust system. Under steady-state conditions exhaust gas temperatures at the inlet to the converter of 673 K to 773 K can be expected.

4. 1981 Emission Control Technology

A. Converters

The catalytic converters used on most 1981 model year vehicles fall into two general categories: three-way converters and dual-bed converters. (Some manufacturers employ dual converters. The distinction between dual-bed and dual converters is whether the two catalysts are housed in the same container or in separate containers.) Both contain a three-way catalyst, but with the dual-bed converter (and dual converter) the three-way catalyst is followed by an oxidation catalyst to provide increased oxidation capability. Supplemental air is added to the exhaust ahead of the oxidation catalyst. The pellet type dual-bed catalytic converter used by General Motors is shown in Figure 2. Both the three-way catalyst and oxidation catalyst are enclosed in the same converter, separated by the air plenum. Operation of the three-way catalyst requires that the exhaust A/F be controlled close to the stoichiometric composition. A diagram which illustrates the relationship between conversion efficiency and A/F for the two converter types is presented in Figure 3. A closed-loop feed-back control system holds the A/F in a narrow region near stoichiometry.

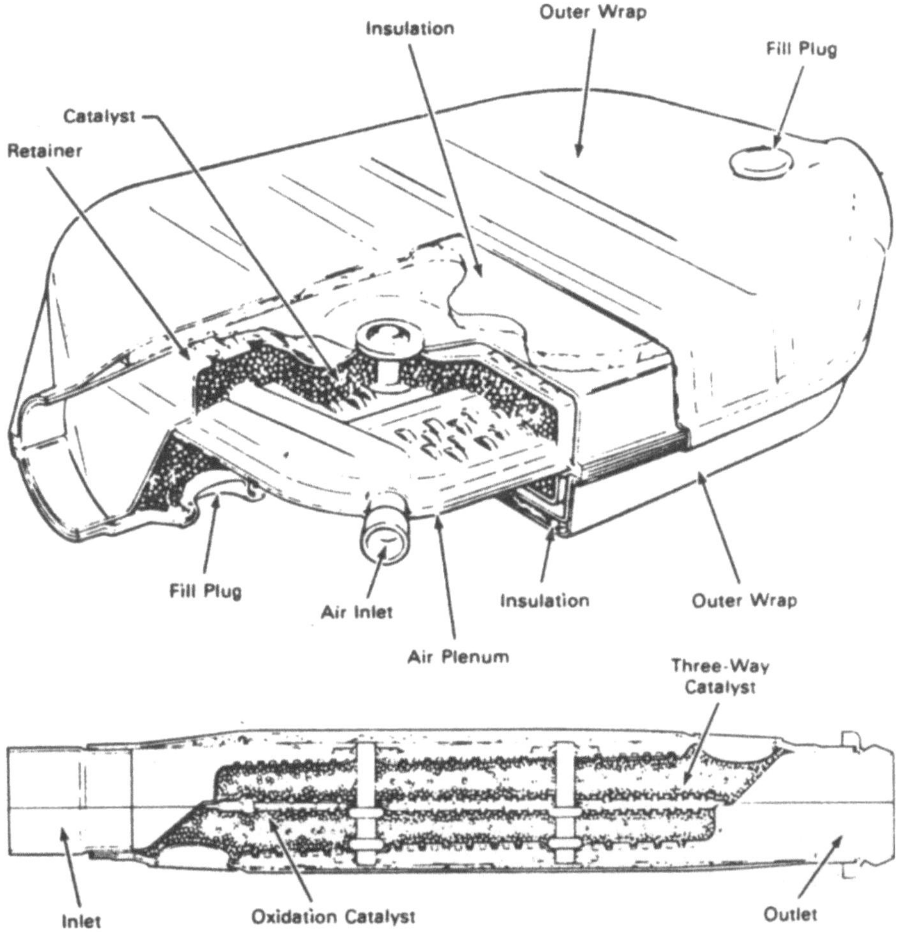

Figure 2. Cross-section of General Motors dual-bed converter. (Reproduced with permission from ref. [18a])

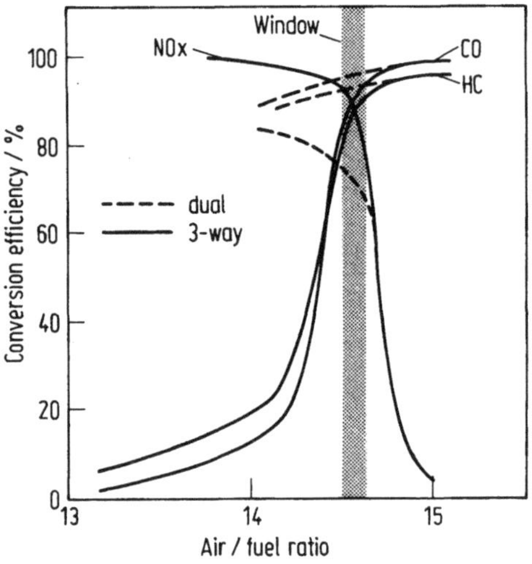

Figure 3. Efficiency scan for a dual-bed catalyst and a three-way catalyst. (Reproduced with permission from ref. [18a])

B. Control System

The basic elements of a closed-loop emission control system are shown in Figure 4. In addition to the catalytic converter (pellet or monolithic type) the system has an exhaust oxygen sensor, electronic control module, carburetor or fuel injector, and air pump [11].

The components of the 1981 General Motors micro-computer engine control system described in a report by Grimm et al. [18, see also 11] will be reviewed here. Reports on other closed-loop A/F control systems with three-way converters are listed in the references [3, 19–22].

The exhaust oxygen sensor indicates whether excess oxygen is present in exhaust. The sensor behaves like a switch in that its output voltage goes through a step change at the stoichiometric point with an A/F change of less than 0.1 A/F unit. The voltage signal from the oxygen sensor passes to the electronic control module and is compared to a preset voltage. The electronic control module then generates a correction control signal which

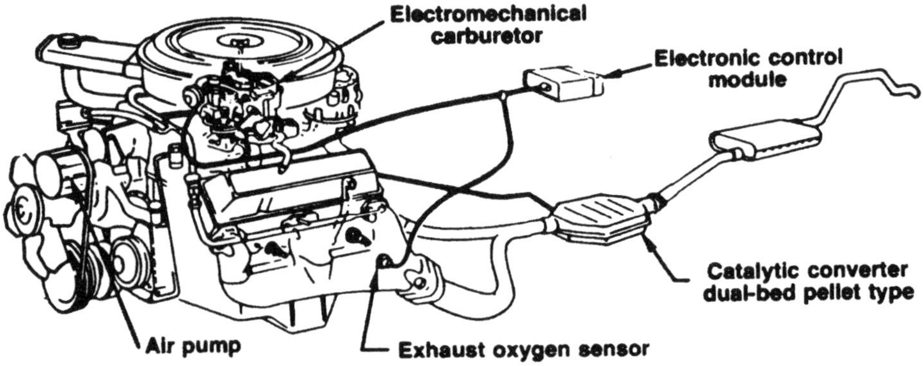

Figure 4. Closed-loop emission control system. (Reproduced with permission from ref. [11])

passes to the carburetor. The control signal increases or decreases at a fixed rate (integral correction) until the oxygen sensor shows that the A/F has crossed the stoichiometric composition; the control signal then reverses following a step change (proportional correction) in the new direction.

The integral and proportional corrections are not fixed but change in value based on input from other signals which follow engine speed and load. The electronic control module therefore receives information on engine coolant temperature, engine speed, throttle position, and manifold pressure. This state-of-the-art system has the potential capability of controlling spark timing, idle speed, exhaust gas recirculation, and transmission converter clutch. It is also able to recognize malfunctioning systems and provides an indication to the driver that a malfunction exists. Appropriate values are then used by the module to permit the vehicle to be operated until the necessary repairs can be made.

The control signal oscillations inherent in the closed-loop system produce oscillations in the exhaust A/F. The amplitude and frequency of the oscillations depend on the transport delay in the engine (manifold and cylinder residence time) and system calibrations. The transport delay decreases as engine speed and/or engine load increases. One cycle per second is a typical oscillation frequency. The closed-loop system is currently the most effective way to maintain the exhaust A/F in the range in which the three-way catalyst simultaneously promotes the reactions of HC, CO, and NO_x.

Not all 1981 and 1982 model year vehicles have the closed-loop control system; some are operated "open loop". Open loop systems lack the oxygen sensor and feed-back mechanism but retain the three-way or dual catalyst feature. Papers on the development of three-way catalyst systems have been published by several automobile manufacturers. Topics covered include methods for fuel delivery [23], secondary air control [24], catalyst durability [25], and influence of system parameters on catalyst performance [23, 26–30]. The development of closed-loop systems with the research objective of keeping the NO_x emissions below 0.4 g mi^{-1} has been reported by Cederquist *et al.* [31] and Zemke and Gumbleton [18a].

5. Catalyst Screening

The automobile manufacturers have developed dynamometer-controlled engine facilities which they use to screen catalysts before certification [30, 32–35]. Aging of catalysts by tests on a dynamometer allows catalyst durability to be measured under precisely controlled conditions and possibly completed in fewer days (by running 24 h days) than for vehicle aging. Dynamometer-controlled engine facilities are used for accelerated aging in which the catalysts are exposed to poisons and high temperatures as well as for short emissions tests. A short description of catalyst screening tests as described in published reports follows.

The AC Spark Plug Division of General Motors has developed a catalytic converter testing facility which consists of a dynamometer-controlled engine, a gas flow control system, emission analyzers, and a computer interface [32, 33]. Engine operation may be varied according to a simulated driving schedule in order to produce a desired exhaust composition and/or catalyst bed temperature. (The catalyst temperature is of course dependent upon the amount of heat generated by reaction.) Conversely, the exhaust characteristics can be fixed by utilizing a constant schedule, e.g. 80 $km \cdot h^{-1}$. Exhaust temperatures and fuel compositions are chosen to be representative of vehicle certification for a given year. For some studies catalytic converters are rotated among a series of engines to ensure that the converters have seen similar conditions [33]. Emission tests are run for these dynamometer aged converters at regular intervals (e.g. 100 h) according to a dynamometer converter performance test [32]. The catalytic conversions of NO_x, HC, and CO are compared at the start and throughout the 1000 h dynamometer durability test. These durability tests are used to compare catalysts, to test the response of catalysts to fuel compositions (additional Pb, P, or S), and to test the response of catalysts to different aging schedules (exhaust composition, temperature, etc.).

Another catalyst property examined in these dynamometer-controlled engine facilities is the "transient conversion efficiency" of the catalyst, that is, the time required for the catalytic converter initially at ambient temperature to start functioning following a step input of a steady-state exhaust gas [32]. Just CO and HC conversions are monitored in this test because in practice converters are usually warmed up under oxidizing conditions (by addition of air to the exhaust).

With the advent of three-way catalysts durability tests have been developed for which catalyst aging using a dynamometer-controlled engine is followed by a converter performance test in which the A/F is deliberately cycled [30, 34]. For the converter performance tests carried out at the AC Spark Plug Division of General Motors the A/F is cycled ± 0.5 A/F units at a frequency of 1 Hz about a fixed value. This fixed value is continuously moved very slowly from lean to rich and back over an A/F band of about ± 0.3 around stoichiometry. Tests of converter performance give information on NO_x, CO, and HC conversions for fresh and aged catalysts as a function of A/F. Typical conditions are a converter inlet gas temperature of 750 K, exhaust gas flow rate about 33,000 $cm^3\ s^{-1}$ (293 K, 1 atm), and a corresponding space velocity of 27,900 h^{-1} [34].

The Ford Motor Company has developed an engine dynamometer test procedure for three-way catalyst screening which like the GM procedure allows for perturbations of the A/F [35]. The Ford aging procedure is *accelerated* compared with the AMA schedule as the test is completed in only 25% of the time of the AMA. Compared with the AMA schedule the number of high speed acceleration (peak temperature) modes is increased fourfold. The fuel lead and phosphorus levels are increased to four times the 1977 certification fuel levels. Just 200 h of dynamometer aging was assumed equivalent to 40,225 km on the standard AMA cycle; a direct comparison of a dynamometer aged catalyst with the same catalyst formulation aged on a vehicle showed the dynamometer aging cycle to be slightly more severe than vehicle aging. The engine dynamometer catalyst screening procedure involved the measurement of NO_x, CO, and HC conversions while the A/F was scanned by applying a DC voltage from a function generator to the electronic control unit. Deliberate oscillations of the A/F around the mean operating point were made by imposing a periodic signal on the DC voltage. Comparison of emission tests with and without A/F oscillations showed that oscillations are necessary to ensure simulation of vehicle operation. For example, for tests of NO_x emissions done with A/F oscillations, catalyst rankings agreed with FTP results, whereas screening tests done without perturbations gave different rankings.

6. Laboratory Testing

Laboratory testing facilities used for the characterization and preliminary evaluation of three-way catalysts have been described in several reports. Descriptions of several of these reactor systems are given here.

A laboratory reactor used in many studies of automobile emission control catalysts is an integral plug-flow reactor with continuous gas analysis [36–38]. This reactor system allows one to simulate typical exhaust conditions in the laboratory but at the same time provides greater reproducibility and control of conditions than available with dynamometer-controlled engine tests.

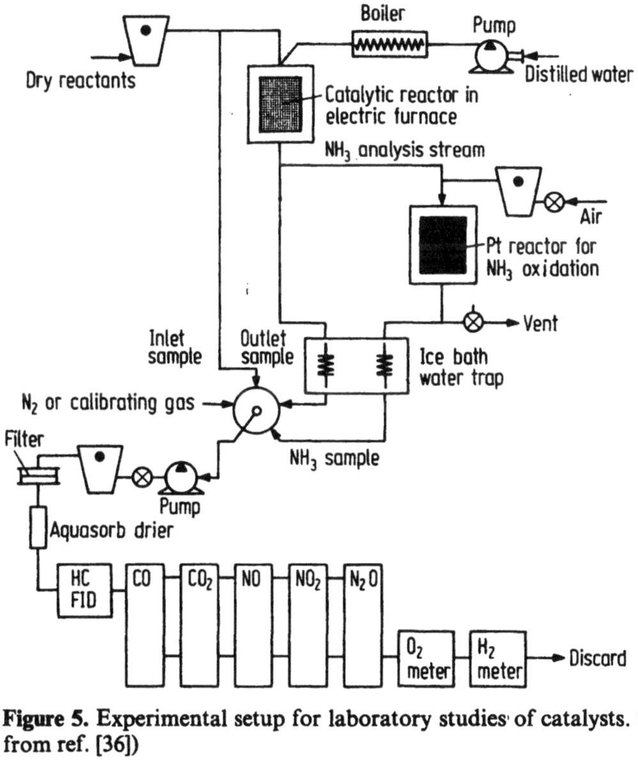

Figure 5. Experimental setup for laboratory studies of catalysts. (Reproduced with permission from ref. [36])

An exhaust-like feedstream blended for laboratory experiments typically contains NO (0.05 to 0.2 vol%), CO (0.5 to 2.0 vol%), H_2 (one-third the amount of CO), propylene (0.1 vol% and also propane or methane to simulate less reactive hydrocarbons), CO_2 (10 vol%), H_2O (10 vol%) and balance N_2. The integral reactor used by Klimisch and Barnes [36] is shown in Figure 5. The reactor itself is a 3/4 inch I.D. stainless steel pipe. The reactor is situated in an electrical furnace which externally heats the catalyst. A proportional controller regulates the furnace temperature. The gases are preheated as they pass through a 30 cm section of silicon carbide pellets at the inlet to the reactor just ahead of the catalyst. Water vapor is formed by pumping water through a mini pump to a stainless steel coil situated in a furnace close to the reactor inlet. The catalyst temperature is monitored by one or more thermocouples in the catalyst bed. The catalyst bed size is typically 15–30 cm^3. The gas space velocity can be varied over a wide range but is limited by the size of the reactor at one extreme and the channeling of the gases at the other. Typical values are 10,000 to 50,000 h^{-1} (STP). After the gases pass through the reactor the gases are passed through a condenser to remove water vapor. A portion of the exit gases are pumped through the analysis train by a throttled diaphragm pump. The composition of the inlet and outlet gases is determined by comparison with calibration gases.

The continuous gas analyzers used with these laboratory reactors are nondispersive IR for CO, CO_2, NO, and H_2O, nondispersive UV for NO_2, paramagnetic detection for O_2, and flame ionization detection for hydrocarbons. Either nondispersive IR or chemiluminescence detection is used for NO. Klimisch and Barnes [36] have reported on a convenient method for ammonia determination whereby the gases leaving the reactor are diverted, oxygen added, and passed over a platinum catalyst which quantitively oxidizes the ammonia back to nitric oxide. The ammonia reading is determined by difference using the NDIR NO analyzer. Breitenbach and Shelef [39] developed a similar technique with a copper oxide catalyst to convert NH_3 to NO. Accurate measurement of H_2 in the effluent is best done by (noncontinuous) gas chromatography.

When sulfur dioxide is included in the laboratory feedstream, certain modifications in the integral reactor system are necessary [40–42]. The reactor tube is changed to Pyrex, and a Goksøyr-Ross coil [43] is added after the catalyst to condense sulfuric acid which is subsequently analyzed by titration of the condensate. A continuous SO_2 analyzer is added to the analysis train. Alternatively SO_2 may be analyzed by sampling through an impinger containing hydrogen peroxide; the sulfate formed is determined by titration [44]. For studies in which the SO_2 is reduced to H_2S, the H_2S can be trapped in a buffered zinc acetate solution and analyzed as methylene blue [45, 46].

Otto *et al.* [47] developed a "pulse-flame reactor" which utilizes combustion products from a laboratory burner for the feedstream passed over the catalyst. This reactor is suitable for aging catalysts for durability studies because either contaminant-free fuel or fuel doped with lead, phosphorus, and sulfur compounds can be used. A diagram of the pulse-flame apparatus

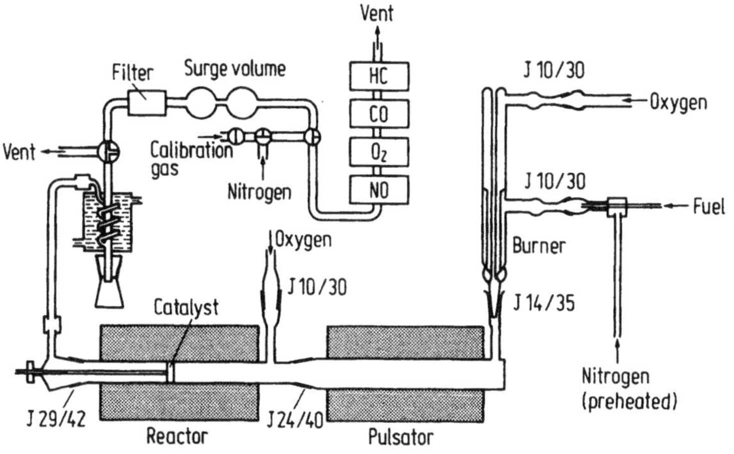

Figure 6. Schematic of pulse flame combustor. (Reproduced with permission from ref. [47])

is shown in Figure 6. The main parts of the system are the feed system, burner, pulsator, catalytic reactor, and analytical instruments for gas analysis. The Pyrex burner is 28 cm in length and 1.7 cm in diameter. Oxygen enters the top of the burner. Preheated nitrogen is mixed with isooctane fuel and enters through another tube so that a combustible mixture forms only near the burner tip. The Vycor pulsator tube which follows is 40 cm in length and 3 cm in diameter. The tube is held at 1023 K. The combustible mixture is heated as it passes through the tube, ignites, burns back to the burner face, and is quenched. This process then repeats itself. The resulting concentration of CO and HC are determined by the oxygen-fuel ratio. The oxygen concentration in the exhaust gases can be adjusted by adding supplemental oxygen just before the reactor. Typical compositions during aging and poisoning are 1.0 vol% CO, 600 ppm HC (as C_6) 3.5 vol% O_2, and 0.005 to 0.01 vol% NO. The catalytic reactor is a 55 cm long by 2.4 cm diameter Vycor tube. The catalyst is a small core of a monolith (1.2 cm length by 1.9 cm diameter) precisely cut to fit the reactor. The catalyst temperature is monitored by a thermocouple at its outlet. The space velocity is a nominal 40,000 h^{-1} (293 K, 1 atm), though higher during a pulse surge.

Schlatter *et al.* [48] have developed a laboratory reactor system in which the composition of the feedstream can be made to oscillate. The frequency and amplitude of the oscillations can be varied independently. This laboratory reactor is used to investigate the behavior of three-way catalysts in an environment which resembles the exhaust A/F fluctuations in a closed-loop emission control system. A schematic diagram of the system appears in Figure 7. Two gas blending systems, each complete with flow controller and mass flow meters, are used to establish the composition of the individual feedstreams. Two fast-acting solenoid selector valves allow one to cycle between the two feedstreams at a switching frequency which can

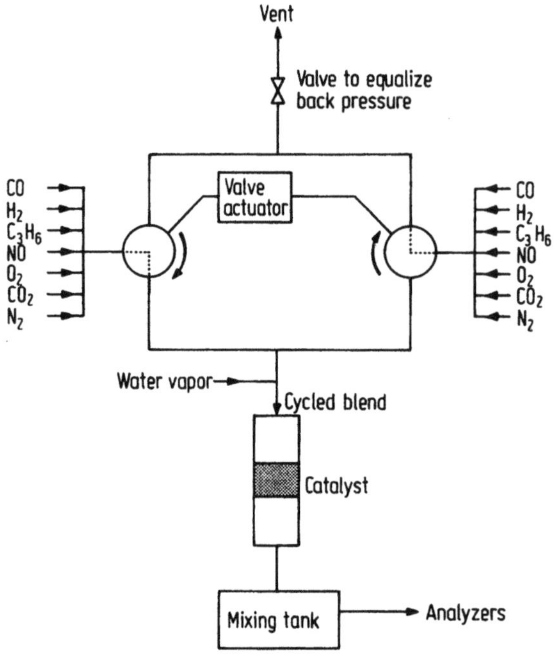

Figure 7. Schematic diagram of system for exposing catalyst to an oscillating feedstream composition. The apparatus was used to make the measurements shown in Figures 19, 20, and 22. (Reproduced with permission from ref. [48])

be varied from 0.03 to 10 Hz. The choice of the two gas blends determines the amplitude and average composition of the oscillations. Mixing of the two feedstreams in the small volume between the switching valves and the catalyst attenuates the amplitude from the square wave shape, especially when the switching frequency is fast. Schlatter *et al.* [48] have examined this characteristic of their laboratory reactor in some detail. The continuous gas analyzers see only average conversions independent of cycling conditions due to the addition of a 2 l mixing volume to the system after the catalyst.

7. The Chemical Reactions

The essential requirement for an effective three-way catalyst is high conversions of NO_x, CO, and hydrocarbons at and near the stoichiometrically balanced exhaust composition. In general, performance is limited by low conversions of CO and hydrocarbons as the A/F is changed in the reducing (rich) direction and decreased conversion of NO(NO_x) as the A/F is changed in the oxidizing (lean) direction. The narrow range of A/F around the stoichiometric point where conversions are high is commonly referred to as the operating "window". A wide A/F range or window of high simultaneous NO_x, CO, and HC conversions is a desirable catalyst characteristic because it lessens the need for tight A/F control.

The overall catalytic reactions which are important for controlling exhaust emissions are given by the following stoichiometric equations:

$$CO + 1/2\,O_2 \rightarrow CO_2 \quad (1)$$

$$\text{hydrocarbons} + O_2 \rightarrow H_2O + CO_2 \quad (2)$$

$$H_2 + 1/2\,O_2 \rightarrow H_2O \quad (3)$$

$$NO + CO \rightarrow 1/2\,N_2 + CO_2 \quad (4)$$

$$NO + H_2 \rightarrow 1/2\,N_2 + H_2O \quad (5)$$

as well as

$$\text{hydrocarbons} + NO \rightarrow N_2 + H_2O + CO_2 \quad (6)$$

$$NO + 5/2\,H_2 \rightarrow NH_3 + H_2O \quad (7)$$

$$CO + H_2O \rightarrow CO_2 + H_2 \quad (8)$$

$$\text{hydrocarbons} + H_2O \rightarrow CO + CO_2 + H_2 \quad (9)$$

and possibly

$$3\,NO + 2\,NH_3 \rightarrow 5/2\,N_2 + 3\,H_2O \quad (10)$$

$$2\,NO + H_2 \rightarrow N_2O + H_2O \quad (11)$$

$$2\,N_2O \rightarrow 2\,N_2 + O_2 \quad (12)$$

$$2\,NH_3 \rightarrow N_2 + 3\,H_2 \quad (13)$$

The desired products are N_2, CO_2, and H_2O. These reaction products are thermodynamically favored at typical exhaust temperatures, *e.g.* 770 K. The overall conversion is determined not only by the activity of the catalysts to promote reactions 1 to 5 but also by the availability of oxidizing and reducing agents. Oxidizing agents are necessarily in short supply at rich A/F so HC, CO, and H_2 may be viewed as competing with each other for NO and O_2. Reducing agents are in short supply at lean A/F so NO and O_2 may be viewed as competing for the available H_2, CO, and HC. The *selectivity* of the three-way catalyst refers to its tendency to promote CO and HC oxidation (reactions 1 and 2) rather then H_2 oxidation (reaction 3) at rich A/F and to promote NO reduction (reactions 4, 5, and 6) rather than H_2 oxidation (reaction 3) and to some extent O_2 reduction by CO (CO oxidation, reaction 1) at lean A/F. In addition, selectivity refers to the tendency of the three-way catalyst to reduce NO to N_2 (reactions 4 and 5) rather than to NH_3 (reaction 7) at rich A/F. Reactions of H_2O with CO (water-gas shift) and hydrocarbons (steam reforming) may contribute to the conversions of CO and HC under net reducing conditions. In exhaust, sufficient reactants are present to completely react NO, CO, and HC over the entire A/F range if we could catalyze particular reactions while suppressing others.

One goal of research and development work on three-way catalyst formulation is to improve catalyst selectivity. As an example, Schlatter and Taylor [49] have demonstrated that three-way catalysts which contain Pd can be less selective for reducing NO to N_2 than are catalysts which contain only Rh or Pt and Rh after exposure of the catalysts to higher temperatures. The Pd containing catalyst reduced more NO to NH_3 under rich conditions and less NO to N_2 under lean conditions than did the Rh and Pt-Rh catalyst [49]. The results may be explained by the greater tendency of Pd to reduce NO to NH_3 compared with Rh, the consumption of CO by reaction 1 over Pd at the expense of reaction 4, and the sintering of Pt during the high temperature oxidation treatment so that its contribution to reactions 1 and 7 was diminished.

The competition among reactants and the selection of reactants by the catalyst largely determines the effectiveness of the three-way catalyst. Laboratory studies of a single reaction such as $2\,NO + 2\,H_2 \rightarrow N_2 + 2\,H_2O$ tell us whether a particular catalyst can catalyze this reaction but do not tell us whether this reaction is solely responsible for the reduction of NO to N_2 in exhaust. Also no single reaction has been identified which can be used for catalyst evaluations in order that a simple feed may substitute for the exhaust-like feed. Laboratory studies of three-way catalysts are therefore carried out using feeds of varying complexity and at a range of concentrations and temperatures in order to obtain a view of the relative contribution of the catalytic reactions 1 through 13 to the overall conversions. Differences in the detailed chemical mechanisms represented by these overall reactions are to be expected as the catalyst composition is changed. The reader should also note that the conversion of NO, CO, and HC in exhaust will be influenced by the promoting or inhibiting effects of all species in exhaust, not just reactant concentrations. Also two sequential chemical reactions may contribute to the overall conversion. For example, ammonia formed by the reaction of NO with inlet H_2 (or with H_2 produced by the water-gas shift reaction) may decompose in a second reaction to yield N_2. Such a mechanism has been shown to contribute to NO reduction over alumina supported Pt—Ni catalysts under reducing conditions [50].

A brief discussion of NO reduction chemistry follows. The literature on fundamental studies of CO and HC oxidation will not be reviewed here.

A complete and detailed chemical mechanism by which NO is reduced to N_2 over Rh containing three-way catalysts in exhaust is not known. Several recent studies of the reduction of NO over Rh single crystal surfaces, over polycrystalline Rh wires, and over supported Rh have made an important contribution to our understanding of the mechanism [51–52]. For example, mechanistic studies of the NO—CO reaction over polycrystalline Rh wire [52] and a Rh(331) single crystal [51] have led to the proposed mechanism for NO reduction to N_2 whereby NO dissociation on Rh is followed by N_2 desorption and the reaction of an intermediate surface oxygen specie with CO to form CO_2. Campbell and White [52] reported that thermal desorption of co-adsorbed NO and CO from polycrystalline Rh produced simultaneous CO_2 and N_2 peaks at 480 K and concluded from this result that the reaction is rate limited by the dissociation of NO. From steady-state kinetic measurements of the NO—CO reaction at constant reactant pressures of 3.2×10^{-6} Pa, Campbell and White [52] observed a larger apparent activation energy below 425 K, which they attributed to inhibition by adsorbed CO.

The ease of removal of surface oxygen from Rh(331) at 700 K by reaction with adsorbed CO was reported by Dubois *et al.* [51] to be dependent upon whether the oxygen had resulted from O_2 or NO dissociation. Surface oxygen produced by O_2 was removed more easily than surface oxygen produced by NO. Taylor and Schlatter [53, see also 54] had observed a somewhat similar selectivity dependence during steady-state reaction of NO, CO, and O_2 under lean conditions over Rh/Al_2O_3: a greater percentage of the initial O_2 than initial NO was reacted at 823 K. This selectivity varied with temperature such that the removal of NO was favored below 550 K and the removal of O_2 was favored above 550 K [53]. Another interesting finding was that the NO/O_2 selectivity of the $NO—H_2—O_2$ reaction and the $NO—CO—O_2$ reaction were not the same [53]. Future research might explore in detail how adsorbed H_2 and CO influence the selectivity and how differences in the surface structure of adsorbed oxygen arising from O_2 and NO influence the NO/O_2 selectivity. Kim *et al.* [55] demonstrated in a recent study that the reactivity of preadsorbed oxygen with gaseous CO on polycrystalline Rh depended on the temperature of exposure to oxygen and possibly the surface ordering of oxygen.

More information is needed on the factors which control NO dissociation and on the detailed kinetics of N_2 desorption from Rh. Campbell and White [52] using a polycrystalline Rh wire observed partial dissociation of NO upon adsorption at 330 K such that only 70% of the surface NO was molecularly adsorbed at saturation coverage. The thermal desorption spectrum obtained with this NO covered surface showed a peak for molecular NO near 430 K, two N_2 peaks at 480 and 650 K, and a broad O_2 peak above 1000 K. Baird *et al.* [56] observed dissociative chemisorption of NO on Rh (110) at low coverages and non-dissociative chemisorption at high coverages. In contrast, Dubois *et al.* [51] using high resolution electron energy loss spectroscopy [EELS] observed only molecularly adsorbed NO or Rh(331) at 300 K. Heating the Rh(331) surface with adsorbed NO to 450 K produced a band characteristic of chemisorbed oxygen, but no separate band due to a surface nitrogen species was observable by EELS.

Studies such as those cited above are beginning to make an important contribution to our understanding of the detailed chemical mechanisms by which NO is reduced to nitrogen. Several excellent fundamental studies of CO oxidation and other chemical reactions important in the control of exhaust emissions have appeared in the recent scientific literature [57–60].

8. Composition of Three-Way Catalysts

Three-way catalysts being used to meet 1981 and 1982 emission control requirements in the United States contain the noble metals rhodium, platinum, and often palladium. Many catalyst compositions are in use. Properties which distinguish these catalysts are the noble metal loadings, the identity of base metal additives, and the supports.

The high cost of noble metals encourages manufacturers to find means to use low noble metal loadings while simultaneously maintaining needed catalytic activity and durability. Methods which have been investigated for improving catalyst performance and durability include the selection of supports with optimized porosity and surface area, attention to the size of the support pellets, the addition of base metals to improve durability and to supplement the activity of noble metals, and attention to the location of the noble metal within the support.

Numerous descriptions of potential and in-use three-way catalysts have appeared in the scientific literature and in patent disclosures during the past ten years. Yaverbaum [61] has compiled a summary of patents on nitric oxide control and removal for the period 1975 to 1979. The patent literature for automobile exhaust catalysts will not be summarized in this review. The major catalyst suppliers to the automobile industry in the U.S. have been Air Products, The Catalyst Company, Degussa, Engelhard, W. R. Grace, Johnson Matthey, and UOP. The patents of these firms as well as those of the manufacturers of catalyst supports (such as Corning and Rhône-Poulenc) and the automobile manufacturers may be of interest to the reader.

In the following sections the composition of three-way catalysts will be considered from the viewpoint of the contribution of each of the noble metals for controlling emissions.

A. Rhodium

Rhodium is widely recognized as the catalyst chosen to promote the reduction of NO to N_2 in three-way catalysts [62–65]. Rhodium exhibits excellent activity for this reaction so only a small amount of Rh is required. Typical rhodium usage is 0.18 to 0.30 g per converter. The low temperature activity of rhodium for promoting the reaction of NO with CO and H_2 and the effects of both oxygen concentration in the feed and catalyst temperature on the selectivity of the reduction to N_2 rather than to NH_3 have been recognized for some time [62, 63, 66–68]. Rhodium catalysts have been studied extensively in recent years because of its important role in exhaust emission control (as well as its catalysis of the CO + H_2 reaction). Many observations of the behavior of rhodium catalysts have been reported which contribute to our understanding of its chemistry, nonetheless our picture is incomplete and many questions remain unanswered.

Over rhodium NO is reduced to N_2 rather than to ammonia provided that the exhaust A/F is close to the stoichiometric composition as shown in Figure 8 [62, 63, 67]. The closed-loop engine control system was designed to control the A/F at the stoichiometric oxygen concentration in order to exploit this property of rhodium. Rhodium catalyzes the reduction of NO to NH_3 at net reducing A/F when the catalyst temperature is low; for example, using 0.1 wt% Rh/Al_2O_3 and without any oxygen in a simulated exhaust feed, more NH_3 than N_2 was formed from NO reduction at 700 K, but the product was mostly N_2 at 923 K [62]. The CO in exhaust most likely inhibits the reaction by which NH_3 could form N_2 at low temperatures [62]. Rhodium catalyzes the reduction of NO with either CO or H_2, though the reaction with H_2 goes more readily at low temperatures [66]. Yao *et al.* [69] have found the turnover frequency for the NO—H_2 reaction to depend on the metal loading (particle size) of Rh supported on alumina.

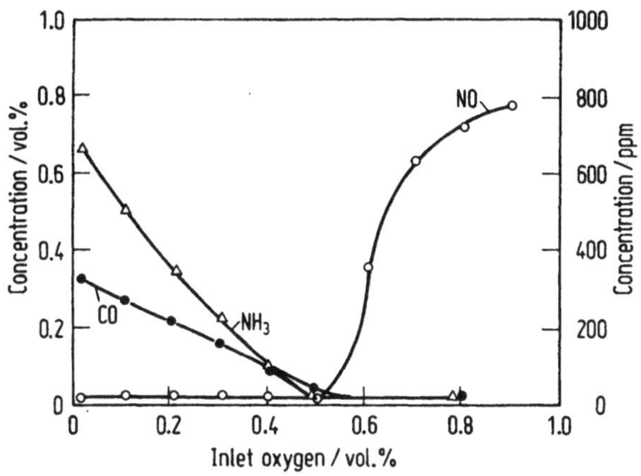

Figure 8. Effect of oxygen concentration on nitric oxide reduction over 0.1 wt% Rh-Al_2O_3 at 811 K. The NO reduced to N_2 is the difference between the initial NO and the NO unreacted (lean conditions) or NH_3 formed (rich conditions). Feedstream: 0.1% NO, 1.0% CO, 0.3% H_2, 10% CO_2, 10% H_2O, and 0–1.0% O_2 in N_2. GHSV = 38,000 (STP). (Reproduced with permission from ref. [62])

At net oxidizing A/F NO in exhaust is not reduced completely. (In contrast, the reaction of NO with NH_3 can be catalyzed in the presence of excess oxygen [70].) The decrease in NO conversion at lean A/F with increasing temperature is explained by the reaction of CO with oxygen at the expense of the reaction of CO with NO as shown in Figure 9 [54]. These selectivity curves were generated by changing the amount of CO in the feedstream at constant temperature and space velocity. The only product of NO reduction observed at net oxidizing A/F is N_2. The temperature dependence of the reduction of NO to N_2 is reversed from what we observed at rich A/F; more

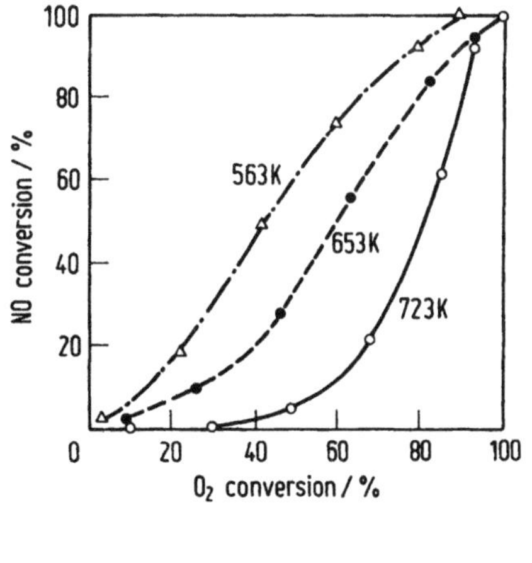

Figure 9. Conversion of NO and O_2 by reaction with CO (added stepwise) over Rh (0.002 wt%)/Al_2O_3 at three catalyst inlet temperatures. Feedstream: 0.05 vol% NO, 0.5 vol% O_2, 10 vol% H_2O, 10 vol% CO_2, 0 to 1.05 vol% CO, balance N_2. (Reproduced with permission from ref. [54])

NO is reduced at 700 K than at 923 K [62]. Greater knowledge of the detailed chemical mechanisms which are responsible for the selectivity characteristics of the catalytic reactions of NO over the Rh catalyst is of interest for the development of catalysts which can promote the reduction of NO to N_2 in the presence of oxygen.

The loading of Rh on a three-way catalyst influences catalyst performance for NO reduction. Mooney *et al.* [71] reported that increasing the Rh/Pt ratio from 5 to 50% by wt but holding the total precious metal loading on a monolithic support constant at 1.77 g l^{-1} lead to an improvement in NO_x conversion when tested in engine dynamometer tests following dynamometer aging with both 1975 FTP specification fuel and 1977 unleaded commercial fuel. These results are shown in Figure 10 [71]. Gandhi *et al.* [72] observed that increasing the Rh/Pt ratio while holding the precious metal loading constant improved catalyst selectivity as demonstrated by a decrease in NH_3 formation. Of course, low Rh availability prohibits wide use of Rh at high Rh/Pt ratios.

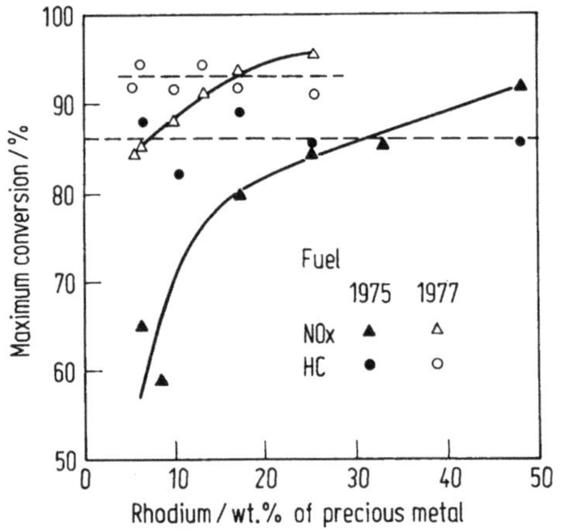

Figure 10. Maximum conversion of HC or NO_x after 8000 km of aging. Evaluation on engine dynamometer at 1600 rpm, 10–11 ins. manifold vacuum, 893 K inlet temperature and VHSV = 112,000. (Reproduced with permission from ref [71])

An important property of rhodium containing three-way catalysts is their ability to catalyze the steam reforming of exhaust hydrocarbons [72, 73]. Schlatter [73] notes, however, that a major product of this reaction over dilute Rh/Al_2O_3 catalysts is CO so the catalyst's burden for CO emission control may be increased.

As an oxidation catalyst for CO, the specific activity of Rh compares favorably with Pt and Pd [68]. In fact, Rh may improve the low temperature performance of oxidation catalysts. One drawback to the use of an all Rh catalyst is that the loadings of Rh would have to be increased beyond the loadings used in current three-way catalysts in order to obtain the oxidation performance needed. In fact, Rh is already being used at loadings above the mine ratio in some three-way catalysts. Another drawback to the use of Rh as an oxidation catalyst could be the sensitivity of Rh to preconditioning in oxygen. As an example, Schlatter *et al.* [54] observed that calcining an alumina-supported rhodium catalyst at 773 K for 4 h lead to a loss in activity for CO oxidation which was reversible upon exposure to a net reducing or a slightly oxidizing $CO—O_2$ mixture above 573 K as shown in Figure 11. Studies of oxygen inhibition of CO oxidation on polycrystalline Rh wire by Kim *et al.* [55] may provide an explanation for these observations. During steady-state reaction CO_2 production over the Rh wire was inhibited by chemisorbed oxygen above 550 K [55]. This temperature agrees well with the temperature Kim *et al.* [55] observed for the onset of oxygen inhibition of CO adsorption. The titration of chemisorbed oxygen with CO at 320 K revealed that the reactivity of adsorbed O formed at 773 K was lower than the reactivity of adsorbed O formed at 340 K [55]. These studies with the Rh wire are also of interest because the Rh is unsupported so no Rh-support interactions are present.

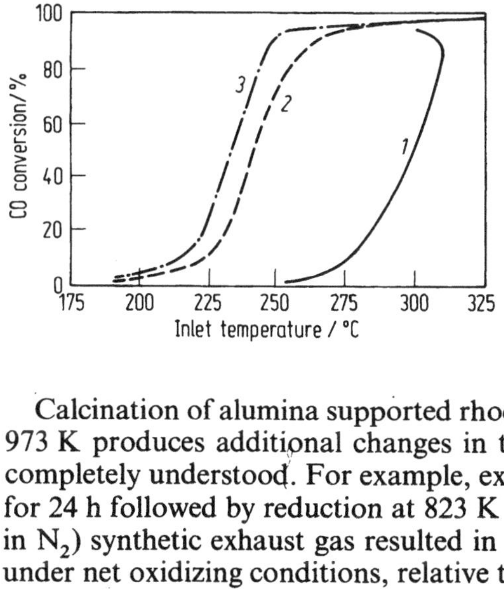

Figure 11. CO conversion during three successive temperature run-ups with a slightly lean feedstream. Curve 1 started with a fresh, air-calcined Rh/Al_2O_3 catalyst. (Reproduced with permission from ref. [54])

Calcination of alumina supported rhodium catalysts at temperatures above 973 K produces additional changes in the catalytic behavior which are not completely understood. For example, exposure of Rh/Al_2O_3 to air at 1173 K for 24 h followed by reduction at 823 K in a net reducing (1 % CO, 0.3 % H_2 in N_2) synthetic exhaust gas resulted in improved activity for NO reduction under net oxidizing conditions, relative to the fresh catalyst [49]. The activity

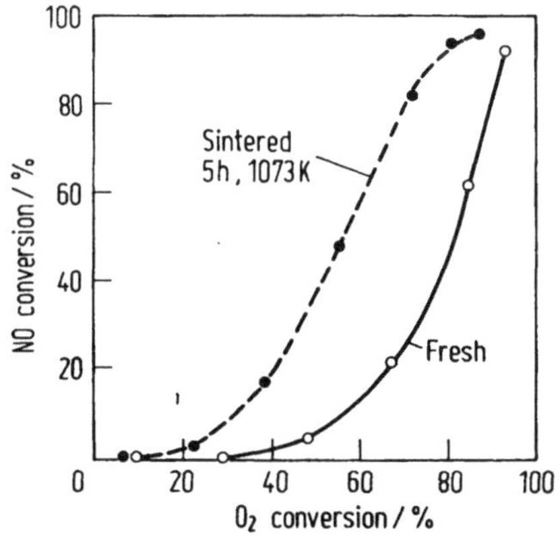

Figure 12. The effect of calcination at 773 K (fresh) and 1073 K on the partitioning of CO between O_2 and NO for a 0.002 wt% Rh/Al_2O_3 catalyst measured at 723 K and various CO levels. (See Figure 9 for conditions). (Reproduced with permission from ref. [54])

of Rh/Al_2O_3 for the $CO—O_2$ reaction is depressed by high temperature oxidation to a greater extent than is the CO—NO reaction [54]; therefore, when O_2 and NO conversions are measured under net oxidizing conditions the selectivity of the catalyst changes in the direction of greater NO conversion as a result of the calcinations (see Figure 12).

Yao *et al.* [74] have described the interaction of rhodium oxide with alumina and the diffusion of rhodium into the bulk when Rh dispersed on α-Al_2O_3 and γ-Al_2O_3 is calcined at 873 K or above. This interaction can only be partially reversed by reduction with H_2 at 823 K. Temperature programmed reduction of Rh(5.51 wt%)/γ-Al_2O_3 calcined at 973 K for 12 h showed two distinct temperature regimes of hydrogen uptake as shown in Figure 13. Catalysts calcined at only 673 K and 773 K did not exhibit this difficult-to-reduce state of rhodium [74]. The catalytic activity of Rh supported on both γ-Al_2O_3 and on α-Al_2O_3 for promoting the NO + H_2 reaction was characterized by greater selectivity for N_2 and N_2O formation (*versus* NH_3 formation) following calcination above 973 K [75]. However, under the experimental conditions used in this study the total NO conversion was depressed by the high temperature calcination [75]. Changes in the performance of three-way catalysts following exposure to lean exhaust at high temperature will be discussed in the section on catalyst deterioration.

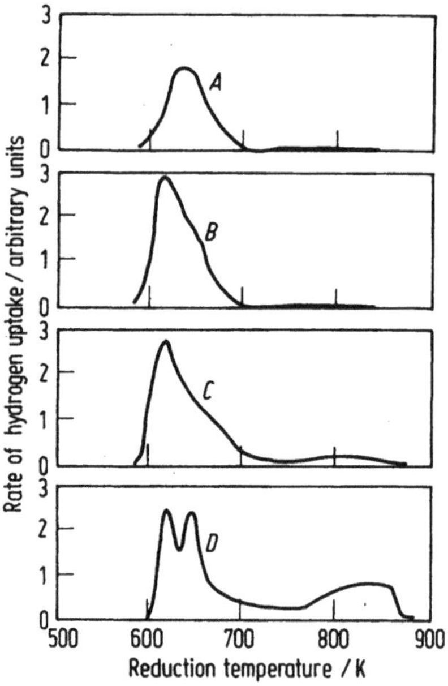

Figure 13. Temperature-programmed reduction of Rh (5.51 wt%)/γ-Al_2O_3. Calcination treatment: (A) 673 K, 16 h; (B) 773 K, 16 h; (C) 873 K, 16 h; (D) 973 K, 12 h. (Reproduced with permission from ref. [74])

B. Platinum

Platinum is used in three-way catalysts for its contribution to CO and HC conversions, especially during the warm-up of converters from a cold start. During warm-up air can be injected into the exhaust before the three-way catalyst so that it functions as an oxidation catalyst. The contribution of Pt to CO and HC oxidation during warm-up becomes more important as the Rh loading is decreased and in the event that Rh suffers thermal damage.

With a dual catalyst or dual converter system in which an oxidation catalyst follows the three-way catalyst, the oxidation catalyst generally contains Pt, Pd, or Pt—Pd. This oxidation catalyst contributes to CO and HC conversion both during the initial warm-up period and during warmed-up operation. The Pt and Pd usage in such systems contributes significantly to the total noble metal usage. The distribution of noble metals between the three-way and oxidizing catalyst beds is a consideration in the design of durable three-way catalysts. Several reports have been written on the philosophy of distribution of noble metals between the front and rear bed of a dual bed converter system as well as the effect of relative noble metal loading on three-way catalyst performance [71–72, 76]. Many different three-way catalysts are in use, and no single noble metal distribution has been established as best for all applications. The Pt content of single bed three-way catalysts is generally in the range of 0.9–2.3 g per converter. Dual bed or dual converter systems have a wide range of Pt distributions between the two catalysts.

Platinum has been known for some time to catalyze the reduction of NO_x to N_2 in an exhaust-like feed which contains the stoichiometric oxygen concentration [77]. Pt, however, probably makes only a minor contribution to the conversion of NO_x in exhaust. Platinum is not as effective as rhodium for reducing NO in the presence of SO_2 and at the high CO concentrations near the stoichiometric air-fuel ratio [78]. Pt has a smaller A/F range (window) than Rh around the stoichiometric A/F where the activity and selectivity for reducing NO to N_2 are favorable [62]. Pt readily catalyzes the reduction of NO to NH_3 under net reducing conditions in laboratory experiments. Whether N_2, NH_3 or N_2O are formed as products of NO reduction over Pt depends on the reaction conditions [79].

C. Palladium

Palladium like Pt is used for the conversion of CO and HC in some three-way catalysts. Possible detrimental interactions between Pd and Pt or Rh when they are used together have been reported [80]. Laboratory studies suggest that Pd may interfere with catalyst selectivity for NO reduction and thereby narrow the air-fuel ratio range for NO conversion [49]. Palladium is frequently used to promote CO and HC oxidation in the second (oxidation) catalyst in dual catalyst systems. Rhodium is too costly to use widely in these oxidation catalysts.

D. Iridium

Iridium is not used in current commercial three-way catalysts. Iridium deserves mention in this review, however, because of its remarkable activity for reducing NO and in particular its ability to reduce NO to N_2 under net oxidizing conditions [53, 81–82]. For example, using an iridium catalyst a greater fraction of the inital NO than the initial O_2 in an exhaust-like feed was removed by reaction with CO (in laboratory experiments at 773 K), whereas the opposite was true for the noble metals Pt, Pd, and Rh [53]. Unfortunately, iridium is not a practical choice for a three-way catalyst because its availability is lower than rhodium. In addition, iridium forms a volatile oxide and is depleted during use. Iridium surface area stabilization has been achieved at low temperatures in laboratory studies described by McVicker *et al.* [83]. Stabilized Ir catalysts are not likely to be used widely as substitutes for Rh in three-way catalysts.

E. Ruthenium and Nickel

Ruthenium and nickel have been studied extensively as NO reduction catalysts and deserve mention in view of their recognized catalytic properties. These metals are not currently being used as stand-alone catalysts but are potential additives in three-way catalysts.

Ruthenium promotes the conversion of NO_x to N_2 under *net reducing* conditions with very little simultaneous NH_3 production [50, 66, 84–86, 86a]. Catalysts which require net reducing exhaust A/F (*e.g.* 13.5) are undesirable because of the associated fuel economy penalty. (Rh in contrast to Ru promotes the reduction of NO to NH_3 under net reducing conditions.) Ruthenium is not very durable due to its tendency to form a volatile oxide and is not generally used in commercial three-way catalysts. Ruthenium stabilization has been the subject of several publications [87–90].

Alumina supported platinum-nickel, palladium-nickel, and copper-nickel (as well as monel which is unsupported) catalysts promote NO_x reduction under net reducing conditions [50, 68, 91–96]. These catalysts are less active and durable than rhodium and cannot replace rhodium in order to meet the 1.0 g mi^{-1} emission requirement for NO_x. Klimisch and Taylor [50] have demonstrated that Ni combined with Pt and Ni combined with Pd exhibit a synergism for promoting NO reduction compared with the singly supported metals. The bimetallic Pt—Ni/Al_2O_3 and Pd—Ni/Al_2O_3 catalysts were shown to be much more active than Ni/Al_2O_3, Pt/Al_2O_3, and Pd/Al_2O_3 alone [50].

Nickel is a constituent of some current Pt-containing three-way catalysts and probably contributes NO reduction activity, at least while the catalyst is fresh [97]. The addition of NiO to three-way catalysts has been examined for its contribution to CO conversion via uptake and release of oxygen or via water-gas shift activity [72, 97]. The water-gas shift activity of nickel and ruthenium exceeds that of platinum between 623 and 773 K and for practical noble metal loadings [72]. Nickel-containing catalysts are susceptible to deactivation of the nickel by aluminate formation when exposed to oxidizing conditions at high temperature, especially in the presence of water vapor. Sulfur dioxide which is present in exhaust may poison shift activity [98].

F. Cerium Oxide

The claimed benefits of Ce addition to catalysts include stabilization of the alumina support against surface area loss and increased dispersion of Pt

[99]. CeO_2 is generally added to three-way catalysts to promote the water-gas shift reaction and to store oxygen under lean conditions for use under rich conditions [98–100]. The role of CeO_2 in influencing the dynamic behavior of three-way catalysts will be described in the section on the transient behavior of three-way catalysts.

Several studies have dealt with the oxygen content of Ce-containing catalysts as well as their activity and stability. For example, experiments with a Pt/Pd/Rh/Ce/Al_2O_3 catalyst showed a higher oxygen content after exposure to lean exhaust than to rich exhaust [101]. Interaction of the noble metal with the CeO_2/Al_2O_3 increases the rate of oxygen uptake by the catalyst but does not change the amount of oxygen which reacts [99]. In steady-state experiments the addition of small quantities of Ce(0.6–1.3 wt% Ce) to Al_2O_3 was found to enhance the CO oxidation activity of a Pt/CeO_2/Al_2O_3 catalyst [100]. Further increase in the quantity of Ce deteriorated the CO oxidation activity, presumably due to a decrease in the exposed Pt [100].

Heating cerium-containing catalysts in air at high temperature leads to a deterioration of the beneficial effect of the Ce. Summers and Ausen [100] have observed that the CO oxidation activity of Pt—Ce/Al_2O_3 catalysts is lower than that of a Pt/Al_2O_3 catalyst following aging *via* heating in air at 1173 K. Yao [99] reported that oxygen storage in a precious metal/CeO_2/Al_2O_3 catalyst decreased with increasing aging temperature up to 1173 K.

G. Search for Alternatives to Noble Metals

No non-noble metal catalysts have been identified to date which can substitute for Rh and Pt in three-way catalysts. (The reader is referred to several interesting reports on base metal oxidation catalysts [102–105]). Base metals are susceptible to poisoning by sulfur compounds in exhaust and generally do not perform as well as noble metals when compared at similar space velocities, temperatures, and oxygen concentration [105]. In fact, base metal additives to noble metals may degrade catalyst performance. Kummer [106] reported that a Cu-treated Pt/Al_2O_3 catalyst had lower activity for oxidizing CO in the presence of SO_2 than the untreated catalyst. Over base metal catalysts oxidation reactions such as CO and H_2 oxidation have been established to proceed *via* a two step mechanism which involves uptake of oxygen during oxidation of the catalyst in one step and removal of oxygen during reduction of the catalyst by CO or H_2 in the second step. Both of these reaction steps must proceed readily and repeatedly below 773 K in a typical exhaust atmosphere.

Two recent studies have explored the substitution of noble metals with base metals in three-way catalysts. Fu *et al.* [107] have evaluated in laboratory experiments the suitability of Fe(II) Y Zeolite as an exhaust gas cleanup catalyst. The catalyst was shown to reversibly change between Fe(II) Y Zeolite and Fe(III) Y Zeolite during repetitive reaction of O_2 and H_2 on a flow microbalance [107]. Using an NO—CO—O_2 feed, the preferred oxidizing agent was O_2, but NO did react when sufficient CO was available [107]. Reduction of the catalyst was found to be limiting in the presence of O_2; the influence of water and SO_2 was not examined [107]. Studies such as this one with the aim of understanding the detailed chemical behavior of base metal catalysts during oxidation-reduction reactions provide an important contribution towards the substitution of noble metals with less costly, more available materials.

A study by Gandhi *et al.* [108] explored the use of MoO_3 in combination with Pt/Al_2O_3 as a replacement for Rh in three-way catalysts. Improved selectivity for the reduction of NO to N_2 and decreased CO poisoning was demonstrated relative to pure Pt/Al_2O_3 catalysts in laboratory experiments. Further research is needed to identify other noble metal-base metal combinations for the goal of lowering noble metal usage in three-way catalysts.

9. Catalyst Supports

Supports for automobile exhaust catalysts are of two general types: alumina pellets (spheres and extrudates) and ceramic monoliths coated with a thin alumina washcoat (Figure 14). The reasons for choosing alumina supports for the noble metals include high surface area, attrition resistance, stable structure under typical exhaust conditions, favorable pore structure, and adequate supply [109]. Both support types have perceived advantages and disadvantages which have influenced their selection for use by automobile manufacturers [110]. Factors considered include converter size and cost,

Figure 14. Supports for automobile emission control catalysts.

catalyst performance and durability on a particular vehicle, availability of support material from suppliers, and ease of replacement. In this section the recent literature on both support types will be reviewed.

A. Pellets

Alumina pellets used in automobile catalytic converters are composed of thermally stable transitional alumina phases often stabilized by incorporation of metal oxides [111]. Since the introduction of catalytic converters, the physical properties of pelleted supports have undergone several changes. The alumina support used for pelleted catalysts from 1975–1979 had lower BET surface area and higher density than supports introduced for use on certain vehicles starting late in 1978 [112]. Support properties were changed to meet the need for improved converter performance dictated by a tightening of the emission control requirements for CO and hydrocarbons [112]. The bulk density of the supports was decreased from approximately 0.69 g cm^{-3} to 0.48 g cm^{-3}; the BET surface area of the new supports was 115 m^2 g^{-1} versus 94 m^2 g^{-1} for the earlier supports [112, 113]. The performance of three-way catalysts made on the low density supports was improved; the Rh usage was simultaneously reduced from 0.018 to 0.004 Troy oz per converter [112].

The relationship between support density and catalyst cold start performance had been demonstrated by Barnes and Klimisch [118] and attributed to the lower catalyst mass with lower support density. Hegedus and Summers [115] demonstrated that selection of an alumina support with an open pore structure and high surface area can improve the poison resistance of oxidation catalysts. The relationship among support pore structure, surface area, and noble metal impregnation profiles of automobile emission control catalysts has been the subject of several engineering studies [109, 115–118].

Further improvements in pelleted catalyst supports were needed in order to meet 1980 and 1981 emission requirements. This time catalyst efficiency was improved by going to supports with high geometric surface area. The geometric surface area was increased through either a change in the physical shape or a reduction in the size of the pellets [112]. For example, bead radius was decreased from 1.55 mm to 1.15 mm [112].

The distribution of noble metals or noble metal profiles in pellets must be controlled if catalyst performance is to be maintained during use and noble metal loadings are to be kept at a minimum [113, 116–117]. In general, noble metals are best placed within the first 150 µm from the outer edge of the catalyst pellet where they are readily accessible to the reacting gases but with sufficient subsurface penetration in order to avoid rapid poisoning [112–113, 117].

B. Monoliths

Monolithic supports are thin walled ceramic honeycombs composed mostly of cordierite ($2\,MgO \cdot 2\,Al_2O_3 \cdot 5\,SiO_2$) [119, 120]. A thin washcoat of alumina applied to the surface of the cordierite serves to increase the area of the surface to which the noble metals are applied [119, 121]. Characteristics of monoliths which make them attractive catalyst supports are the low thermal expansion of the ceramic, their small size, fast heat-up, low back pressure, low vibration attrition, and design flexibility [119].

The characteristics of ceramic monolithic catalyst supports underwent changes in geometry between 1978 and 1981 which provided improved physical and catalytic properties [123]. Specifically, the cell density of monolithic supports was increased from 46 to 62 cells per cm^2 while the wall thickness was reduced from 0.30 to 0.15 mm [123]. Benefits cited for the changed design include higher geometric surface area, improved thermal shock resistance, faster light-off time, and lower pressure drop across the converter [123].

Studies of supports for automobile emission control catalysts should be pursued actively in coming years in order to meet the simultaneous need for smaller and lighter catalytic converters, lower noble metal usage, and improved durability. New direction should be provided by design studies which incorporate information on support properties together with catalytic properties, exhaust conditions, and poisoning characteristics [124, 125].

10. The Transient Behavior of Three-Way Catalysts

A typical closed-loop control system causes the A/F to cycle rapidly about the stoichiometrically balanced composition with a frequency of about 1 Hz. This operating condition has generated considerable interest in how the A/F perturbations influence catalyst performance compared with a non-cycled stoichiometric feedstream and how catalysts may be formulated to perform optimally in cycled feeds. The recent literature aimed at understanding the transient behavior of three-way catalysts will be reviewed here.

In 1976 Gandhi *et al.* [65] described the concept that a three-way catalyst should contain an "oxygen-storage" component in order to remove excess oxygen under lean conditions and thereby extend the selectivity of the catalyst for reducing NO. The oxygen picked up by the catalyst under oxidizing conditions would be removed under slightly reducing conditions by reaction with the reducing agents, thereby increasing the conversion of CO and HC. Thus, immediately after switching to net oxidizing or to net reducing conditions, a stoichiometric oxidation-reduction reaction involving a component of the catalyst supplements the catalytic reaction. Research aimed at understanding the contribution of this phenomenum to catalyst performance and at understanding catalyst response to feedstream oscillations in general has included measurements of the oxidation-reduction capacity of catalysts [72, 98], the oxygen content of catalysts as a function of A/F [29, 126, 127], the amount of reducing chemical species on catalysts as a function of A/F [29, 126, 127], the rate at which the catalyst is oxidized and reduced [29, 72, 126], the rate at which the catalyst performance responds to a step A/F change [27, 128, 129], and catalyst performance in a cycled feed and as a function of the amplitude and frequency of the A/F perturbations [23, 48, 72, 130–132]. Supported noble metal catalysts and noble metals in combination with base metals have been examined [34, 65, 97–98, 126, 128, 130]. Exhaust A/F perturbations have been simulated with laboratory reactors using synthetic exhaust gas and with engine-dynamometers using either deliberate A/F switching or closed-loop control. The results of these investigations are summarized below.

Three-way catalysts can take up oxygen via interaction of the noble metal with oxygen and *via* interaction of base metal additives with oxygen. The *capacity* of various three-way catalysts for taking up and releasing oxygen has been determined by titrating the catalysts with O_2 and monitoring O_2 uptake, followed by monitoring CO_2 evolution during catalyst reduction with CO at 773 K [72, 98]. Schlatter and Mitchell [98] reported that 5 wt% Ce on a 0.045 wt% Pt/Ce/Al_2O_3 catalyst greatly increased the oxygen adsorption capacity compared with the platinum catalyst without Ce, both measured at 298 and 773 K. Not all of the oxygen in the catalyst would be expected to be removed upon cycling the exhaust A/F. In laboratory experiments about half of the oxygen added to the Pt/Ce/Al_2O_3 catalyst could be removed by reduction with CO at 773 K, and three-fourths could be removed with H_2 [98]. The oxygen content of three-way catalysts has also been measured as a function of exhaust A/F [29, 126]. Herz [126] measured the change in the oxygen content of a 0.09 wt% Pt/0.03 wt% Pd/0.006 wt% Rh/2.6 wt% Ce/Al_2O_3 catalyst as the exhaust A/F was varied between 14.1 and 15.1 (Figure 15). He observed a dramatic change in oxygen content with A/F; however, the catalyst was neither depleted of oxygen nor saturated with oxygen at any time.

In order for stored oxygen to contribute to the performance of a three-way catalyst as the A/F is rapidly changing, the catalyst must take up and release oxygen at a *rate* which is fast compared with the 1 Hz cycling frequency.

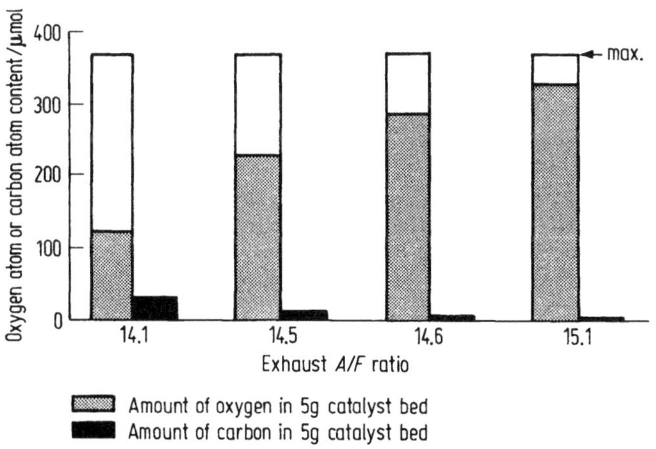

Figure 15. The oxygen and carbon atom contents of the 5 g catalyst bed after 30 min exposure to exhaust at four air-fuel ratios (A/F). The maximum oxygen content of the bed was 370 µmol of oxygen atoms. The amount of carbon is shown on the same scale in carbon atom equivalents. The three-way catalyst contained 0.09 wt% Pt, 0.03 wt% Pd, 0.006 wt% Rh, and 2.6 wt% Ce on alumina. (Reproduced with permission from ref. [126])

Three-way catalyst formulations have, therefore, been compared in terms of the time dependence for O_2 uptake and for CO_2 formation by the catalysts [29, 72, 98, 126]. These measurements have shown that the rate of catalyst oxidation (oxygen storage) is faster than the rate of catalyst reduction [29, 72, 98, 126]. The faster oxidation rate implies that the catalyst oxygen content will be high during symmetric cycling of the A/F about the stoichiometric point. Herz [126] suggests that the reaction of the reduced catalyst with water vapor contributes to the rapid oxidation because in the first 0.5 s after changing the exhaust A/F from 14.1 to 15.1 the oxygen atom content of his catalyst increased by a greater amount than the oxygen supplied as O_2 and NO.

During A/F cycling the capacity of a three-way catalyst to store oxygen is determined by the amount of oxygen removed during the slower reduction process. Figure 16 shows Herz's results for oxygen removal from a three-way catalyst containing 2.6 wt% Ce during the first 0.5 s and 1.0 s after a switch of the exhaust A/F setting from 15.1 to 14.1. In 1.0 s the oxygen content decreased by 36% of the difference between the 15.1 and 14.1 steady-state amounts [126]. The total amount of oxygen held in a catalytic converter depends upon the rate of reduction, the volume of catalyst in the converter, the space velocity, the catalyst composition, and the catalyst temperature [29, 126, 127].

Three-way catalysts are able to retain some reducing chemical species during exposure to rich exhaust gas [29, 126, 127]. Kaneko *et al.* [29] reported that at 823 K a Rh containing three-way catalyst retained reducing species somewhat more easily during exposure to C_3H_6 than during exposure to C_3H_8, but CO was not retained. (The chemical nature of the retained reducing species was not identified, also see Figure 15.) The retained reducing

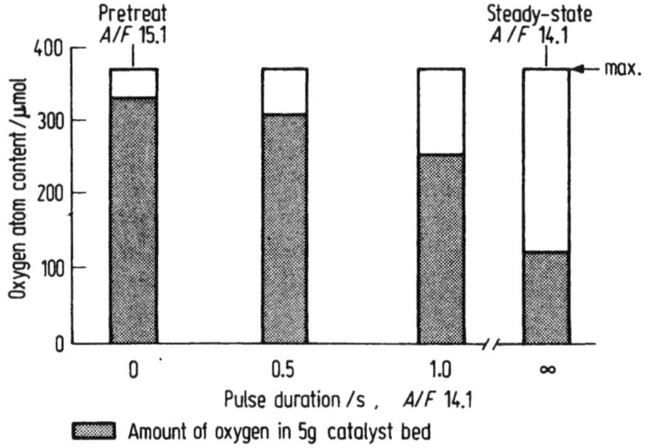

Figure 16. Rate of decrease of oxygen content of a pelleted $Pt/Pd/Rh/Ce/Al_2O_3$ catalyst in rich exhaust at a space velocity of 110000 h^{-1} (STP) and a catalyst inlet temperature of 680 K. (Reproduced with permission from ref. [126])

species reacted with oxidizing species when the gas composition was changed to net oxidizing.

At low temperatures (*e.g.* less than 573 K) or at net reducing exhaust A/F black carbonaceous deposits can build up on the surfaces of three-way catalysts [133]. The deposition and reaction of the deposit is another "storage" phenomena which can attenuate the influence of A/F perturbations.

The distinction between a catalytic reaction and reduction of a metal oxide component on a three-way catalyst is not readily apparent during A/F cycling. For example, both catalysis of the water-gas shift reaction and the reduction of cerium oxide by CO can give enhanced CO removal during the net reducing part of the cycle. Schlatter and Mitchell [98] observed that a 0.005 wt% $Rh/0.7$ wt% CeO_2/Al_2O_3 catalyst converted more CO than did a 0.005 wt% Rh/Al_2O_3 catalyst when the composition of a simulated exhaust gas was rapidly cycled between net reducing and net oxidizing at 1 Hz frequency. Because the enhancement disappeared when water was removed from the feed, the reaction of CO with H_2O is a likely explanation for this observation. Similarly Shulman *et al.* [129] calculated that the amount of oxygen storage on their three-way catalyst during oscillation of the exhaust A/F was insufficient to account for the CO oxidation above that which reacted with inlet oxygen during the rich portion of each cycle.

Experiments conducted by Sell *et al.* [27] using an engine mounted on a dynamometer and fast response infrared diode laser spectroscopy for the

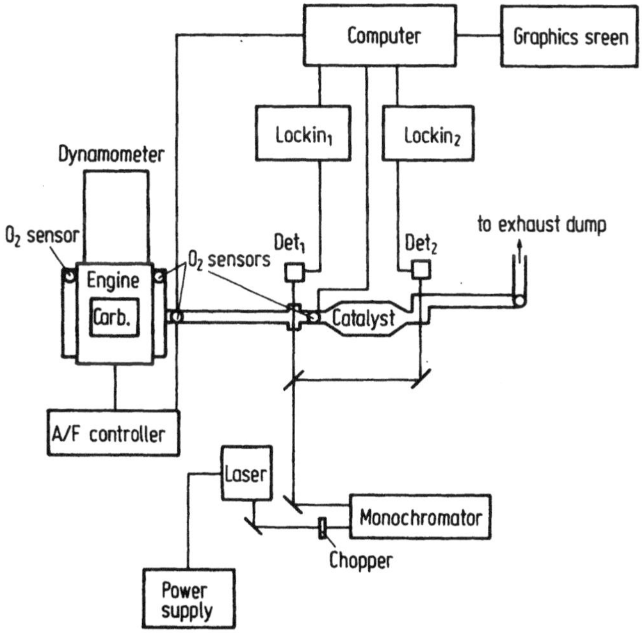

Figure 17. Schematic diagram of the apparatus used to study the dynamic behavior of three-way catalysts in engine exhaust. This apparatus was used to make the measurements shown in Figures 18 and 21. (Reproduced with permission from ref. [27])

dynamic measurement of CO concentrations in exhaust provided the first real-time measurements of CO concentration during A/F oscillations (Figure 17). This technique was used to monitor the instantaneous response of a three-way catalyst to a step A/F change [27]. Figure 18 shows plots of the catalyst inlet and outlet CO concentrations versus time for a switch in the exhaust A/F from rich to lean (A) and from lean to rich (B). The outlet CO level quickly reached a new steady-state value following the rich to lean step change; whereas, a new outlet CO level was reached only after several seconds following the lean to rich step change [27]. This result is in agreement with the CO_2 and O_2 titration experiments [72, 98, 126] in which the catalyst took up O_2 more readily than the catalyst was reduced.

Measurements of catalyst performance while cycling the composition of the laboratory feedstream or cycling the A/F of exhaust delivered to the catalyst have been used to gain understanding of the dynamic response of catalysts in closed-loop emission control systems [23, 29, 34, 48, 65, 129–132].

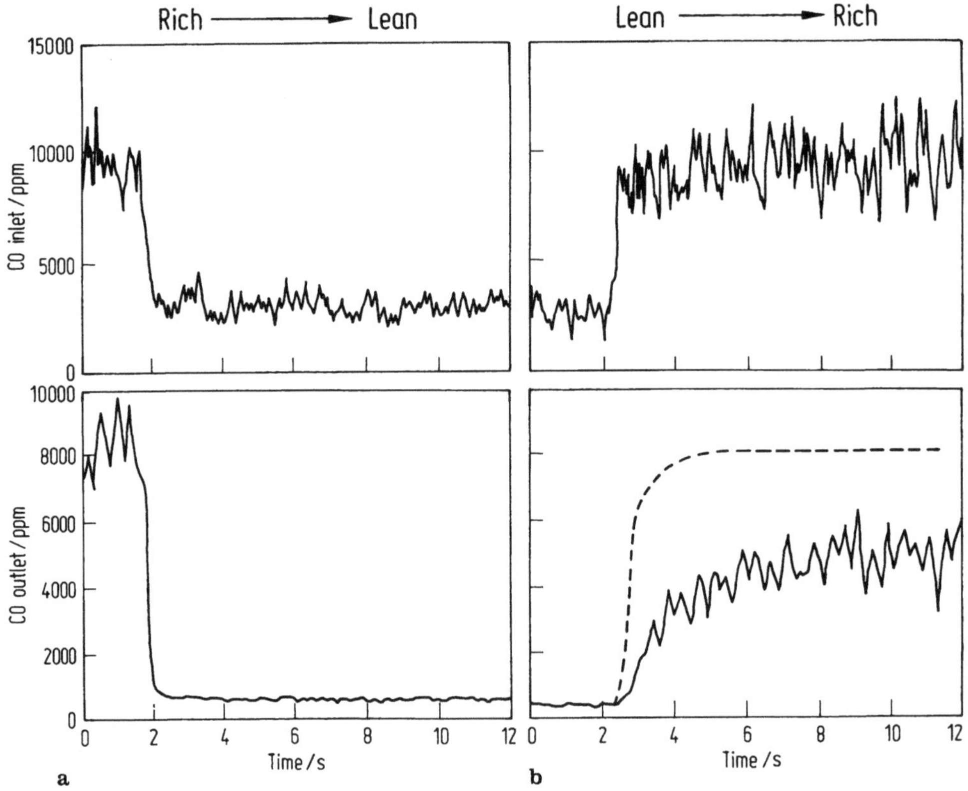

Figure 18. a CO concentrations measured over a pelleted $Pt/Pd/Rh/Ce/Al_2O_3$ catalyst at a space velocity of 50,000 h^{-1} (STP) as the engine A/F control setting was stepped from rich (A/F = 14.1) to lean (A/F = 15.5). **b** CO concentrations measured as the A/F control setting was stepped from lean (A/F = 15.5) to rich (A/F = 14.1). The dashed curve gives the approximate, smoothed result that would have been obtained if the catalyst response were instantaneous. (Reproduced with permission from ref. [27])

Symmetrical cycling of the A/F at 1 Hz about the stoichiometric composition leads to lower conversion of NO and CO compared with an uncycled stoichiometrically balanced feed [29, 48, 65]. If the A/F is symmetrically cycled at 1 Hz about an A/F which is more reducing by 0.5 A/F unit than the stoichiometrically balanced composition, the time-averaged conversion of CO exceeds the conversion obtained using a steady feed with a composition characteristic of the mean A/F of the cycled feed [23, 29, 48, 65]. If the A/F is symmetrically cycled at 1 Hz about an A/F which is more oxidizing by 0.5 A/F unit than the stoichiometrically balanced composition, the NO conversion exceeds the conversion obtained with the corresponding uncycled feed [23, 29, 48, 65]. In general, the conversions of NO and CO at net oxidizing and net reducing A/F respectively, are poor. Operation close to the stoichiometrically balanced A/F is required in order to convert all three pollutants simultaneously [29, 65, 98]. Figure 19 is an illustration of the effect of cycling frequency on the time-averaged conversion of NO and CO taken from Schlatter *et al.* [48]. The simulated exhaust gas was cycled ± 0.25 A/F units about the mean value at 0.25, 1.0, and 8 Hz [48]. Close to the stoichiometric A/F, conversions are seen to decrease as the frequency is decreased. The 8 Hz cycling frequency is equivalent to a steady flow because mixing decreased the amplitude of the inlet composition oscillations to zero. A very rapid cycling frequency of the control system has been suggested as a way to improve performance and therefore save on noble metal usage [130]. Practically speaking, increasing the cycling frequency means eliminating the delays in the feedback control system.

The amplitude of the A/F oscillations has a significant influence on the NO and CO conversions. Increasing the cycling amplitude during symmetric

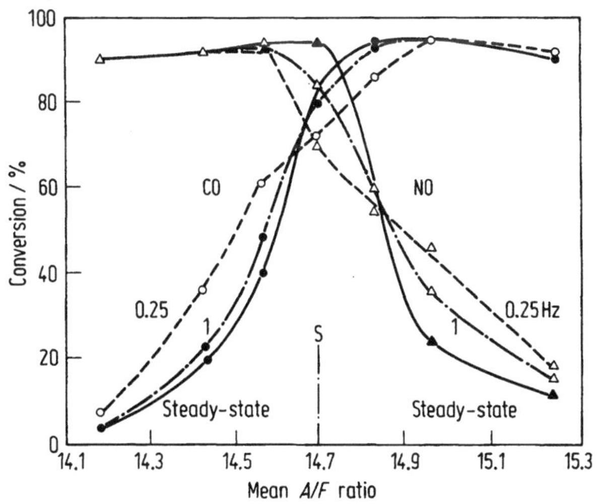

Figure 19. Time-averaged CO and NO conversions measured using the laboratory reactor system shown in Figure 7. A fresh, pelleted $Pt/Rh/Al_2O_3$ catalyst was operated at a middle-bed temperature of 820 K and a space velocity of 52000 h^{-1} (STP). The feedstreams simulated exhaust that would be obtained with various engine air-fuel ratios (A/F) but did not contain SO_2. The feedstream compositions were cycled at 0.25 and 1 Hz at an amplitude of ± 0.25 A/F about the mean A/F. For the curves labeled steady-state, conversions were measured with feedstreams at the mean A/F values. (Reproduced with permission from ref. [48])

perturbations of the A/F about the stoichiometric point leads to a decrease in NO and CO conversions [29, 48, 130]. Figure 20 shows the well-known result that perfect stoichiometric A/F control (no cycling) gives the highest NO and CO conversions. Developmental work on three-way control systems is aimed at minimizing the amplitude of the A/F perturbations in order to obtain high NO and CO conversions. One way to reduce the cycling amplitude would be to develop an exhaust sensor that responds linearly to changes in A/F in contrast to current sensors which operate like a switch. Sell and Chang [134] have demonstrated this relationship.

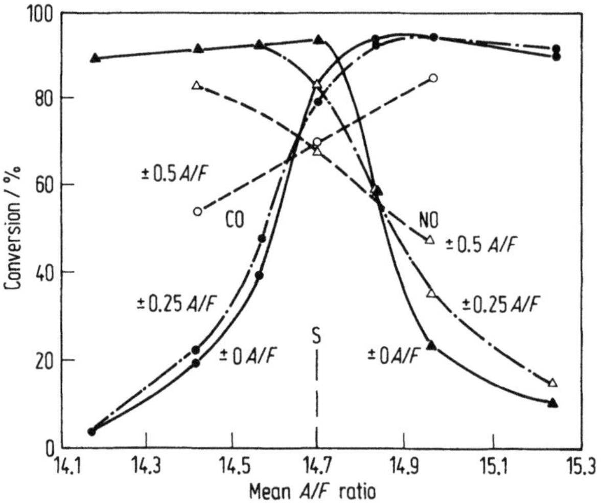

Figure 20. CO and NO conversions measured at various cycling amplitudes. Frequency = 1 Hz. Space velocity = 52,000 h^{-1} (STP), catalyst temperature = 823 K. The catalyst contained 0.024 wt % Pt and 0.018 wt % Rh on alumina. (Reproduced with permission from ref. [48])

Figure 20 also shows that as the mean A/F of the synthetic exhaust gas is shifted to a value more reducing than the stoichiometric composition the CO conversion does not fall off so sharply with A/F with the ± 0.5 A/F cycling amplitude as with the ± 0.25 A/F cycling amplitude. Likewise, as the mean A/F is shifted to a value more oxidizing than the stoichiometric composition, the NO conversion increases with cycling amplitude. Similar trends have been observed with exhaust [29, 130].

Falk and Mooney [130] have examined the relationship among the amplitude of the A/F perturbations, noble metal loading, and conversion efficiency for a series of exhaust-aged monolithic Pt—Rh three-way catalysts. Increasing the noble metal loading from 0.35 to 1.41 g l^{-1} resulted in an increase in the CO, NO, and HC conversions during 1 Hz perturbations of the A/F when the amplitude of the A/F perturbations was ± 0.1 and ± 0.5 A/F [130]. At a fixed noble metal loading the conversion was increased by a decrease in the amplitude of the A/F perturbations which is in agreement with the laboratory studies [130]. The best conversion and smallest dependence on noble metal loading were observed with the lowest A/F perturbation amplitude tested (± 0.3 A/F) [130].

Taylor and Sinkevitch [132] have explored the dependence of catalyst performance on the symmetry of the perturbations of the feedstream stoichiometry. An asymmetric cycle is defined as a cycle for which the flow times for the exposure of the catalyst to the rich and lean parts of a complete cycle are not equal. Examination of the conversions of NO and CO using both symmetric (equal flow times) and asymmetric cycling revealed that asymmetric cycling benefited conversion at average A/F away from the stoichiometric composition, that is a wider operating window was found with asymmetric cycling than with symmetric cycling. However, no one cycling scheme was identified which improved NO and CO conversions simultaneously [132].

The foregoing observations of the influence of cycling parameters on the conversion efficiency of three-way catalysts have monitored only the time-average conversion and not the instantaneous response of the catalyst. Herz et al. [132] have recently used fast response infrared diode laser spectroscopy to follow the instantaneous conversion of CO over a three-way catalyst during 1 Hz perturbations of the exhaust A/F. These results provide additional details of the relationship between catalyst composition and conversion efficiency under dynamic A/F conditions. Figure 21 is a plot of the outlet CO concentrations from the catalyst for two differently formulated catalyst samples. The experimentally determined outlet CO concentration is compared with the outlet CO concentration predicted if the catalyst responded instantaneously to the lean-to-rich A/F change. We observe for both catalysts complete CO conversion immediately following lean-to-rich A/F changes and lower CO emissions overall than predicted. The Ce-containing catalyst emitted less CO than the non-Ce catalyst. This experiment demonstrates *in*

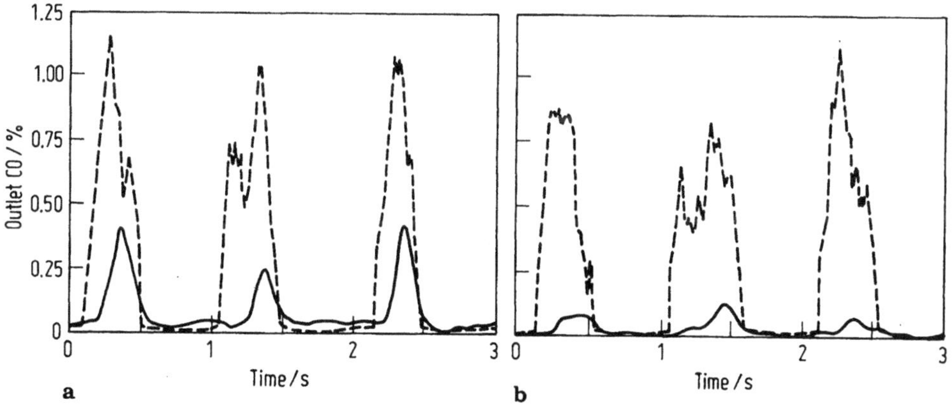

Figure 21. Outlet CO concentration from two catalysts during exhaust A/F cycling at 1.0 Hz between 14.1 and 15.1. Catalyst **a** contains Pt—Pd—Rh—Al_2O_3. Catalyst **b** contains Pt—Pd—Rh—Ce—Al_2O_3. Dashed line is the estimated instantaneous response to the cycling; solid line is the measured response. Of the CO that would have come out of the converter if the catalyst had responded instantaneously only 30% an 10% were observed with catalysts a and b, respectively. (Reproduced with permission from ref. [131])

real-time the improved CO conversion with cycling which has been attributed to storage of oxygen in the catalyst and water-gas shift activity.

Using a timed sampling method (similar to the technique described by Kaneko *et al.* [23]) Shulman *et al.* [129] have determined the time duration of the enhanced instantaneous conversions of all three exhaust constituents during oscillation of the exhaust A/F. Exhaust samples were collected by tapping the exhaust through rapid response solenoid valves to evacuated sample bags during the same 0.1 s intervals during repetitive cycles. With a 0.25 Hz cycling frequency, hydrocarbon conversion and NO conversion exceeded their steady-state values during the entire 2 s rich part of the 4 s cycle; CO conversion exceeded its steady-state value for almost 2 s following the lean-to-rich A/F transition. Other experiments showed that 50 s of rich operation was required for NO_x conversion to return to the steady-state value. The enhanced HC conversion can be rationalized on the basis of steam reforming and HC storage; the enhanced CO conversion is explained by water-gas shift and oxygen storage. The enhanced NO conversion observed during A/F perturbations is not completely understood.

11. Deterioration of Three-Way Catalysts

The deterioration of the activity of three-way catalysts during use is an important characteristic because the Federal emission standards require that performance be maintained for 80,000 km, as described earlier. Three-way catalysts lose activity both due to thermal effects and due to poisoning by contaminants in exhaust, namely phosphorus, lead, and sulfur. Combined poisoning and thermal effects and interaction among poison species complicate deterioration studies. A considerable incentive exists to minimize catalyst deterioration. If deterioration were eliminated, catalysts could have lower noble metal loadings and vehicles might be calibrated for greater fuel economy.

A. Thermal Effects

Exhaust catalysts and catalytic converters are designed to withstand occasional high temperature operation. However, prolonged and repeated exposure to temperatures in excess of 1000 K leads to loss of noble metal surface area and even deterioration of the catalyst support.

Exposure of three-way catalysts to oxidizing conditions at high temperatures results in changes in activity for NO reduction and CO oxidation which are generally attributed to Rh deactivation [63, 64, 135, 136]. The exposure of oxidized three-way catalysts to reducing conditions restores some activity, but repeated oxidation-reduction cycling may permanently damage three-way catalysts. Summers *et al.* [136] reported that a fresh Pt—Rh/Al_2O_3 catalyst exposed to oxidizing exhaust gas (A/F = 15.1) at 1023 K lost activity for reducing NO at A/F = 14.1. The lost activity was recovered by treatment with reducing exhaust gas (A/F = 14.1) at 1023 K [136]. The NO reduction activity of a $Pt/Rh/Pd/Ce/Al_2O_3$ catalyst used for 10,000 km on a vehicle was likewise improved following exposure to rich exhaust (A/F = 13.8) for 15 min at 1023 K. Comparison of the CO oxidation of this reduced catalyst with a fresh sample of the same catalyst indicated that activity had been only partially restored by the reduction at 1023 K [136]. Koberstein [135] found that a rhodium-containing three-way catalyst heated at 1073 K in air for 0.5 h was less active at low temperature for oxidizing CO and HC than a similar catalyst treated with H_2 for 0.5 h at 873 K. Gandhi *et al.* [65] reported that 168 h aging of a three-way catalyst (noble metal composition not reported) in air at 1073 K drastically narrowed the selectivity "window". This catalyst did not exhibit any shift in the air-fuel ratio point of best simultaneous NO and CO conversions, which implies that both the NO and CO conversions were influenced by the aging [65].

The transformation which Rh undergoes upon exposure to oxidizing conditions at high temperatures has important implications for the use of Rh in three-way catalysts. Transformation to the hard-to-reduce form and the loss of catalytic activity which has been found for these pretreatments limits converter operating conditions. A temperature of 900 K where these transformations start to occur may be reached during normal operation of a catalytic converter; however, much higher temperatures (*e.g.* 1300 K) are unlikely. Surface interactions between Rh and alumina whereby Rh reportedly diffuses into the subsurface region and the bulk during heat treatment above 813 K was discussed earlier [74].

Several research groups have reported on surface segregation of Rh in Pt—Rh containing three-way catalysts and in supported Pt—Rh alloys following exposure to oxidizing conditions [64, 137]. For example, Cooper *et al.* [64] reported that a Pt—Rh three-way catalyst aged 50 h in oxidizing exhaust showed both crystal growth of the Pt—Rh phase and surface enrichment of Rh in the crystallites. Wang and Schmidt [137] have explained the surface enrichment of Rh on silica supported Pt—Rh alloy particles during oxidation-reduction cycling as due to segregation of Rh by the formation of rhodium oxide followed by reduction of the oxide at temperatures where bulk diffusion is minimal. While surface enrichment may explain changes in the activity of three-way catalysts following a mild oxidation (*e.g.* 900–1000 K), surface enrichment of Rh probably does not provide an explanation for severe thermal deterioration. Furthermore, the interpretation of the XPS results of alumina supported catalysts may be influenced by the greater tendency of Pt than Rh to sinter. Clearly more work is needed to unravel the relationship between surface analysis and three-way catalyst deactivation.

Many current catalysis studies are aimed at finding ways to prevent oxidative deterioration of Rh containing three-way catalysts. One approach which has met with success is first to impregnate Rh on ZrO_2 to form a well dispersed catalytic phase followed by supporting the Rh/ZrO_2 on a high surface area γ-Al_2O_3 washcoated monolith [138]. This approach combined thermal stability with poison resistance in laboratory tests [138].

Pt and Pd used in three-way catalysts are susceptible to sintering and loss of active metal area by analogy to earlier studies reported for oxidation catalysts. In general, Pd is less susceptible to thermal deterioration than is Pt [139]. Barnes and Schlatter [139] have reported that both the HC and CO emissions were increased during cycle 1 of FTP tests as a result of sintering Pt/Al_2O_3 and Pd/Al_2O_3 catalysts at 900 °C for only 5 h. The activity of three-way catalysts used as oxidation catalysts during warm-up could be similarly impaired by exposure to high temperatures.

B. Phosphorus Poisoning

Both phosphorus present in fuel and phosphorus derived from engine oil additives are potential sources of phosphorus contamination of three-way catalysts. The phosphorus level in unleaded regular gasoline (0.02 mg l^{-1}) is very low compared with the phosphorus in engine oil ($\sim$1.2 g l^{-1}) [140, 141]. Monroe [141] calculates that in 80,000 km of driving a catalyst might be exposed to 13 g of P of which 93% is derived from engine oil. Phosphorus derived from engine oil was established as a poison of oxidation catalysts; in fact, a direct relationship between catalyst poisoning by phosphorus accumulation and HC emissions was determined [130]. Spearot and Caracciolo [142] determined that 0.1 g mi^{-1} of the HC emissions from a vehicle aged 80,000 km on an AMA driving schedule with a high phosphorus fuel was attributable to catalyst poisoning by phosphorus. No such direct correlation between P poisoning and emissions has been reported to date for three-way catalyst systems. Differences in the poisoning action of fuel and engine oil derived P, possible differences in the susceptibility of pelleted versus monolithic three-way catalysts to poisoning by P, and the effect of aging conditions and additives on P poisoning have generated lively discussions.

Much of the current research on phosphorus poisoning of three-way catalysts in some way relates to questions of fuel-versus engine-oil-derived phosphorus. McArthur [143] has reported that when operating a catalyst under net reducing conditions the retention of phosphorus on the catalyst was much greater for phosphorus derived from engine oil than for phosphorus derived from gasoline. Williamson *et al.* [144] reported that fuel derived phosphorus did not poison a monolithic three-way catalyst in a laboratory pulsator test which simulated 28,000 km of operation on isooctane containing 2.2 mg P l^{-1} added as cresyl diphenyl phosphate (CDP). In a similar study the removal of phosphorus from the fuel made the poisoning effect of Pb and S on NO and HC activities worse [145]. A monolithic three-way catalyst aged in fuel doped with much higher levels of CDP in order to simulate P throughput levels typical of oil derived P did result in loss of catalytic activity [146].

For pelleted Pt—Pd—Rh/Al_2O_3 three-way catalysts the addition of 15.8 mg P l^{-1} as the engine oil additive zinc dialkyldithiophosphate (ZDP) to the fuel lowered the conversions of NO, CO, and HC compared with clear fuel following aging in exhaust for 17.5 h [141]. The catalytic reactor was divided into four sections, and the ability of the catalyst to scavenge P was shown by greater P collection by catalyst pellets in the front section (0.32 wt%) than by catalyst pellets in the rear section (0.09 wt%). Less P was retained by the catalyst in all four sections of the reactor and catalytic activity was greater when Ca was added to the fuel with the ZDP as calcium sulfonate in order to simulate an alkaline metal oil additive. Here the P retention on the catalyst was 0.08 wt% for the front section and 0.06 wt% for the rear [141].

Spearot and Caracciolo [147] reported that phosphorus derived from ZDP contributed significantly to the loss of performance of a three-way catalyst system. A 200 h test using a dynamometer-mounted engine, a closed-loop emission control system, and a pelleted three-way catalyst was conducted with both a high phosphorus oil (0.20 mass percent P) and a phosphorus-free ashless oil. For the high phosphorus oil a large increase in NO_x emissions following the aging was attributed to deterioration of both oxygen sensor and catalyst performance. The aged oxygen sensor controlled the A/F at a lean value (14.94 versus 14.73) where NO conversion is low. The effect of the aging with the phosphorus containing oil on the conversion efficiency of the catalyst revealed a greater loss in CO than NO conversion at the stoichiometric A/F. In general, when the oil contained alkaline earth metal additives (Mg, Ca) less phosphorus was retained on the catalyst and the poisoning of HC conversion activity by ZDP was less [147, 148]. The chemical identity of the phosphorus species which accumulate on three-way catalysts during vehicle operation has not been characterized in great detail. Aluminum phosphate has been identified on vehicle-aged and dynamometer-aged three-way catalysts [146].

To date the poisoning susceptibility of pelleted and monolithic three-way catalysts by phosphorus has not been compared in a well controlled test. For monolithic oxidation catalysts phosphorus has been found to accumulate in a sharp axial profile with most of the phosphorus uptake at the front end of the monolith [149]. The distribution of phosphorus in a pelleted three-way catalyst bed is not known; complete mixing of the pellets in the bed is probably a good assumption. We know from electron microprobe measurements of sectioned pellets and monoliths that phosphorus penetrates the alumina surface of oxidation catalysts in a sharply defined shell-like profile [150, 151]. Most likely phosphorus accumulates similarly on three-way catalysts.

C. Lead Poisoning

So-called unleaded gasoline is required for use with catalyst equipped vehicles. The lead content in unleaded regular gasoline is currently about 1 mg l^{-1} Pb [140]. Any increase in lead usage would accelerate the deterioration of three-way catalysts.

The effect of lead on the performance of oxidation catalysts has been an important and popular area of study for several years. Shelef *et al.* [10] reviewed the literature on lead poisoning of automobile catalysts through 1976. Lead poisoning of oxidation catalysts is brought about by several different lead containing species (oxides, halides, and sulfides). Lead is added to gasoline as Motor Mix which contains halide scavengers as well as tetraethyl lead. Exhaust gas temperature and A/F influence the identity and quantity of the lead species accumulated. Lead poisoning of oxidation catalysts will be reviewed here in more detail in the section on catalysts for Europe.

Evidence of the detrimental effect of lead on the performance of three-way catalysts has been reported by several groups. Gandhi *et al.* [72] observed a direct relationship between fuel contaminant level and durability for two differently formulated three-way catalysts following aging on a laboratory pulsator test apparatus. The fuels used were lead free isooctane, 1976/1977 simulated certification fuel (isooctane with 0.026 g l^{-1} Pb and 0.003 g l^{-1} P), and 1975/1976 simulated certification fuel (isooctane with 0.11 g l^{-1} Pb and 0.008 g l^{-1} P). The performance of one three-way catalyst as measured by the width of the selectivity window in A/F units (at 80% net conversion of NO, CO, and HC at 823 K and 60,000 h^{-1} space velocity) fell from 0.16 before the aging to zero A/F units with increasing fuel Pb level following 25,000 simulated miles. Gandhi *et al.* [65] observed that the point on the A/F scale of maximum activity was not shifted following aging. Also, hydrocarbon conversion rather than CO conversion was the limiting reaction under rich conditions following aging [65]. In general, Gandhi *et al.* [72] found that catalysts containing higher amounts of Rh showed better overall performance. Schlatter *et al.* [48] observed the accumulation of 0.10 wt% Pb, 0.21 wt% P, and 0.1 wt% S during a 20 h exposure of a three-way catalyst to exhaust from an engine operated on fuel containing high levels of lead and phosphorus (20 mg l^{-1} each) in an accelerated aging test. The conversions of NO, CO, and propylene at the stoichiometric A/F were all below 50 percent following the aging (see Figure 22). While these studies do not strictly distinguish between the poisoning effects of lead and phosphorus, they do demonstrate the need for contaminant-free fuels. Monroe [141] has reported on a study aimed at distinguishing the poisoning effects of lead and phosphorus additives. A Pt—Pd—Rh—Ce/Al_2O_3 three-way catalyst was aged 17.5 h in exhaust from an engine operated with a cycled A/F (20 s at 0.5 A/F units rich and 20 s at 0.5 A/F units lean of the stoichiometric composition) on fuel doped with 0.027 g Pb l^{-1} [141]. Comparison with a similar catalyst aged on

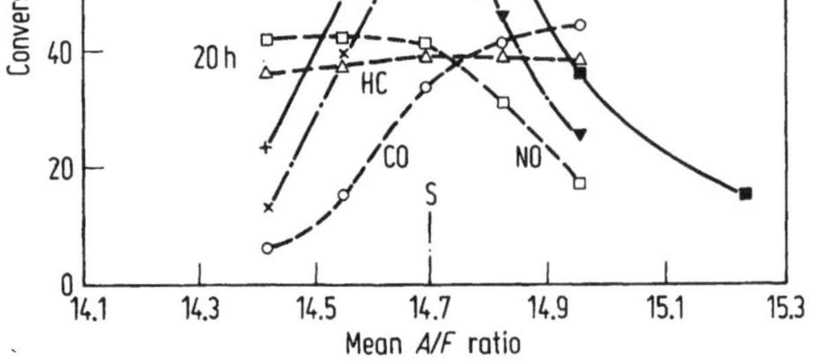

Figure 22. Effect of exhaust aging on catalyst efficiency of a Pt—Rh three-way catalyst (measured at 1 Hz, ±0.25 A/F ratio). (Reproduced with permission from ref. [48])

clear fuel showed that HC and NO conversions were lowered by exposure to the lead-containing fuel whereas CO conversion was not. In contrast, HC, CO, and NO conversions were all lowered by aging with fuel doped with 0.015 g l^{-1} [141]. Monroe [141] concludes that lead is a less severe poison of three-way catalyst activity than is phosphorus at the contaminant levels typical of 1000 h operation on certification fuel but cautions that interaction between different species (*e.g.* P and Pb [142, 151]) may affect the poisoning process.

D. Catalyst Poisoning by Sulfur

Sulfur dioxide is present in exhaust gas at a concentration of about 20 ppm, arising from 0.03 wt % (208 mg l^{-1}) sulfur in unleaded regular gasoline [140]. The interaction of SO_2 with oxidation catalysts received considerable attention when exhaust emission control catalysts were first introduced because of the concern that SO_2 would be catalytically oxidized to sulfuric acid. Little sulfuric acid was emitted by oxidation catalysts, but a large research effort was committed to the study of this subject. This literature on SO_2 oxidation will not be reviewed here. The lower activity of three-way catalysts for SO_2 oxidation in vehicle tests compared with oxidation catalysts has been attributed to the use of Rh [152, 153]. Furthermore, the low oxygen content and the high CO concentration close to the stoichiometric A/F suppress SO_2 oxidation [152]. Likewise, control of the A/F at the stoichiometrically balanced composition restricts SO_2 reduction to H_2S [154]. Reduction of exhaust SO_2 to H_2S is favored by rich A/F, low exhaust flow rates, and high catalyst temperatures [154, 154a].

Sulfur dioxide at levels normally present in exhaust has been shown to affect the activity of three-way catalysts [80, 122, 145, 155–157]. Williamson *et al.* [156] demonstrated in laboratory studies that increasing the sulfur content of gasoline from 0.0 to 0.03 wt % resulted in lower CO, hydrocarbon, and gross NO_x conversions, but a further increase in fuel sulfur content to 0.09 wt % did not result in any additional drop in activity. The effect of sulfur on three-way catalyst activity has been found to be more pronounced during fuel-rich operation than during A/F modulation [156, 157]. This result was attributed to a smaller coverage of catalyst sites by sulfur under dynamic conditions than under rich conditions [157].

A vehicle study of the effect of gasoline sulfur concentration on the performance of aged (8000 km and 80,000 km) three-way catalysts was reported by Furey and Monroe [122]. These results are shown in Table 4. Increasing the fuel sulfur content to 0.03 and 0.09 wt % compared with 0.01 wt % was found to reduce the conversion of CO, hydrocarbons, and NO_x in FTP tests [122]. In these tests which examined the immediate rather than long term effect of sulfur, catalyst poisoning by sulfur was reversible. Laboratory durability studies which compared catalysts aged with fuel containing 0.03 wt % sulfur and sulfur free fuel resulted in a faster drop in activity with time with sulfur-containing fuel [145, 156].

Table 4. Conversion of Hydrocarbons, CO, and NO_x During Vehicle FTP Emission Tests [122]

Catalyst Age	Fuel Sulfur Content/Mass %	Average Conversion/%		
		HC	CO	NO_x
8,000 km	0.01	83.3	64.5	72.4
	0.03	83.9	61.8	71.4
	0.09	81.4	56.2	67.6
80,000 km	0.01	80.3	51.0	66.6
	0.03	79.7	48.2	64.6
	0.09	76.9	44.0	62.8

Sulfur dioxide has been found to influence the selectivity of three-way catalysts, and moreover these influences have been traced to the interaction between SO_2 and specific catalyst components. For example, SO_2 suppresses the reduction of NO to NH_3 by Pt catalysts under net reducing conditions [77, 140]. During A/F cycling, SO_2 was observed to suppress the enhanced CO conversion attributed to Ce in a Rh/Ce/Al_2O_3 catalyst [98]. SO_2 does not significantly poison the NO_x reduction activity of Rh, so high Rh use in three-way catalysts helps to maintain NO reduction [80, 145, 156]. Others have observed with Pt catalysts that SO_2 suppressed the oxidation of propylene [80, 156, 158] but under certain conditions not propane oxidation [156, 158]. The influence of SO_2 on CO oxidation over noble metals has been recognized for some time, but SO_2 has not been viewed as a severe poison of Pt—Pd oxidation catalysts [159].

Joy *et al.* [157] have noted that under fuel-rich operation and in the presence of SO_2 the water-gas shift and steam reforming reactions do not occur over three-way catalysts. These reactions might otherwise make a significant contribution to the removal of CO and unburned HC. The need for dual catalyst and dual converter systems could probably be reduced if the detrimental effects of SO_2 on CO and HC oxidation over three-way catalyst were elminated.

While the presence of sulfur in the fuel is often cited as one of the reasons why noble metal catalysts must be used for exhaust emission control, it is unlikely that current emission control requirements could be met using non-noble metal catalysts even if sulfur were to be removed from gasoline.

12. The 0.4 NO_x Research Objective

Two vehicle manufacturers have published papers in which they describe their research aimed at the development of emission control systems which meet the Federal research objective emission levels of 0.41 g mi^{-1} HC, 3.4 g mi^{-1} CO, and 0.4 g mi^{-1} NO_x [18a, 31]. The development programs are aimed at achieving these emission levels at low mileage and at establishing the necessary durability of the emission control systems using the 80,000 km certification durability schedule [18a, 31]. Of course, vehicle emission levels far below the research objective are sought at low mileage in order to allow for vehicle-to-vehicle variation in emissions and for catalyst deterioration during use. The published reports of these programs indicate that further development is required before the experimental systems can be considered viable [18a, 31].

13. Control of Diesel Particulate Emissions

Federal emission control requirements for diesel-fueled automobiles include a particulate standard of 0.6 g mi^{-1} in 1982 and 0.2 g mi^{-1} in 1985. Development programs aimed at meeting these requirements include both engine modifications and the use of aftertreatment devices. Only aftertreatment will be considered in this review.

Diesel particulates are typically of the order of 0.1 μm diameter and contain both a solid fraction (mostly carbon) and a soluble organic fraction (typically 20 wt%) [160]. The mass emission rate and the organic content of the diesel particulates vary with driving conditions. The mass emissions also vary depending upon engine size and vehicle make.

Control devices currently under development involve collection of the particulates in traps followed by periodic incineration. The low temperature of diesel exhaust (473–673 K) at the trap location precludes continuous incineration. The development of a reliable aftertreatment system will require good collection efficiency, reliable periodic incineration, and durable trapping materials.

Trapping materials which have been described in published reports include alumina-coated metal wool [161], catalyzed alumina-coated metal wire mesh [162], a cellular ceramic trap [163], and fiberglass cloth [164]. Oh *et al.* [165] have developed a mathematical model which describes the initial performance of fibrous filter traps and considers important filter design parameters such as filter face area, filter thickness, packing density, and fiber size.

The potentially large volume of collected particulates requires a method for periodic clearing of the trap. The thermal regeneration of diesel particulate filters has been described by a mathematical model developed by Shadman and Bissett [166]. Experimental studies by Otto *et al.* [167] have explored the requirements for rapid oxidation of trapped diesel particulates. Oxygen pressure, bulk density of the soot, and organic fraction all influenced soot combustion [167, 168]. Reliable periodic ignition may require use of catalysts or an external ignition source. The temperature at which trapped particulates ignite has been shown to be lowered by the addition of inorganic catalysts [167, 169, 170]. Incorporation of the catalyst into the soot (*e.g.* by the addition of metal additives to the fuel) may be preferred to a catalyzed trap material. Other ignition techniques being explored include fuel addition to the exhaust, burners [161, 162] and combinations of these techniques.

At this writing trapping systems which make use of the concepts described above are under development. No single system has demonstrated adequate performance, durability, and reliability for all applications.

14. Exhaust Gas Purification for Europe

Exhaust emissions are currently controlled without the use of catalysts in Europe; however, exhaust emission standards have been proposed which might lead to catalyst systems which differ significantly from those used in the United States and Japan [171]. Exhaust emission standards being discussed are aimed at 90% reductions in emissions compared with 1969 levels [171]. Koberstein *et al.* [171] state that the proposed CO emission standard varies with the inertial weight of the vehicle, while the HC and NO_x emissions are added together to meet a single standard. No durability schedule has yet been adopted. The emissions test method (called the ECE-test after the Economic Commission for Europe) differs from the CVS-test used in the United States, and emissions measured with the two tests cannot be directly compared. These two driving schedules are shown in Figure 23. A main difference between the European and U.S. approach to exhaust emission control is that lead-free gasoline will not be available in Europe (except perhaps in Switzerland). Accordingly catalyst systems must be developed which are compatible with lead content in the range 0.15 to 0.4 g l^{-1} [171]. No catalyst schemes have been identified which can be used to meet U.S. exhaust emission standards at such high lead content.

The catalyst system which has received the most attention for application in Europe is a platinum-containing oxidation catalyst [171]. Platinum is less susceptible to poisoning by lead than is palladium; in fact, palladium increases the susceptibility of platinum to poisoning when the two metals occur together

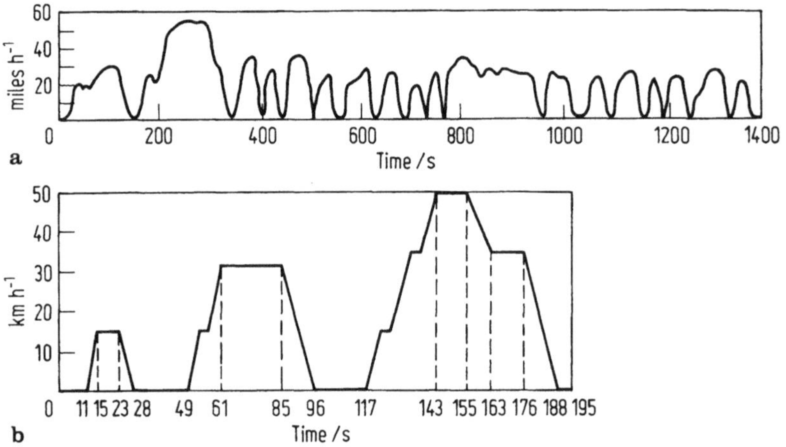

Figure 23. U.S. driving schedule of the CVS Test (FTP 72) **a**; and European ECE Test **b**. (Reproduced with permission from ref. [171])

on the same catalyst [172–175]. The catalyst would be aimed at lowering CO and HC emissions under net oxidizing conditions. NO_x emissions would not be controlled catalytically under these conditions. Both ceramic and metallic monolithic catalyst supports are being considered for this application [171, 174].

Lead poisoning of automobile oxidation catalysts has been studied extensively because Pb is a severe poison of catalysts at levels typical of leaded fuels. The scientific literature on lead poisoning of automobile emission control catalysts prior to 1981 primarily addressed the poisoning of oxidation catalysts as used to meet U.S. regulations. The literature contains much information currently of interest for the application of catalysts in Europe. Here we review some of the highlights of the early work together with the recent literature aimed specifically at a European catalyst system.

The oxidation catalyst used with leaded fuel must be able to withstand considerable poisoning by several different lead compounds [143, 174, 176] which, depending upon the conditions of exposure, can both poison catalytic sites via pore mouth poisoning [141, 173] and restrict access to the catalyst *via* pore blockage [5, 177]. At temperatures below 873 K Pt and Pd oxidation catalysts are severely poisoned by lead halides and halide scavengers [172, 174, 178]. Activity loss due to the scavenger alone (ethylene dibromide and ethylene dichloride) is reversible upon removing the scavengers whereas the effects of tetraethyl lead are not fully reversed on changing to unleaded fuel [16, 172]. Ethylene dibromide was a more severe catalyst poison than ethylene dichloride when the effect of scavengers on catalytic activity was examined without tetraethyl lead in the fuel [172, 178]. At catalyst temperatures above 873 K halide compounds are not stable in the solid phase and lead accumulates in an oxidation catalyst as $PbSO_4$ and $PbO \cdot PbSO_4$ [179]. Steady-state lead poisoning studies have shown a relationship between fuel Pb content and catalyst deterioration [173, 180]. For example, Klimisch *et al.* [173] report that exposure of 1000 cm^3 of a 0.1 % Pt on Al_2O_3 catalyst at 823 K to exhaust at $A/F = 16$ for 60 h did not result in deactivation of activity for HC oxidation when the fuel contained 0.026 g Pb l^{-1}, whereas the catalyst was severely deactivated in 5 h when the fuel contained 0.77 g Pb l^{-1}. As expected, increasing the volume of catalyst beyond 1000 cm^3 improved the poison tolerance [173]. CO oxidation activity loss at high temperature is in general initially less severe than HC oxidation activity loss [174].

A complex reaction scheme which describes the accumulation and volatilization of lead compounds on oxidation catalysts has been proposed by White and Zekrajsek [174]. These authors speculate that cyclical rather than steady-state exposure of catalysts to lead-containing exhaust is less detrimental to the catalysts because halide compounds present in the fuel to scavange lead can react with stable lead deposits (*e.g.* $PbO \cdot PbX_2$ and $PbO \cdot PbSO_4$) to form lead halide compounds which are subsequently volatilized from the catalyst during a hot cycle [174]. This scheme incorporates the model for the mechanism of poisoning described by McArthur [143] which predicts that lead retention is temperature dependent, and for operating temperatures greater than 973 K Pb is removed and catalytic activity rejuvenated. Furthermore, this scheme explains the observed influence of driving mode or test conditions on the accumulation of lead [171].

Diwell and Harrison [181] have recently reported on a development program on lead tolerant catalysts for Europe at Johnson Matthey. The effect of inlet temperature to the catalyst on efficiency for methane oxidation was investigated by pulse flame reactor tests between 823 and 1023 K using fuel doped with 0.4 g l^{-1} lead (Motor Mix). After eight hours of aging, the best conversion was demonstrated by the catalyst aged and tested at 923 K. These authors considered the appearance of an optimum temperature of operation to reflect a window where the lead halides are unstable and the low vapor pressure of PbO limits transport of lead to the catalysts [181]. Studies of the effect of A/F on the rate of loss of catalyst activity suggested that engine operation at lean A/F ratios produces a lower rate of poisoning than do rich A/F ratios [181]. Using an AMA driving schedule and ECE-test Diwell and Harrison [181] achieved 50% conversion after 37,000 km of driving with a fuel containing 0.4 g l^{-1} lead; however, the channels of the monolithic catalyst became blocked by deposition of particulate lead after 48,000 km of driving. Both Diwell and Harrison [181] and Koberstein *et al.* [171] have demonstrated in independent vehicle tests the application of a noble metal catalyst and monolithic metal substrate to emission control using an AMA durability cycle and the ECE emissions test. Further development work is required before these systems can be adopted. Catalyst efficiency is low, and the platinum usage is likely to be high (2.5–2.8 g l^{-1}). Catalyst system reliability under all possible driving conditions has not been demonstrated. Controlled aging cycles probably simulate customer usage poorly. These studies [171, 174, 180] have all helped to make known the extreme complexity of the catalyst-poison interaction and have identified several catalyst design and operating parameters which appear important for the successful development of lead tolerant catalysts.

15. Concluding Remarks

The research challenges in emission control catalysis have not been exhausted with the introduction of the three-way catalyst. Four areas of continuing need are:

1) *To reduce dependence on noble metals.*
   The annual consumption of Pt in the manufacture of catalytic converters is the largest single use of Pt in the United States and exceeds all other uses combined [182].
   The current usage of Rh and Pt in three-way catalysts exceeds the Rh/Pt mine ratio. Reduction of noble metal usage or replacing noble metals would help conserve the limited noble metal supply.
2) *To reduce catalyst deterioration.*
   Catalyst deterioration during use results in catalysts and catalytic converter systems being designed to exceed performance requirements throughout

all 80,000 km of use. Elimination of deterioration would permit use of smaller converters and lower noble metal usage.

3) *To optimize converter performance under transient A/F conditions.* The chemical engineering community has only limited field experience with nonsteady-state catalytic reactor operation. Evidence to date on the three-way converter system suggests that further design studies and development could lead to improved three-way performance and elimination of secondary air supply and oxidation catalysts.

4) *To plan for future application of alternate fuels and fuel additives.* Catalytic control of aldehyde emissions from alcohol fueled vehicles was not covered in this review but is an area of current interest in view of the recent introduction of alcohol fuels and fuel blends.

16. References

1. Quader, A. A.: SAE Transactions **80**, (1971) Paper No. 710009
2. Gumbleton, J. J.; Bolton, R. A.; Lang, H. W.: Paper No. 740104, presented to Society of Automotive Engineers, February, 1974
3. Canale, R. P.; Winegarden, S. R.; Carlson, C. R.; Miles, D. L.: Paper No. 780205, presented to Society of Automotive Engineers (1978)
4. Kummer, J. T.: in: Progress in Energy and Combustion Science, Vol. 6, pp. 177–99, Pergamon Press Ltd., 1980
5. Ebel, R. H.: Advances in Environ. Sci. **1**, 237 (1969)
6. Dwyer, F. G.: Catalysis Rev. **6**, 261 (1972)
7. Wei, J.: Advances in Catalysis **24**, 57 (1975)
8. Shelef, M.: Catalysis Rev.-Sci. Eng. **11**, 1 (1975)
9. Wei, J.: Adv. Chem. Ser. **148**, 1 (1975)
10. Shelef, M.; Otto, K.; Otto, N. C.: Advances in Catalysis **27**, 311 (1978)
11. Hegedus, L. L.; Gumbleton, J. J.: Chemtech **10**, 630 (1980)
12. Egelhoff, W. G. Jr.: in: The Chemical Physics of Solid Surfaces and Heterogenous Catalysis, Vol. 4, p. 397 (D. A. King and D. P. Woodruff, eds.) Elsevier, New York 1982
13. Harrison, B.; Cooper, B. J.; Wilkins, A. J. J.: Platinum Met. Rev. **25**, 14 (1981)
14. Agnew, W. G.: Proc. Roy. Soc. **A307**, 53 (1968)
15. McDonald, G. C.: Technometrics **23**, 239 (1981)
16. Jackson, M. W.: Paper No. 780624, presented to Society of Automotive Engineers, 1978
17. 1978 Annual Report of General Motors on Advanced Emission Control System Development Progress, submitted to Environmental Protection Agency, June, 1979
18. Grimm, R. A.; Bremer, R. J.; Sonestreet, S. P.: Paper No. 800053, presented to Society of Automotive Engineers, February, 1980. Reprinted in Automotive Engineering **88**(8) 33 (1980)

18a. Zemke, B. E.; Gumbleton, J. J.: Paper No. 800398, presented to Society of Automotive Engineers, February, 1980

19. Creps, W. D.; Spilski, R. A.: Paper No. 750368, presented to Society of Automotive Engineering, February, 1975
20. Seiter, R. E.; Clark, R. J.: Paper No. 780203, presented to Society of Automotive Engineers, February, 1978
21. Masaki, K.; Akaeda, M.; Minami, H.: Paper No. 780204, presented to Society of Automotive Engineers, February, 1978
22. Zechnall, R.; Baumann, G.; Eisele, H.: Paper No. 730566, presented to Society of Automotive Engineers, February, 1973
23. Kaneko, Y.; Kobayashi, H.; Komagome, R.; Nakagami, T.; Fukui, T.: Paper No. 770197, presented to Society of Automotive Engineers

24. Toyoda, T.; Yamakawa, Y.; Inoue, T.; Oishi, K.; Hattori, K.: Paper No. 800399, presented to Society of Automotive Engineers, February, 1980
25. Wotring, W. T.; Meguerian, G. H.; Gandhi, H. S.; McCuiston, F. D.; Piken, A. G.: Paper No. 780608, presented to Society of Automotive Engineers, June, 1978
26. Oser, P.: Paper No. 790306, presented to Society of Automotive Engineers, February, 1979
27. Sell, J. A.; Herz, R. K.; Monroe, D. R.: Paper No. 800463, presented to Society of Automotive Engineers, February, 1980
28. Sell, J. A.; Herz, R. K.; Perry, E. C.: Paper No. 810276, presented to Society of Automotive Engineers, February, 1981
29. Kaneko, Y.; Kobayashi, H.; Komagome, R.; Hirako, H.; Nakayama, O.: Paper No. 780607, presented to Society of Automotive Engineers, June, 1978
30. Hammerle, R. H.; Wu, C. H.: Paper No. 810275, presented to Society of Automotive Engineers, February, 1981
31. Cederquist, A. L.; Devlin, S. S.; Hart, D. L.; Moon, R.: Paper No. 790508, presented to Society of Automotive Engineers, February, 1979
32. Herod, D. M.; Nelson, M. V.; Wang, W. M.: Paper No. 730557, presented to Society of Automotive Engineers, February, 1973
33. Casassa, J. P.; Beyerlein, D. G.: Paper. No. 730558, presented to Society of Automotive Engineers, February, 1973
34. Hegedus, L. L.; Summer, J. C.; Schlatter, J. C.; Baron, K.: J. Catal. **56**, 321 (1979)
35. Heyde, M. E.; Adawi, M. K.; Andersen, D. H.; Schroeder, M. P.; DeLosh, R. G.: Paper No. 770371, presented to Society of Automotive Engineers, February, 1977
36. Klimisch, R. L.; Barnes, G. J.: Environ. Sci. Technol. **6**, 543 (1972)
37. Shelef, M.; Gandhi, H. S.: Ind. Eng. Chem. Prod. Res. Develop. **11**, 2 (1972)
38. Mannion, W. A.; Aykan, K.; Cohn, J. G.; Thompson, C. E.; Mooney, J. J.: in: Catalysts for the Control of Automobile Pollutants, Adv. in Chem. Ser. **143**, American Chemical Society, Washington, D.C., 1975
39. Breitenbach, L. P.; Shelef, M.: J. Air Pollution Control Assoc. **23**, 128 (1973)
40. Taylor, K. C.: Ind. Eng. Chem. Prod. Res. Develop. **15**, 264 (1976)
41. Barnes, G. J.; Summers, J. C.: Paper No. 750093, presented to Society of Automotive Engineers, February, 1975
42. Hammerle, R. H.; Mikkor, M.: Paper No. 750097, presented to Society of Automotive Engineers, February, 1975
43. Goksøyr, H.; Ross, K.: J. Inst. Fuel **35**, 177 (1962)
44. West, P. W.; Gaeke, G. C.: Anal. Chem. **28**, 1816 (1956)
45. Flamm, D. L.; James, E. R.: Environ. Sci. Technol. **9**, 159 (1975)
46. Dietzmann, H. E.; Smith, L. R.; Porness, M. A.; Fanick, E. R.: Analytical Procedures for Characterizing Unregulated Pollutant Emissions from Motor Vehicles. EPA Publication EPA-600/2-79-017, February, 1979
47. Otto, K.; Dalla Betta, R. A.; Yao, H. C.: J. Air Pollution Control Assoc. **24**, 596 (1974)
48. Schlatter, J. C.; Sinkevitch, R. M.; Mitchell, P. J.: Ind. Eng. Chem. Prod. Res. Develop. **22**, 51 (1983). General Motors Research Publication GMR-2911
49. Schlatter, J. C.; Taylor, K. C.: J. Catal. **49**, 42 (1977)
50. Klimisch, R. L.; Taylor, K. C.: Environ, Sci. Technol. **7**, 127 (1973)
51. Dubois, L. H.; Hansma, P. K.; Somorjai, G. A.: J. Catal. **65**, 318 (1980)
52. Campbell, C. T.; White, J. M.: Appl. Surface Sci. **1**, 347 (1978)
53. Taylor, K. C.; Schlatter, J. C.: J. Catal. **63**, 53 (1980)
54. Schlatter, J. C.; Taylor, K. C.; Sinkevitch, R. M.: Paper presented at the Advances in Catalytic Chemistry Symposium, Snowbird, Utah, October 3–5, 1979. To be published in Symposium Proceedings
55. Kim, Y.; Shi, S.-K.; White, J. M.: J. Catal. **61**, 374 (1980)
56. Baird, R. J.; Ku, R. C.; Wynblatt, P.: Surface Sci. **97**, 346 (1980)
57. Gland, J. L.; Kollin, E. B.: J. Catal. **68**, 349 (1981)
58. Engel, T.; Entl, G.: Advances in Catalysis **28**, 2 (1979)
59. Gland, J. L.; Kollin, E. B.: J. Chem. Physics to be published
60. Niwa, M.; Lunsford, J. H.: J. Catal. **75**, 302 (1982)

61. Yaverbaum, L. H.: Nitrogen Oxides Control and Removal. Noyes Data Corp., New Jersey, 1979
62. Taylor, K. C.: in: The Catalytic Chemistry of Nitrogen Oxides (R. L. Klimisch and J. G. Larson, eds.) Plenum Press, New York, 1975
63. Ohara, T.: in: The Catalytic Chemistry of Nitrogen Oxides (R. L. Klimisch and J. G. Larson, eds.) Plenum Press, New York, 1975
64. Cooper, B. J.; Harrison, B.; Shutt, E.; Lichtenstein, I.: Paper No. 770367, presented to Society of Automotive Engineers
65. Gandhi, H. S.; Piken, A. G.; Shelef, M.; Delosh, R. G.: Paper No. 760201, presented to Society of Automotive Engineers, February, 1976
66. Kobylinski, T. P.; Taylor, B. W.: J. Catal. **33**, 376 (1974)
67. Ashmead, D. R.; Campbell, J. S.; Davies, J. S.; Farmery, K.: Paper No. 740249, presented to Society of Automotive Engineers, February, 1974
67a. Bauerle, G. L.; Service, G. R.; Nobe, K.: Ind. Eng. Prod. Res. Develop. **11**, 54 (1972)
68. Taylor, K. C.: Ind. Eng. Chem. Prod. Res. Develop. **15**, 264 (1976)
69. Yao, H. C.; Yao, Y.-F.; Otto, K.: J. Catal. **56**, 21 (1979)
70. Markvart, M.; Pour, V. L.: J. Catal. **7**, 279 (1967)
71. Mooney, J. J.; Thompson, C. E.; Dettling, J. C.: Paper No. 770365, presented to Society of Automotive Engineers, February, 1977
72. Gandhi, H. S.; Piken, A.; Stepien, H.; Shelef, M.; Delosh, R.; Heyde, M.: Paper No. 770196, presented to Society of Automotive Engineers, February, 1977
73. Schlatter, J. C.: Paper No. 780199, presented to Society of Automotive Engineers, February, 1978
74. Yao, H. C.; Japar, S.; Shelef, M.: J. Catal. **50**, 407 (1977)
75. Yao, H. C., Gandhi, H. S.: J. Catal. **61**, 547 (1980)
76. Summers, J. C.; Monroe, D. R.: Ind. Eng. Chem. Prod. Res. Develop. **20**, 23 (1981)
77. Jones, J. H.; Kummer, J. T.; Otto, K.; Shelef, M.; Weaver, E. E.: Environ, Sci. Technol. **5**, 790 (1971)
78. Summers, J. C.; Hegedus, L. L.: J. Catal. **51**, 185 (1978)
79. Otto, K.; Yao, H. C.: J. Catal. **66**, 229 (1980)
80. Summers, J. C.; Baron, K.: J. Catal. **57**, 380 (1979)
81. Tauster, S. J.; Murrell, L. L.: J. Catal. **41**, 192 (1976)
82. Lester, G. R.; Joy, G. C.; Brennan, J. F.: Paper No. 780202, presented to Society of Automotive Engineers, February, 1978
83. McVicker, G. B.; Garten, R. L.; Baker, R. T. K.: J. Catal. **54**, 129 (1978)
84. Taylor, K. C.; Klimisch, R. L.: J. Catal. **30**, 478 (1973)
85. Taylor, K. C.; Sinkevitch, R. M.; Klimisch, R. L.: J. Catal. **35**, 34 (1974)
86. Klimisch, R. L.; Taylor, K. C.: Ind. Eng. Chem. Prod. Res. Develop. **14**, 26 (1975)
86a. Shelef, M.; Gandhi, H. S.: Ind. Eng. Chem. Prod. Res. Develop. **11**, 393 (1972)
87. Shelef, M.; Gandhi, H. S.: Platinum Met. Rev. **18**, 2 (1974)
88. Kobylinski, T. P.; Taylor, B. W.; Young, J. E.: Paper No. 740250, presented to Society of Automotive Engineers, February, 1974
89. Gandhi, H. S.; Stepien, H. K.; Shelef, M.: Paper No. 750177, presented to Society of Automotive Engineers, February, 1975
90. Pepper, M. W.; DeLuca, J. A.; Rohdes, R. P.; Bernstein, L. S.: Paper No. 750684, presented to Society of Automotive Engineers, June, 1975
91. Bernstein, L. S.; Kearby, K. K.; Raman, A. K. S.; Vardi, J.; Wigg, E. D.: Paper No. 710014, presented to Society of Automotive Engineers, February, 1971
92. Lunt, R. S.; Bernstein, L. S.; Hansel, J. G.; Hoh, E. L.: Paper No. 720209, presented to Society of Automotive Engineers, February, 1972
93. Bernstein, L. S.; Lang, R. J.; Lunt, R. S.; Musser, G. S.: Paper No. 730567, presented to Society of Automotive Engineers, May, 1973
94. Lee, C. H.; Makowski, M. P.: Paper No. 741090, presented to Society of Automotive Engineers, October, 1974
95. Fedor, R. J.; Dunlevey, F. M.; Henry, R. W.; Steiner, R. R.: Paper No. 741081, presented to Society of Automotive Engineers, October, 1974
96. Fedor, R. J.: in: The Catalytic Chemistry of Nitrogen Oxides (R. L. Klimisch and J. G. Larson, eds.) Plenum Press, New York, 1975

97. Cooper, B. J.; Keck, L.: Paper No. 800461, presented to Society of Automotive Engineers, February, 1980
98. Schlatter, J. C.; Mitchell, P. J.: Ind. Eng. Chem. Prod. Res. Develop. **19**, 288 (1980)
99. Yu Yao, Y.-F.: Paper presented at 6th No. Am. Cat. Soc. Mtg., Chicago, 1979
100. Summers, J. C.; Ausen, S. A.: J. Catal. **58**, 131 (1979)
101. Herz, R. K.: Ind. Eng. Chem. Prod. Res. Develop. **20**, 451 (1981)
102. Kummer, J. T.; Yao, Y.; McKee, D.: Paper No. 760143, presented to Society of Automotive Engineers, February, 1976
103. Yu Yao, Y.-F.; Kummer, J. T.: A.C.S. Division of Petroleum Chemistry Preprints **21**, 807 (1976)
104. Yu Yao, Y.-F.: J. Catal. **39**, 104 (1975)
105. Barnes, G. J.: in: Catalysts for the Control of Automotive Pollutants, Adv. in Chem. Ser. 143, American Chemical Society, Washington, D.C., 1975
106. Kummer, J. T.: J. Catal. **38**, 166 (1975)
107. Fu, C.-M.; Deeba, M.; Hall, W. K.: Ind. Eng. Chem. Prod. Res. Develop. **19**, 299 (1980)
108. Gandhi, H. S.; Yao, H. C.; Stepien, H. K.: in: Catalysis Under Transient Conditions, A.C.S. Symposium Ser. 178, American Chemical Society, Washington, D.C., 1982
109. Osment, H. E.: Paper No. 730276, presented to Society of Automotive Engineers, January, 1973
110. Summers, J. C.; Thompson, C. E.: Paper No. 800842, presented to Society of Automotive Engineers, June, 1980
111. Gauguin, R.; Graulier, M.; Papee, D.: in: Catalysts for the Control of Automotive Pollutants, Adv. in Chem. Ser. 143, American Chemical Society, Washington, D.C., 1975
112. Adomaitis, J. R.; Smith, J. E.; Achey, D. E.: Paper No. 800084, presented to Society of Automotive Engineers, February, 1980
113. Ernest, M. V.; Kim, G.: Paper No. 800083, presented to Society of Automotive Engineers, February, 1980
114. Barnes, G.; Klimisch, R.: Paper No. 730570, presented to Society of Automotive Engineers, February, 1973
115. Hegedus, L. L.; Summers, J. C.: J. Catal. **48**, 345 (1977)
116. Wei, J.; Becker, E. R.: in: Catalysts for the Control of Automotive Pollutants, Adv. in Chem. Ser. 143, American Chemical Society, Washington, D.C., 1975
117. Hegedus, L. L.; Summers, J. C.; Schlatter, J. C.; Baron, K.: J. Catal. **56**, 321 (1979)
118. Oh, S. H.; Cavendish, J. C.; Hegedus, L. L.: AIChE J. **26**, 935 (1980)
119. Bagley, R. D.; Doman, R. C.; Duke, D. A.; McNally, R. N.: Paper No. 730274, presented to Society of Automotive Engineers, January, 1973
120. Pentecost, J. L.; Cochran, J. K.: Paper No. 740198, presented to Society of Automotive Engineers, February, 1974
121. Briggs, W. S.; Graham, J. R.: Paper No. 730275, presented to Society of Automotive Engineers, January, 1973
122. Furey, R. L.; Monroe, D. R.: Paper No. 811228, presented to Society of Automotive Engineers, October, 1981
123. Howitt, J. S.: Paper No. 800082, presented to Society of Automotive Engineers, February, 1980
124. Lester, G. R.; Marinangeli, R. E.: Paper No. 800844, presented to Society of Automotive Engineers, June, 1980
125. Oh, S. H.; Cavendish, J. C.: Ind. Eng. Chem. Prod. Res. Develop. **21**, 29 (1982)
126. Herz, R. K.: Ind. Eng. Chem. Prod. Res. Develop. **20**, 451 (1981)
127. Meitzler, A. H.: Paper No. 800019, presented to Society of Automotive Engineers, February, 1980
128. Herz, R. K.: in: Catalysis Under Transient Conditions, A.C.S. Symposium Ser. 178, American Chemical Society, Washington, D.C., 1982
129. Shulman, M. A.; Hamburg, D. R.; Throop, M. J.: Paper No. 820276, presented to Society of Automotive Engineers, Feburary, 1982
130. Falk, C. D.; Mooney, J. J.: Paper No. 800462, presented to Society of Automotive Engineers, February, 1980

131. Herz, R. K.; Kiela, J.; Sell, J. A.: Ind. Chem. Prod. Res. Develop. **22**, 381 (1983)
132. Taylor, K. C.; Sinkevitch, R. M.: Ind. Eng. Chem. Prod. Res. Develop. **22**, 45 (1983).
133. Oh, S. H.: Private communication
134. Sell, J. A.; Chang, M.-F.: Paper No. 820388, presented to Society of Automotive Engineers, February, 1982
135. Koberstein, E.: Paper No. 770366, presented to Society of Automotive Engineers, March, 1977
136. Summers, J. C.; Monroe, D. R.; Chang, C. C.; Gaarenstrom, S. W.: Paper presented at Materials Research Society Meeting, Boston, MA 1979. Available as General Motors Research Publication GMR-3147
137. Wang, T.; Schmidt, L. D.: J. Catal. *71*, 141 (1981)
138. Stepien, H. K.; Williamson, W. B.; Gandhi, H. S.: Paper No. 800843, presented to Society of Automotive Engineers, June, 1980
139. Barnes, G. J.; Schlatter, J. C.: American Chemical Society Division of Petroleum Chemistry, Inc. Preprints **22**, 1258 (1977)
140. MVMA National Gasoline Survey, Fall 1980, Motor Vehicle Manufacturers Association, January, 1981
141. Monroe, D. R.: Paper No. 800859, presented to Society of Automotive Engineers, June, 1980
142. Spearot, J. A.; Caracciolo, F.: Paper No. 770637, presented to Society of Automotive Engineers, February, 1977
143. McArthur, D. P.: in: The Catalytic Chemistry of Nitrogen Oxides (R. L. Klimisch and J. G. Larson, eds.) Plenum, New York, 1975
144. Williamson, W. B.; Stepien, H. K.; Watkins, W. L. H.; Gandhi, H. S.: Environ. Sci. Technol. **13**, 1109 (1979)
145. Williamson, W. B.; Stepien, H. K.; Gandhi, H. S.: Environ. Sci. Technol. **14**, 319 (1980)
146. Williamson, W. B.; Perry, J.; Gandhi, H. S.; Bomback, J. L.: Paper No. 810084, presented to Society of Automotive Engineers, February, 1981
147. Caracciolo, F.; Spearot, J. A.: Paper No. 790941, presented to Society of Automotive Engineers, February, 1979
148. Caracciolo, F.; Spearot, J. A.: Paper No. 760562, presented to Society of Automotive Engineers, 1976
149. Weaver, E. E.; Shiller, J. W.; Piken, A. G.: AIChE Symposium Series **72**, (156) 369 (1976)
150. Hegedus, L. L.; Summers, J. C.: J. Catal. **48**, 345 (1977)
151. Acres, G. J. K.; Cooper, B. J.; Shutt, E.; Malerbi, B. W.: in: Catalysts for the Control of Automobile Pollutants, Adv. in Chem. Ser. 143, American Chemical Society, Washington, D.C., 1975
152. Gandhi, H. S.; Otto, K.; Piken, A. G.; Shelef, M.: Environ. Sci. Technol. **11**, 170 (1977)
153. Krause, B.; Bouffard, R. A.; Karmilovich, T.; Kayle, E. L.: Paper No. 760091, presented to Society of Automotive Engineers, February, 1976
154. Barnes, G. J.; Summers, J. C.: Paper No. 750093, presented to Society of Automotive Engineers, February, 1975
154a. Cadle, S. H.; Mulawa, P. A.: Paper No. 780200, presented to Society of Automotive Engineers, February, 1978
155. Gandhi, H. S.; Yao, H. C.; Stepien, H. K.; Shelef, M.: Paper No. 780606, presented to Society of Automotive Engineers, 1978
156. Williamson, W. B.; Gandhi, H. S.; Heyde, M. E.; Zawacki, G. A.: Paper No. 790942, presented to Society of Automotive Engineers, 1979
157. Joy, G. C.; Lester, G. R.; Molinaro, F. S.: Paper No. 790943, presented to Society of Automotive Engineers, 1979
158. Yao, H. C.; Stepien, H. K.; Gandhi, H. S.: J. Catal. **67**, 231 (1981)
159. Hunter, J. E.: Paper No. 720122, presented to Society of Automotive Engineers, January, 1972
160. Amann, C. A.; Stivender, D. L.; Plee, S. L.; MacDonald, J. S.: Paper No. 800251, presented to the Society of Automotive Engineers, 1980
161. McMahon, M. A.; Faist, C. H.; Virk, K. S.; Tierney, W. T.: Paper No. 820183, presented to Society of Automotive Engineers, February, 1982

162. Enga, R. E.; Buchmann, M. F.; Lichtenstein, I. E.: Paper No. 820184, presented to the Society of Automotive Engineers, February, 1982
163. Mogaka, Z. N.; Wong, V. W.; Shahed, S. M.: Paper No. 820272, presented to Society of Automotive Engineers, February, 1982
164. Tessier, L. P.; Sullivan, H. F.; Bragg, G. M.; Hermance, C. E.: Paper No. 800338, presented to the Society of Automotive Engineers, February, 1980
165. Oh, S. H.; MacDonald, J. S.; Vaneman, G. L.; Hegeus, L. L.: Paper No. 810113, presented to Society of Automotive Engineers, February, 1981
166. Shadman, F.; Bissett, E. J.: Paper presented to Combustion Inst., Spring Meeting, Warren, MI, March, 1981
167. Otto, K.; Sieg, M. H.; Zinbo, M.; Bartosiewicz, L.: Paper No. 800336, presented to the Society of Automotive Engineers, February, 1980
168. Wade, W. R.; White, J. E.; Florek, J. J.: Paper No. 810118, presented to Society of Automotive Engineers, February, 1981
169. Murphy, M. J.; Hillenbrand, L. J.; Trayser, D. A.; Wasser, J. H.: Paper No. 810112, presented to Society of Automotive Engineers, February, 1981
170. Hillenbrand, L. J.; Trayser, D. A.: Paper No. 811236, presented to Society of Automotive Engineers, Tulsa, OK, October, 1981
171. Koberstein, E.; Pletka, H. D.; Volker, H.: Paper No. 800394, presented to Society of Automotive Engineers, February, 1980
172. Barnes, G. J.; Baron, K.; Summers, J. C.: Paper No. 741062, presented to Society of Automotive Engineers, October, 1974
173. Klimisch, R. L.; Summers, J. C.; Schlatter, J. C.: in: Catalysts for Control of Automotive Pollutants, Adv. in Chem. Ser. 143, American Chemical Society, Washington, D.C., 1975
174. White, J. E.; Zakrajsek, C. E.: Paper No. 810086, presented to Society of Automotive Engineers, February, 1981
175. Gracomazzi, R. A.; Holmfeld, M. F.: Paper No. 730595, presented to Society of Automotive Engineers, January, 1969
176. Hirschler, D. A.; Gilbert, L. F.; Lamb, F. W.: Niebylski, L. M.: Ind. Eng. Chem. **49**, 1131 (1957)
177. Hegedus, L. L.; Baron, K.: J. Catal. **37**, 127 (1975)
178. Otto, K.; Montreuil, C. N.: Environ. Sci. Technol. **10**, 154 (1976)
179. McArthur, D. P.: in: Catalysts for the Control of Automobile Pollutants, Adv. in Chem. 143, American Chemical Society, Washington, D.C., 1975
180. Weaver, E. E.: Paper No. 690016, presented to Society of Automotive Engineers, January, 1969
181. Diwell, A. F.; Harrison, B.: Platinum Met. Rev. **25**, 142 (1981)
182. Chaston, J. C.: Platinum Metals Rev. **26**, 3 (1982)

Chapter 3

Infrared Spectroscopy in Catalytic Research

J. B. Peri

Amoco Oil Company, Research and Development Department,
Amoco Research Center, P.O. Box 400 Naperville, Illinois 60566, USA

Contents

1. Introduction

After nearly thirty years of intensive application, infrared spectroscopy remains the most widely used, and usually most effective, spectroscopic method for characterization of the surface chemistry of high-area heterogeneous catalysts and of many other high-area solids of major commercial importance. Starting from the work of Terenin and other Russian scientists in the 1940's and early 1950's [1] and important pioneering work by Eischens and coworkers [2, 3] in the 1950's on supported metal catalysts, the use of infrared spectroscopy in surface studies generally, and in catalytic research particularly, has grown rapidly. Hundreds of articles, reviews [3–11], and three books [12–14] presently attest to its popularity. Infrared spectroscopy [5] and closely related vibrational spectroscopies of other kinds [15–17] have also played, and continue to play, a very important role in research on the surface chemistry of low-area solids of many kinds.

Reasons for this success are not hard to find. Infrared spectra of surface groups and adsorbed molecules on high-area catalysts and catalyst supports are usually easily obtainable using relatively inexpensive commercial spectrometers and conventional high vacuum or simple flow systems. Infrared cells adequate for many applications can be constructed at modest expense by the research worker, and sample preparation does not usually present major difficulties. Spectra of surface species can usually be obtained under conditions not too far removed from those of typical catalytic reactions. Infrared spectroscopy also provides a fairly gentle, nondestructive probe of the surface, in contrast to the use of electron or ion bombardment or ionizing radiation which may cause important changes in surface chemistry. Interpretation of the infrared spectral data, although often very difficult, is aided by a large background literature, and the connection between the data and its interpretation is generally more immediately obvious than with certain other applicable techniques which require elaborate mathematical treatment of the data to produce a final result.

Because of its intensive past application, infrared spectroscopy in surface studies is often regarded as a fairly mature technique [6]. When, however, consideration is taken of the questionable quality of much of the literature in this field, the great complexity of solid surfaces, and the very significant recent improvements in spectrometers and in related techniques, it seems more accurate to regard infrared study of surfaces as simply in adolescence. In comparison with other techniques used for surface research, which are often still in infancy, infrared spectroscopy has attained a size and self-assurance suggesting maturity, but much additional learning, and unlearning, will be needed for true maturity.

The availability of better tools, plus growing realization by most investigators that better work is needed, augurs well for future development. We also increasingly find valuable informational inputs from other types of surface studies, ranging from "surface science" techniques, such as electron energy loss spectroscopy (EELS), low energy electron diffraction (LEED), Auger electron spectroscopy (AES), photoelectron spectroscopy (XPS, UPS, ESCA) and secondary ion mass spectrometry (SIMS), requiring ultrahigh vacuum, to X-ray absorption edge fine structure (EXAFS), Mössbauer spectroscopy and other techniques which can be used for high pressure studies. Such information often permits more reliable interpretation of the infrared spectra. Most research workers are not equipped to use many tools on a particular problem, however, and should, in any case, use a few well, rather than many badly. Major progress can still be made with simple, relatively inexpensive equipment in infrared characterization of catalyst surfaces.

Catalysis is nearly as broad as chemistry in its scope, and new dimensions are added by the various and changing aspects of surface exposure and support interactions in most "real" catalyst systems. Definitions of catalytic research vary. Any research elucidating the surface chemistry of solids or the structure of coordination complexes can be viewed as contributing ultimately to a better understanding of catalysis. In recent academic research in surface science, the systems studied, although possibly catalytic, have often been simplified to such a degree that they cannot be regarded by most industrial catalytic research workers as even roughly representative of typical commercial catalysts. Obviously it is often necessary to simplify a system to permit the use of special techniques or to adequately interpret the data obtained. Simplification can be self-defeating, however, if it leads to addressing the wrong questions. Justification for most of the extensive research effort currently funded under the heading of catalysis depends on the widely perceived potential importance of improved catalysts for industrial processes ranging from petroleum refining to chemicals production. The information sought in catalytic research should, if it is to be useful, contribute to improvements in "real" catalysts in the not-too-remote future. Infrared spectroscopy has demonstrated an ability to provide such information.

Many reviews already broadly cover the use of infrared spectroscopy in surface and catalytic research, and little purpose would be served by an attempt to summarize this material in this chapter. Some summary is essential, but my intention is to emphasize those techniques and topics which appear to be most useful for gaining a better understanding of "real" catalysts, ideally as they exist under reaction conditions. Because of my own background, I will stress transmission studies of catalysts similar to those used in petroleum refining or of actual commercial catalysts and attempt to show that useful information can be obtained on such catalysts, despite their complexity. Other infrared methods, and related vibrational spectroscopic techniques applicable to high-area catalysts will also be briefly discussed. This is not to deny the intrinsic interest and potential practical importance of good "surface science" research on, for example, single crystals of metals. Such work, using infrared as well as other techniques, continues to contribute information vital to sound interpretation of spectra obtained on the supported-metal catalysts used commercially. Most commercial catalysts, however, are not fully-reduced metals. Interesting research has also been carried out on homogeneous catalysts using infrared characterization, but this will not be covered here.

2. General Background — Infrared Spectroscopic Studies

Both the theoretical and experimental aspects of infrared spectroscopy in general have been adequately covered elsewhere [18–24]. Introductory chapters or sections summarizing the major features of this background have usually been included in books and reviews covering the use of infrared spectroscopy in surface studies [8–14]. Most readers are familiar with the main features of infrared spectroscopy. Briefly, however, infrared spectra arise from transitions between vibrational (and rotational) energy levels with corresponding absorption or emission of energy at discrete frequencies corresponding to certain "normal" modes of vibration of molecules or surface groups. The number of vibrational modes increases with molecular complexity, with $3n$–6 normal modes for a free non-linear molecule composed of n atoms. If a vibrational mode is accompanied by a dipole oscillation, infrared radiation can be adsorbed or emitted at that frequency. In general, all the atoms in a molecule are involved in a vibrational mode, but most of the motion is often concentrated in one bond or atomic grouping. In this case, the masses of the atoms in that group and the "force constant" associated with the bond between them primarily govern the frequency of vibration. Approximately the same frequency is therefore observed for a given group in different compounds. Such "group frequencies" aid the analyst in identifying structural features, such as —OH or —CH groups, where the terminal atom has a small mass, or where the force constant of a bond is much different from those of other bonds. Extensive literature on useful diagnostic frequencies exists [19–24]. These frequencies generally fall in the spectral region from 4000 to 400 cm^{-1}. Interactions of rotational energy level transitions with vibrational transitions can also give rise to discrete bands in this region, but these are usually seen only for molecules rotating freely in the gas phase. Such spectra are not normally of concern in studies of surface groups or adsorbed molecules, but they can be encountered in a gas phase background when infrared cells with a fairly long gas path length are used or even with short path lengths in high pressure studies.

Surface groups and adsorbed molecules are often identified by comparison of observed infrared frequencies with known group frequencies or with those in spectra of known compounds. Spectra of adsorbed molecules tend to resemble those seen in the liquid state, but surface interactions can shift band frequencies to higher or lower values and, by inducing dipoles where none would otherwise exist, give rise to dipole oscillations in vibrations that would not normally exhibit them and thus to the appearance of bands not normally seen in the infrared spectrum [12, 25]. Very often, however, the band frequency alone is not sufficient to permit unambiguous identification of a surface group or adsorbed molecule. This is particularly true in typical studies of oxide-supported catalysts where strong absorption of infrared by the oxide support prevents obtaining spectra below 1000–1100 cm^{-1}. Chemical or other evidence must then be obtained to support a particular assignment. The use of isotopic substitution (*e.g.* the use of exchange with D_2O to convert —OH to —OD groups) is one such approach. Reaction with various reagents (*e.g.* reduction with H_2) followed by analysis of products formed can also aid in positive identification of surface groups. Demonstration of marked alteration of the frequencies of certain bands when adsorbates are added can be used to show accessibility or surface location of particular groups. In general, frequency shifts can be caused by either mechanical (geometric) effects or electronic (force constant) effects although the two are frequently interrelated.

The use of adsorbates as "probe" molecules, besides demonstrating the surface location of hydroxyl or other groups, can also provide information on the nature and environment of atoms and ions exposed (*i.e.* "coordinatively unsaturated") on the surface, even though such atoms or ions cannot be studied directly by infrared spectroscopy. The use of such probe molecules as CO, NO, and NH_3 dates back to the earliest research work in this field, and continues to provide important information today. Changes in both the frequencies and intensities of the infrared bands and in the binding energy of the probe molecules on surface sites can be used to elucidate the number and nature of the adsorptive sites. Subtle differences in surface sites can readily be demonstrated by changes in the spectra of adsorbed molecules, but the exact nature of these differences is often difficult or impossible to determine. As will be shown, several causes are theoretically possible for shifts in band frequencies and/or intensities as well as for changes in binding energies of adsorbed molecules. Factors at present very imperfectly understood can strongly influence the spectra obtained, and conclusions must often be very tentative.

Although infrared spectroscopy provides fairly direct evidence on many surface groups, such as hydroxyl, sulfate, carbonate, etc., the study of exposed metal or other atoms through the use of probe molecules generally provides no *direct* evidence on the oxidation state or even on the chemical identity of an adsorbing atom. If the chemical composition of the catalyst is known certain correlations, accepted by most workers in the field, exist which can aid in identifying certain types of bonding and certain oxidation states of the adsorbing metal site. Experience and analogies with spectra of known compounds have led to assignments for bands from adsorbed CO [11, 12], for example, to "linear" (*i.e.* attached by one end to one metal atom) or "bridged" (attached to two or more atoms) species depending on whether band frequencies are above or below 2000 cm^{-1}. Bands much above 2100 cm^{-1} are generally associated with CO on metal atoms in an oxidation state greater than zero. Differences can certainly be seen between spectra of CO adsorbed on different transition metals on the same support after similar reduction procedures. Such differences, however, usually appear to reflect differences in the chemical reactivities of the metals which result in different support interactions or differences in crystallite size or face exposure rather than intrinsic differences between the binding of CO by atoms of the two metals in equivalent states. Real differences certainly exist in some cases, but many reduced metals show very similar bands for adsorbed CO, and a particular metal may exhibit a wide range of band frequencies depending on various factors, as will be shown below.

Possible reactions of probe molecules with the catalyst surface or with other probe molecules should also be kept in mind. These reactions can in some cases be very helpful in detecting unusually reactive surface sites, present at very low concentrations but potentially important catalytically. In other instances unsuspected reactions of probe molecules used may lead to erroneous interpretation of the data.

The problems confronting infrared characterization of catalyst surfaces are in large measure, however, common to any study of catalyst surface chemistry. They arise from the inherent complexities of the systems studied. Many techniques yield only an averaged value for a particular surface property. Infrared, with greater sensitivity and resolution, discloses variations in detail and anomalies of surface structure which would be invisible with other techniques. This can create major problems in interpretation, but it also permits attacking most directly key questions in relating surface chemistry to catalytic performance.

3. Experimental

A. Sample Preparation

1. High Area Oxide Supports and Supported Catalysts

A major problem in infrared studies of conventional supports and catalysts lies in the very large scattering and absorption losses typically encountered in transmission studies of these finely-divided solids. To obtain useful spectra of adsorbed molecules or surface groups it is necessary to put a detectable amount of the surface species, often held at very low concentration, in the path length of the infrared beam. Sample thickness must usually be sufficient to provide many equivalent surface layers in the beam, although a practical upper limit to sample thickness is normally set by absorption and scattering of infrared radiation by the adsorbent itself [5]. Requirements for mechanical strength, convenience of preparation, and reproducible, uniform sample thickness usually favor moderately thick (*i.e.* 10–40 mg cm^{-2}) samples. Various special, low scattering, oxide supports have been widely used in past infrared studies to minimize scattering losses. These include certain oxides made by flame preparation procedures (Aerosil or Cabosil, Alon C, Kadox 25, etc.), porous glass, and transparent plates of silica, alumina and other aerogels. Sample preparation procedures have included size fractionation of powders to obtain very small particles and the use of standard KBr and oil mull techniques. A good discussion of such techniques has been given by Blyholder [4]. For surface studies of more conventional catalysts, however, most special methods are, in one way or another, too limiting, being applicable only in special cases or giving samples which cannot safely be considered typical of "real" catalysts. The use of KBr disks or oil mulls generally precludes subsequent treatment of the catalyst surface and may also alter the surface chemistry.

Scattering losses of powder samples can usually be considerably reduced by pressing the powder (often obtained by handgrinding pellets or granules of a high-area catalyst), without the use of any binder, to form a thin, self-supporting disk. Such disks, usually of silica, alumina, aluminosilicate, or other oxide-supported catalysts can be prepared easily using stainless steel dies 1.5–2.5 cm in diameter and pressures up to 172 MPa (25,000 psi). These are the samples most commonly used today in infrared studies of catalysts. Pressing can be troublesome on occasion. Disks can crack, crumble, or stick to the dies. Control of humidity or changing the die surfaces may be helpful.

Pressed disks can be used directly as prepared in cylindrical dies or trimmed to fit special holders. They can be pressed in special dies or holders to rectangular or other shapes. They are usually quite stable toward heating and other normal pretreatment in the infrared cell. Unless unusually high pressures are used in their preparation, such disks normally retain most of the surface area of the original powder sample. Although some closed pores and other surface changes may occasionally be produced by the pressing procedure, pressed disk samples are generally regarded as fairly representative of the original material.

2. *Unsupported Metals — Evaporated Films, Foils, Ribbons, and Single Crystals*

Dating from the early work of Eischens and Pliskin [3] infrared study of unsupported metal surfaces [5, 26, 27] has paralleled study of supported metals. For transmission infrared studies, only evaporated metal films or metal deposited in oil [28] or some other matrix are possible samples. Evaporated metal films can provide sufficient metal surface to permit study of adsorbed molecules, without the frequency range restrictions imposed by the presence of high area oxide supports, but they can hardly be considered ideal samples. Such films, which must generally be very thin (< 10 nm), to permit obtaining transmission spectra, are prepared by evaporation of the metal in vacuum and condensation of the vapor on a cool, infrared-transparent substrate such as CaF_2 or NaCl. Evaporation must be carried out under ultrahigh vacuum conditions in a fully bakeable infrared cell if major contamination is to be avoided. These conditions were not present in early work. Evaporated metal films sinter easily, however, and do not have well-defined surfaces even when prepared under ultrahigh vacuum conditions. Evaporation into oil suffers both from contamination and from the usual problem of mulls preventing further surface treatments. Matrix isolation methods in which metal vapor is condensed together with argon on a salt plate and a further layer of CO and argon then condensed on this have also been applied to chemisorption studies [29] and show promise for characterization of small clusters of metal atoms. As with vacuum deposited films, however, surface heterogeneity is here probably almost as great as with supported metals.

For studies of foils, ribbons, and well-defined metal single crystal surfaces, where reflection-absorption spectroscopic methods must be used, sample preparation and surface cleaning often requires special methods which cannot be covered here.

B. Cell Design

The infrared cell in which the catalyst sample is pretreated and subsequently studied is extremely important in surface studies. The perfect, all-purpose cell has yet to be devised, and cell design is normally chosen to suit the purposes of a particular study. Some features are usually of overriding importance in a given application. If catalytic reactions are to be studied the exposure of catalytic metals must be eliminated in cell construction, and bare heating elements within the cell are ruled out. In some surface characterization studies such features may be completely acceptable, but even in such studies it is well to avoid any possibility of Ni or other carbonyls being formed from cell components.

A variety of relatively simple, but effective, cells have been used for studies at subatmospheric pressures. Many of these have been described in the literature [2–17]. Cells are usually constructed of Pyrex, Vycor, or fused quartz. The two latter materials permit sample treatments above 720 K. These cells are of two basic types, depending on whether or not the sample is moved out of the infrared beam for heating and other pretreatments or is held in place in the infrared beam during such pretreatments. Cells suitable for ultrahigh vacuum studies have also been described [6, 30, 31]. Cells for high pressure studies are, of course, constructed of metal. These will be discussed further below.

The main problem in cell design lies in the seals required between the cell body and the infrared-transparent windows, which are generally of calcium fluoride or sodium chloride. Because of the high coefficients of thermal expansion and sensitivity to thermal (and mechanical) shock of such window materials, no simple, completely satisfactory method has yet been devised to make a vacuum-tight seal for windows which can be repeatedly heated much above 570 K and cooled again without failure. There are a few possible exceptions. With care, calcium fluoride windows can be sealed with silver chloride to silver tubes or flanges which can then be attached to the cell body, providing a seal bakeable in principle to 700 K [31]. The use of other less-satisfactory window materials also offers a possible solution to the problem of bakeable window seals. Silica and sapphire windows can be successfully attached to make cells bakeable at very high temperatures, but the use of these materials severely restricts the frequency range of infrared study. Sheet silver chloride can be used for some purposes, being readily sealed to silver or noble metals and bakeable to about 650 K, but such windows are soft, deform easily, and darken under the influence of light, giving decreased transmission. Periclase (MgO) windows can be sealed to soda glass, silicon windows to Pyrex, and various other combinations employed to make cells bakeable to at least 650 K [32], but none of these solutions is ideal. Through use of a double vacuum chamber, with two sets of windows, the window sealing problem can be avoided and a cell suitable for high temperature sample pretreatments as well as ultrahigh vacuum studies constructed [33].

In most cases, however, windows are attached to the cell body with waxes, epoxy, or silicone resin cements or through use of Viton or other gaskets. If the sample is to be heated, but not moved out of the infrared beam, the window seals are usually cooled while the sample is heated. Unless the normal optical path is modified the maximum distance between cell windows with most commercial spectrometers is only 10–12 cm. With such window separation simultaneous heating of the sample to 1100 K or more and cooling of the window seals is quite possible. Absorption spectra of hot samples can readily be obtained without complications from emission if the incident radiation is somehow modulated or "chopped".

Figure 1 shows a simple cell of this type. Many variations exist, having internal heating elements in some cases, constructed of various materials. Addition of a suitable internal cold finger or an external jacket for liquid nitrogen or other coolants permits study at low temperatures with such cells [12–14, 34, 35]. More elaborate cells can combine these features with an attached gas cell for the reference beam to eliminate gas phase contributions [35].

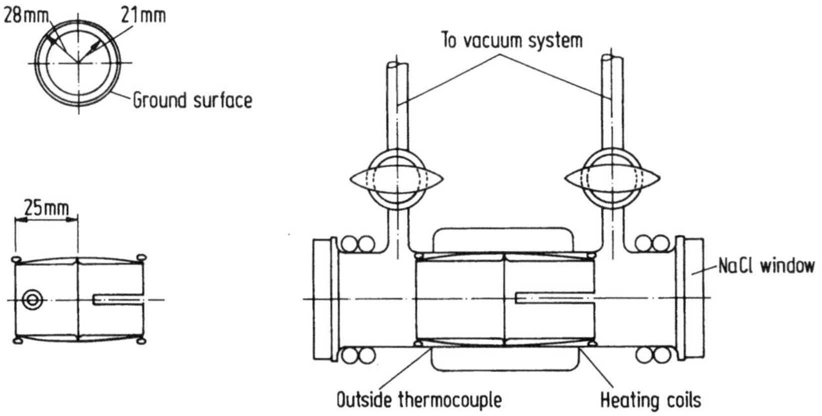

Figure 1. Infrared cell and sample holder. (Reproduced with permission from ref. [4a])

Quite high temperatures can be achieved with such cells, but the temperature uniformity leaves much to be desired. Non-uniform heating is, however, a problem common to most of the cell designs used to date, particularly when heating effects of an undispersed infrared beam are taken into account. Another possible problem arises from condensation of material desorbed from the sample during heating on the cooler windows, giving rise to infrared bands which could be confused with those of surface groups on the catalyst. Finally, because of the long gas path length and the inability to move the sample out of the infrared beam, assessment of gas phase spectral contributions may be difficult.

Cell designs which permit movement of the sample into a separate furnace section eliminate some of these problems. Whether the furnace section is above or to one side of the infrared beam the distance between cell windows can be greatly reduced, if desired, cutting the gas path length. Condensation on the windows can be avoided. The sample can be heated as hot as desired and the furnace section lengthened to increase temperature uniformity. The major problem with this type of cell is that observation of hot (or cold) samples is not usually possible. Figure 2 shows a simple cell of this type which has been used successfully in this laboratory for several years.

Combination of the special features of the two types of cells is obviously feasible to some extent. An additional refinement useful in some studies is the inclusion of a microbalance to measure adsorption gravimetrically [36–38]. Something is generally lost, however, for each gain in versatility.

Cells for *in-situ* studies of catalytic reactions may present special challenges. If high pressure is not required in such studies a rather simple cell can be

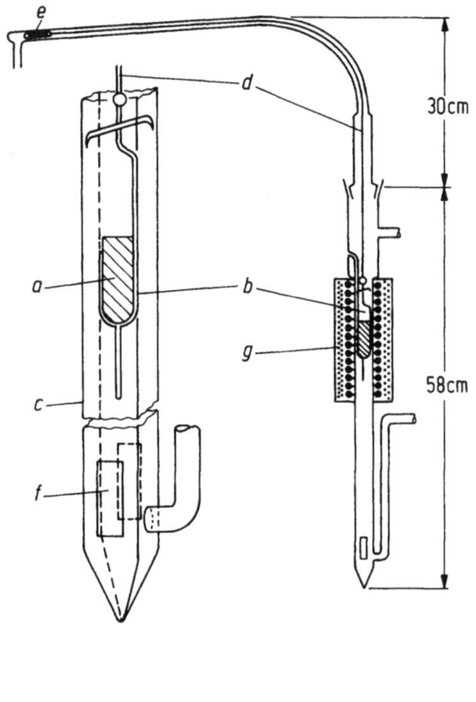

Figure 2. Infrared cell permitting sample movement. *a* Sample. *b* Quartz sample holder. *c* Cell body (quartz, 2.5 cm square cross-section, attached to 34/45 Vycor joint.) *d* Fiber or wire of quartz, stainless steel or gold. *e* Pyrex-enclosed magnet. *f* Window openings. *g* Furnace

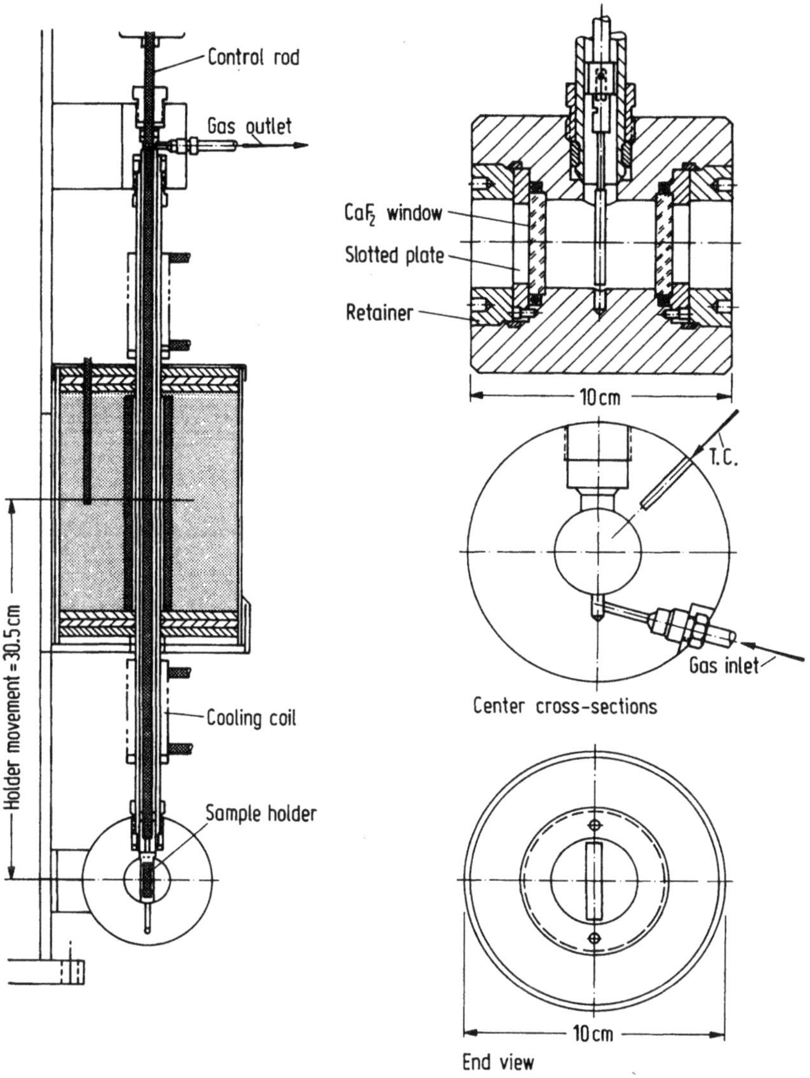

Figure 3. High-pressure infrared cell permitting sample movement

used [39–40]. Studies of many commercially interesting reactions require the use of high pressure infrared cells, however. Several such cells have recently been described [41–44]. Figure 3 shows a cell designed in collaboration with Autoclave Engineers, Inc., and used successfully in this laboratory in several studies including *in-situ* CO hydrogenation [41]. This cell permits heating a catalyst sample to at least 573 K under pressure of 2 to 5 MPa or higher, depending on window material and thickness. Design of the lower windowed section has been described [41]. It mainly follows that of Tinker and Morris [45] but a small internal heating coil has been added to this section which permits heating the sample somewhat hotter than the stainless steel cell body. For pretreatment at still higher temperatures, the sample can be raised by a hydraulically activated control rod to an upper furnace section where treatment under pressure to 850 K is possible. Because of small leaks through the packing around the control rod the cell is not entirely vacuum tight, but this creates no problems in studies under high pressure.

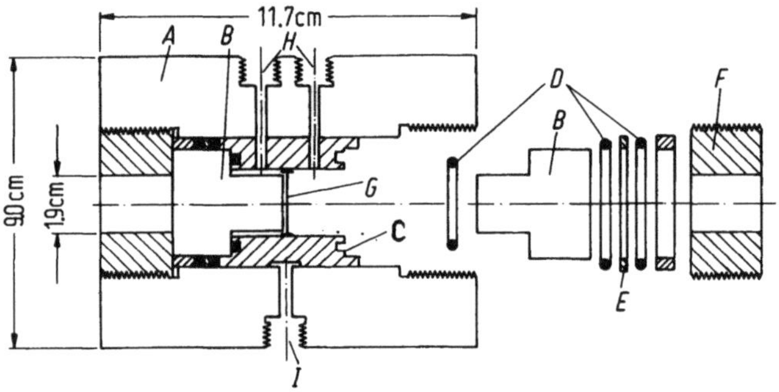

Figure 4. High-pressure infrared cell. (Reproduced with permission from ref. [42])

Another similar cell design [42] for high pressure catalytic studies is illustrated in Figure 4. This cell and those described by Hecker and Bell [43] and Penninger [44] offer the shorter gas path length, desirable for high pressure studies, but because of window seals 573 K is probably a normal upper temperature limit. Penninger [44] has reported, however, that under some conditions 723 K and 30 MPa is feasible with his design, using compressed Rulon-25 (Dixon Corp., Bristol, R.I.) seal rings. Although high pressure cells permit infrared study under more realistic reaction conditions in many cases, additional problems are frequently encountered in such studies. Sometimes their use is indispensable, but much additional work will be needed in this area before their full benefits will be realized.

C. Vacuum and Gas Handling Systems

For effective study of catalyst surfaces, control of ambient conditions is obviously essential. A suitable infrared cell, as described above, can permit control of sample temperature and pressure of adsorbing gases or vapors, but some effective means of ensuring an initially reproducible, and hopefully fairly clean, catalyst surface is needed, along with an effective gas-handling and purification system. Conventional high vacuum systems capable of, at best, about 10^{-4} Pa pressure have been used in most published infrared studies, but a substantial amount of ultrahigh vacuum (10^{-6}–10^{-9} Pa) work has also been reported. In many studies flowing gas streams were used for all necessary catalyst pretreatments. The use of rather modest vacuum in studies of high area catalysts has been justified in most cases by the very large surface area of most such catalysts relative to the potential supply of contaminants [46] and the extremely slow diffusion of contaminants present at very low pressures [47]. Improved vacuum, particularly in studies of reactive metals at low concentrations on supports, would undoubtedly prove advantageous in many cases. At present both oil and mercury diffusion pumped systems are used. Either is satisfactory, but suitable traps must be used to remove H_2O, Hg, or other contaminants.

The use of flow systems, at pressures from subatmospheric to 10 MPa or more may be required in *in-situ* studies of catalytic reactions and investigation of catalyst pretreatment or regeneration procedures. It is beyond the scope of this chapter to discuss the details of such systems. Impurities in flowing gas streams, even at very low levels, can be extremely troublesome if they accumulate on a catalyst surface. The use of upstream traps containing the same catalyst being studied by infrared can often prove useful, but minute leaks at joints in stainless steel flow systems can be difficult to detect and may lead to irreproducible results. As with vacuum system work, careful attention to detail is needed if reliable data are to be obtained.

D. Spectrometers

Although spectrometers have been greatly improved in recent years, as will be discussed, most recent published work in this field has been carried out with spectrometers no better than those available twenty years ago when commercial grating or prism + grating, spectrometers first became readily available. These spectrometers opened a new era in infrared surface studies because, through a trade-off of resolution for energy, they permitted spectra to be readily obtained on a wide variety of catalysts and supports which could not previously be studied. With prism spectrometers, prior to this, special samples, which scattered relatively little infrared radiation had been the key to successful research in this area. Only special supports or catalysts prepared, usually at rather high metal loadings, on these supports had been suitable for infrared studies. The ability to study a wider variety of catalysts, including most commercial catalysts, has since greatly increased the scope and practical importance of the infrared technique. Most of the spectrometers in use today are dispersive with roughly the same capabilities as those of twenty years ago. These instruments are still entirely adequate for obtaining valuable information on a wide variety of catalysts. Much work continues to be done on special supports, as in the past, because of the desirability of high transmission, but it is neither necessary nor desirable to restrict research solely to such materials.

The availability in recent years of low cost minicomputers and the development of commercial Fourier transform infrared spectrometers have once again introduced a new era in infrared surface studies. Improvements in the potential sensitivity of infrared surface studies and in the ability of the research worker to handle the spectral data amount to a major breakthrough.

The potential of Fourier transform spectroscopy was well recognized in the 1960's, and surface studies had been made using it, but major limitations existed. Resolution was poor and reliability was very poor. The introduction of good commercial Fourier transform infrared (FTIR) spectrometers in the 1970's inspired serious interest in this area. The past five years have witnessed widespread use of FTIR spectrometers in catalytic research and the introduction of new instruments by several manufacturers. (Digilab, Bruker, Nicolet, IBM, Analect, Perkin-Elmer, and others currently have instruments on the market.) Applications of FTIR spectroscopy in surface and catalytic studies have recently been reviewed [9, 48, 134].

The theoretical optical advantages of FTIR spectroscopy are impressive. Because all spectral elements are sampled at the same time, a Fourier transform spectrometer can theoretically acquire a spectrum over the region 4000 to 400 cm^{-1} at a resolution of 1 cm^{-1} 3600 times faster than a dispersive spectrometer operating at the same signal-to-noise ratio. Alternatively, if the noise is detector limited, a spectrum acquired by an FTIR spectrometer in the same length of time as taken by a dispersive spectrometer should have a signal-to-noise ratio better by $(3600)^{1/2} = 60$ X. This is called Fellgett's advantage. The absence of slits in an interferometer also yields a potentially greater throughput of infrared energy than with dispersive spectrometers (Jacquinot's advantage). For high resolution work — where narrow slits are required by dispersive instruments — this advantage can be very important. Beyond these important theoretical optical advantages, commercial FTIR spectrometers all offer computer processing of the final spectral data.

Much stress has been laid on the optical advantages unique to FTIR spectroscopy but the advantages of computer data processing, which can equally be applied to spectra obtained with dispersive spectrometers, are probably more important in most research on catalyst surface chemistry. The current availability of low cost mini- and microcomputers which can easily be interfaced to dispersive spectrometers has provided a viable alternative to FTIR where the unique features of FTIR are really unnecessary. With an interfaced minicomputer, signal averaging, effective background correction, and subtraction in absorbance can greatly improve the effective sensitivity of a dispersive spectrometer. Comparison of different spectra is greatly facilitated. Deconvolution becomes a feasible routine operation, when desired, which can sometimes aid considerably in interpretation of the complex spectra often encountered in infrared studies of adsorbed molecules and surface groups on catalysts.

Comparisons of the performance of FTIR spectrometers with that of dispersive spectrometers have been published [49–51], but such comparisons are necessarily of specific instruments on specific types of samples. For any realistic comparison of FTIR with computerized dispersive spectrometers the nature of the research problem should first be accurately defined. The theoretical advantages of FTIR in greater energy and ability to record an entire spectrum quickly can be very important in, for example, gas chromatography "on the fly" or in studies of atmospheric pollutants, but unimportant for studies of condensed phase samples such as KBr disks, oil mulls, or liquids [50]. With such samples some time may be saved recording a spectrum (although the total time needed for obtaining a copy of the final spectrum is often comparable), but unless very many samples are run daily and sample preparation time is not limiting, the overall time saving would be insignificant. Unless the need exists for both speed in capturing a complete spectrum (*i.e.* 4000–200 cm^{-1}) *and* high resolution, FTIR is not necessarily best for a given problem. Where measurements are needed only over narrow spectral regions and where, as in most catalyst research, high resolution is not required, computerized dispersive spectrometers may have the advantage [52]. To quote Cheney and Sheppard [51], "Indeed it could be argued that if spectroscopy had developed historically through interferometry first, then the invention of the dispersion spectrometer would have been hailed as a wonderful invention which allowed the study of limited wavelength regions at a time with great simplicity and substantially lower cost." The cost of FTIR has been steadily decreasing, but it is still high. A minicomputer interfaced to a dispersive spectrometer can also be used for purposes other than simply collecting and processing infrared spectra. Manufacturers of both FTIR and dispersive spectrometers are increasingly providing "stand-alone" minicomputer capability, however.

Published work to date on the use of FTIR in catalyst surface studies has been less impressive than might have been anticipated. Few spectra have been presented which could not have been obtained as well, or better, by computerized dispersive spectroscopy. The theoretical optical advantages of FTIR spectrometers do not usually seem to be fully realized in catalytic studies. With FTIR spectrometers in the past it has been difficult to obtain useful transmission spectra on typical pressed disk catalyst samples in the region above 2500 cm^{-1}. This problem and possible solutions have recently been discussed by Angell [134]. Recent publications [115, 136, 137] show that FTIR can presently give acceptable OH-stretching spectra on sufficiently transparent alumina samples.

Various technical factors can affect the amount of energy actually available in a given spectral region. Griffiths *et al.* [50] have discussed some of these factors. The advantages of FTIR in obtaining high resolution spectra are markedly reduced when low resolution measurements are considered. Severe reduction in the total advantage of an interferometer over a grating spectrometer can also be caused by the relative performance of the detectors commonly used with each type of instrument. The (common) use of triglycine sulfate (TGS) pyroelectric detectors can put FTIR spectrometers at a disadvantage compared to dispersive instruments using thermocouple detectors. Major improvement in FTIR performance at higher frequencies can be achieved, at a price, through use of photoconductive or photovoltaic detectors such as liquid-nitrogen-cooled mercury-cadmium-telluride, indium antimonide, or lead sulfide detectors. Such detectors, along with photoacoustic and diffuse reflectance attachments, can be very useful additions for FTIR studies of catalysts.

The energy actually reaching the sample from the infrared source is also generally lower in FTIR than in dispersive spectrometers, because the interferometer is located between the source and sample. This can be advantageous, because it means less heating of the sample, but may also mean less energy at the detector. A special high-throughput optical bench (Nicolet) appears useful in increasing energy [115]. Heating of the sample can be reduced in dispersive spectrometers either by using suitable filters or by placing the sample between the spectrometer exit slit and the detector, although this is not always easy [12]. Other problems can also exist in optimization of FTIR hardware and software for studies in particular spectral regions because of compromises needed to obtain good spectra over the entire frequency range. Use of electronic or optical high-pass or low-pass filters may improve spectra in a given region, but, of course, at the expense of other regions [134].

Speed in acquiring a spectrum is seldom a requirement in catalyst surface studies, and hardly a major advantage. Many FTIR spectra must usually be averaged to obtain good signal/noise ratio with poorly transmitting samples, and much more time is usually spent pretreating the catalyst than in actually recording the spectrum. An argument has been made for the use of FTIR in time-resolved studies of surface species [9]. In most cases, however, the time required to collect a spectrum by FTIR is too long to permit study of moderately fast transients. It would probably usually be more effective to monitor one or more frequencies by "sitting on a band" or to scan narrow spectral regions with a dispersive spectrometer, as has been past practice, having first determined which bands are likely to be of interest.

Other disadvantages exist. "Previewing" a spectrum as is commonly done with dispersive spectrometers is not usually feasible with FTIR. Positioning cells and accessories for maximum optical throughput is generally more difficult with FTIR spectrometers. FTIR spectrometers are single beam instruments. Some of the problems in quantitative FTIR has been discussed by Hirschfeld [53]. Sample positioning or orientation in the beam is apparently a more critical variable in FTIR than in dispersive spectroscopy, and considerable difficulty has sometimes been experienced in comparing different catalyst samples. Over the past 20 years most of the "bugs" have been worked out of dispersive spectrometers and good commercial spectrometers do their jobs very reliably and well. Improvements are continually being made in FTIR spectrometers, and the theoretical advantages of FTIR may soon be more-fully realized in catalyst studies.

At present either FTIR or computerized dispersive spectrometers are capable of performance markedly surpassing that of unaided dispersive spectrometers. Costs vary; cheaper models of both types of spectrometers are currently being produced and computer performance is steadily improving. Software varies for different systems but usually provides effective routines for background subtraction in absorbance, baseline correction, expansion of bands, replotting of spectra, etc. Provision for the spectroscopist to easily reprogram or to use his own programs in handling spectral data can be a most important consideration, however. FTIR manufacturers have apparently not always been ready to update software for existing instruments and user modification has often been rather difficult.

E. Computer Treatment of Spectra

With computer processing of spectral data many treatments of spectra which would have been much too laborious to have been attempted in the past become feasible. Pitfalls exist for the unwary in some of these procedures, however. Changing background spectra can lead to fictitious bands on subtraction — as can subtraction of broadened or shifted bands. The "black box" aspect of these instruments, combined with inexperienced spectroscopists, can spell trouble. Hirschfeld [53] has discussed some of these pitfalls. They are by no means limited to FTIR. Criteria for presentation of spectra from computerized IR instruments have recently been recommended by a Coblentz society subcommittee [54]. Their recommendations call for extensive information which, although desirable, is probably unrealistic to expect in most publications.

The ability to computer process spectral data fortunately comes at a time when the questions to be answered are becoming increasingly difficult. The tools are better, but such better tools will now usually be needed for further major progress.

4. Typical Applications

A. General — Spectra of Solid Catalysts

Infrared spectroscopy can, of course, be used for study of catalysts just as for study of any powdered solid. Bands arising from the bulk structure of catalysts or supports can readily be observed. Such characterization can be very useful in distinguishing one zeolite from another, in determining the nature of the support used in a particular commercial catalyst, and in other applications. If no need exists for surface characterization standard mull or KBr disk techniques can be used for sample preparation. The pressed disk samples normally used for surface characterization are usually too thick to permit study of the bands arising from the bulk of the solid below 1000 cm^{-1}, but thin powder layers can be deposited on NaCl windows or held without binder in a screen as an alternative to the use of KBr or mull techniques where surface accessibility is also required.

B. Study of Surface Groups on Oxide Supports and Catalysts

1. Existence and Importance of Surface Groups

Commercial catalysts are commonly made with high area oxide supports such as alumina, silica, or silica-alumina gels. In other applications "acidic" oxides such as silica-alumina cracking catalysts, halogenated aluminas, or large-pore crystalline aluminosilicates (zeolites) are important catalysts by themselves. Pure oxides such as alumina and zinc oxide can also show interesting catalytic activity.

In addition to providing a large surface on which a metallic component can be highly dispersed to increase its exposed surface, supports exhibit surface features such as hydroxyl, sulfate, carbonate and other groups which may in some cases be catalytically important. Hydroxyl groups may participate directly in catalytic reactions or, by reacting with supported metals, make metal reduction very difficult. Sulfate groups can produce H_2S during reduction of a catalyst in H_2. This, on further reaction with a supported metal, can change its catalytic behavior. Such effects may or may not be desirable. Obtaining infrared spectra of such groups is fairly simple and, although resolution of individual bands may at times be difficult, interpretation of the spectra, at least as regards identification of particular groups, generally poses no major problems. Many studies of surface groups have been reported.

2. *Hydroxyl Groups on Oxides and Related Supports*

Most "dry" oxides, aluminosilicates, phosphates, and other oxygen-containing solids tenaciously retain some hydroxyl groups on their external surfaces or, in the case of zeolites, at locations in the crystal structure where stoichiometry would otherwise be violated by substitution of a cation for another cation of higher oxidation number. Hydroxyl groups which are not involved in H-bonding interactions give rise to infrared bands in the region from 3600–3800 cm^{-1}, corresponding to the O—H stretching vibrations of "isolated" groups. The exact frequencies of these groups depend on their particular surface or crystalline environments. Hydroxyl groups which are hydrogen-bonded to other groups give bands at lower frequencies.

Spectra of surface hydroxyl groups may be fairly simple, as is the case for pure silica gels [12–14, 55, 56] which exhibit basically one sharp band near 3750 cm^{-1}, corresponding to isolated OH groups and a broad band or "tail" in the 3300–3700 cm^{-1} region caused by H-bonded OH groups. Hydroxyl spectra may also be quite complex. Transition aluminas [57–62] typically exhibit up to five discrete bands arising from the stretching vibrations of isolated hydroxyl groups in the region from 3690 to 3800 cm^{-1} in addition to bands arising from H-bonded hydroxyl groups. These bands evidently correspond to hydroxyl groups held on different types of sites on a crystalline surface. Zeolites also exhibit a variety of isolated hydroxyl-stretching bands [10, 63, 64].

Spectra of hydroxyl (or other) groups raise questions regarding the location of the groups as follows: a) Are they accessible to probe molecules? (*i.e.* are they surface groups?), and b) What is the nature of the local surface environment responsible for their differences in frequency and, often, in chemical behavior? Answers to these questions are sought through study of the effects of adsorbed probe molecules or of chemical reaction with various reagents. Exchange of hydrogen- or deuterium-labelled molecules can show differences in reactivity and/or accessibility of surface hydroxyl groups.

Physical adsorption of probe molecules can also produce the marked shifts and broadening of isolated hydroxyl bands which indicate that the groups are surface groups. Figure 5 shows such effects of N_2 adsorption on the

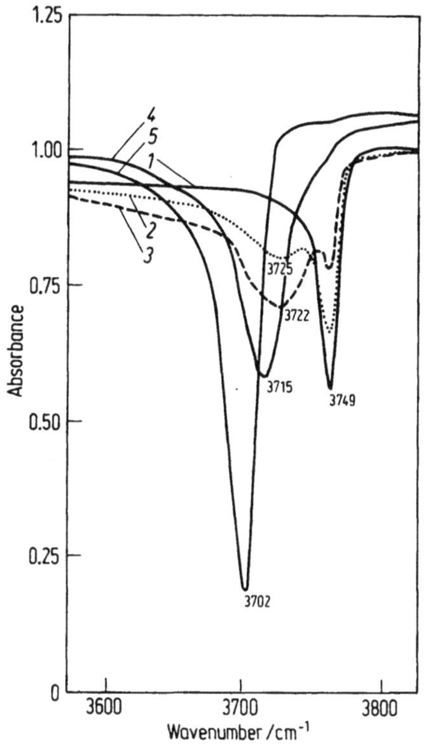

Figure 5. Spectrum of Aerosil 2491 silica in equilibrium with various pressures of nitrogen: *1* Degassed silica; *2, 3, 4* increasing amounts of nitrogen physically adsorbed at 83 K; *5* Sample immersed in liquid nitrogen. (Reproduced with permission from ref. [55])

isolated hydroxyl stretching band of dry silica [55]. The hydroxyl groups on this silica are evidently all surface groups, but this is not invariably the case on other oxides or zeolites. If some or all of the hydroxyl groups remain unaffected by adsorbed molecules even at high surface coverages they are probably not surface groups.

Participation of hydroxyl groups in exchange reactions is not always proof of surface location. In exchange with D_2O, proton migration through the bulk oxide structure cannot always be excluded. Traces of H_2O or D_2O may be present in exchange experiments, perhaps as a result of reduction of metal oxides left after a previous incomplete reduction. Sometimes such H_2O can act as an intermediate in an exchange reaction, producing an overall exchange that might be impossible in its absence. Differences reported in the literature on the relative exchange rates of different types of hydroxyl groups may in some cases merely reflect such effects. Differences in the extent to which H-bonding or other interactions of hydroxyl groups with adsorbed molecules shift the frequencies of the bands corresponding to a given type of hydroxyl group may also sometimes be confused with true exchange. Beyond these and other possible errors in interpretation of spectral changes, there are real differences between different preparations of alumina, silica, etc., and it is not safe to generalize regarding specific surface chemical behavior.

Clearly, however, careful infrared study of a particular support can yield valuable information on the numbers, types, and chemical reactivities of surface hydroxyl groups. Such information can then aid in explaining catalytic properties, in identifying supports, and in establishing differences in surface structure. Aside from demonstrating differences in the accessibility and chemical behavior of different types of hydroxyl groups, the frequencies and intensities of hydroxyl stretching bands should also, in combination with other evidence, eventually show how such groups are attached to the surface.

Figure 6 shows spectra of isolated deuteroxyl groups on dry $AlPO_4$ [65]. This material is completely isomorphous with silica, but is composed of linked oxide tetrahedra whose centers are alternately filled by Al and P atoms rather than by Si atoms. The two major isolated deuteroxyl bands seen in Figure 6 at 2800 and 2712 cm^{-1} (corresponding to hydroxyl bands at 3800 and 3680 cm^{-1}) appear to arise from surface deuteroxyl groups attached to Al and P atoms held in tetrahedral sites. Following usual terminology, Al^{3+} and P^{5+} alternately replace Si^{4+} ions in the tetrahedra holding the hydroxyl groups. This evidently creates a charge "defect" in the local environment of the deuteroxyl or hydroxyl groups which either increases or decreases the frequency of their stretching vibrations relative to the value seen for such groups attached, with no charge defect, to Si atoms ($\sim$3750 cm^{-1} for hydroxyl and $\sim$2760 cm^{-1} for deuteroxyl groups). Usually, however, identification of the sites holding isolated hydroxyl or deuteroxyl groups is not as simple as with $AlPO_4$. The three to five isolated hydroxyl bands seen in spectra of dry high-area aluminas have also been assigned to hydroxyl groups whose vibration frequencies have been raised or lowered by local charge defects

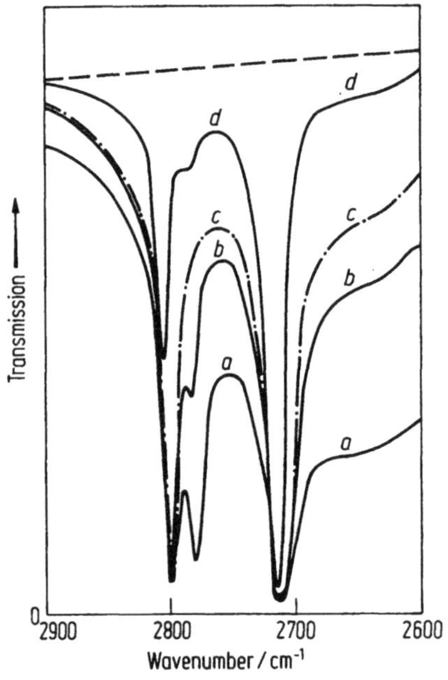

Figure 6. Deuteroxyl stretching bands on $AlPO_4$ aerogel predried for 1 hour at *a* 473 K, *b* 673 K, *c* 873 K, and *d* 30 min at 1073 K (Reproduced with permission from ref. [65])

[59, 62]. The details of the surface environments or site configurations producing the charge defects, if such defects truly exist, are still unclear, as also the absolute magnitude of charge defect needed to cause a given vibrational frequency shift for an isolated hydroxyl group. The assignment of a hydroxyl band near 3800 cm^{-1} to hydroxyl groups held on Al in tetrahedral sites [61, 62] by analogy with $AlPO_4$ hydroxyl groups may be correct, but the essential factor appears to be the charge defect associated with the site rather than the tetrahedral site *per se*. Such a charge defect could result from other than tetrahedral configuration [59]. Assignments of hydroxyl bands must be rather tentative at present. Hydrogen bonding interactions with other hydroxyl groups, local charge defects, and other factors can all shift frequencies appreciably. No sure guide generally exists to show whether a particular band arises from hydroxyl groups attached to a given element. Nickel addition to silica gel, for example, can under some conditions produce a strong new hydroxyl band (near 3620 cm^{-1}) in the spectrum of the dry nickel/silica catalyst. This band is not necessarily caused by Ni—OH groups, however, and may instead reflect inductive effects of neighboring Ni^{2+} ions on Si—OH groups [66].

3. *Hydroxyl Groups in Zeolites*

Zeolites are porous crystalline aluminosilicates widely used as adsorbents and catalysts. Their commercial importance would be hard to overestimate, but they are also of fundamental interest. Ideally, zeolites need contain no hydroxyl groups. In the crystalline aluminosilicate framework of zeolites the (AlO_4) tetrahedra have a net negative charge of -1, because the coordination number of the Al^{3+} ion is 4. To compensate for this charge defect, metal cations are normally held in, or can be exchanged into, the zeolite to occupy positions outside the framework in numbers sufficient to provide electroneutrality (*e.g.* one univalent cation for each $(AlO_4)^-$ tetrahedron or one divalent cation for each two $(AlO_4)^-$ tetrahedra).

No protons, (which are held as hydroxyl groups), need be retained in a perfectly crystalline zeolite which holds enough other extra framework metal cations—except possibly at the (relatively small) external surface of the crystal. According to Ward [10], non-cation-deficient monovalent zeolites apparently contain no hydroxyl groups. Actually most zeolites do contain hydroxyl groups (although those holding monovalent cations contain very few), because some metal cations are usually removed in the course of zeolite preparation and washing. These hydroxyl groups produce a variety of hydroxyl bands in the spectra of zeolites, depending on the metal cations held, pretreatment conditions, and various other factors. The vibrational spectra of structural hydroxyl groups of zeolites, (as well as those on various oxides) have been reviewed by others in some detail. Discussion by Kiselev and Lygin [14] and references cited therein provide further details on studies through 1971. This and more recent work has also been extensively reviewed by Ward [10]. Work continues in this area, but bewildering complexity is everywhere evident and unifying concepts are little in evidence.

The hydroxyl bands observed are generally assigned to hydroxyl groups attached either to Si or Al, and variously influenced by neighboring cations or held as part of a complex metal cation. General agreement exists that a sharp band near 3750 cm^{-1} reflects Si—OH groups. When divalent cations are present the reaction $Zeolite^{2-} + M^{2+} + H_2O \rightarrow M(OH)^+ + Zeolite{-}OH^-$ gives two OH species which may explain the 3650 and 3540 cm^{-1} bands seen on such zeolites [10]. In the case of trivalent cations species such as $M(OH)^{2+}$ or $M(OH)_2^+$ may give rise to OH bands [10].

Many studies of catalytic reactions over zeolites have been reported, but the active sites for these reactions are still uncertain [10]. Acidic hydroxyl groups, particularly those responsible for bands near 3650 cm^{-1} in X and Y zeolites appear to be important in acid-catalyzed reactions, but whether other sites are also important and whether acid site strength is important remain uncertain. Decationized Y zeolites and ultrastable zeolites often show no significant band near 3650 cm^{-1} and yet are catalytically very active. Hydroxyl bands on such zeolites have been tentatively assigned [64] to cations and as $Al(OH)^{2+}$ and $Al(OH)_2^+$ which may coexist with AlO^+. The positive charge on such cations should promote protonic acidity.

The confusion in the literature on zeolites is rather easily understood in light of present knowledge of the extent to which dealumination occurs during steaming, heating and other conditions [67]. Zeolites, even those having apparently identical compositions and crystal structure, can easily differ in the extent to which aluminum has been removed from the framework and exists either as amorphous or microcrystalline alumina or as complex aluminum cations held on cation sites outside the crystalline framework. As is often stated, science can adequately handle either complete disorder or complete order. Unfortunately zeolite chemistry, like most of catalysis, falls somewhere well between these extremes.

Lacking really convincing explanation for many of the hydroxyl bands on zeolites, we can still use them in combination with X-ray, NMR, or other data to show similarities or important differences between batches of presumably identical zeolites. Such differences can be important in practical catalysis, and specifications which include infrared spectra have appeared in the patent literature.

In some published infrared work on zeolites, and also on other subjects, proper precautions were not always taken to eliminate contributions from radiation emitted by the sample. "Proton delocalization" studies sometimes suffer from this problem. As previously mentioned, the infrared literature concerned with surface studies is quite uneven. Many factors, ranging from the techniques used in obtaining spectra to variations in sample preparation and pretreatment, can make it difficult to compare results obtained by different research workers. Possible readsorption of H_2O in zeolite studies is another problem. Most early workers were unaware of the major changes in crystallinity that can easily result from exposure of decationated zeolites to water. A very critical look should be taken at the abundant literature in this field. Zeolite studies will be discussed further below.

C. Acid Sites on Acidic Oxides and Zeolites

1. *Use of Hydrocarbon Probes*

Because of their importance in many hydrocarbon reactions, acid sites on catalysts are of considerable interest. Such sites may be either Lewis or Brønsted in nature and often appear, in fact, to be acid-base sites wherein a coordinatively unsaturated (CUS) anion directly adjoining a Lewis acid (CUS cation) also plays a important role in adsorption. Hydrocarbons provide, in principle, good probes for the study of acid sites and have often been used in such studies [10, 12–14]. Olefins can make useful probes because they are more basic than saturated hydrocarbons, and are more selectively adsorbed on acid sites, but also less basic than ammonia or other nitrogen bases to be discussed below. Successful use of olefin probes usually depends, however, on low activity of the catalyst for olefin reactions. Use of ethylene for the study of a series of cation-exchanged X zeolites by Carter *et al.* [68] and the use of butenes for the study of surface sites on alumina [36] and zinc oxide [69] provide examples where adsorbed olefin probes reacted only slowly, or not at all under the conditions studied. Generally, however, olefins adsorbed on acid sites on active catalysts react quite rapidly to yield oligomers or other complex products. Acetylene has also been used, but it reacts very readily with oxide surfaces.

Aromatics and other liquid hydrocarbons tend to physically adsorb on the catalyst surface as well as being held on acid sites, but have also been successfully used in infrared characterization of catalyst and support surfaces, particularly in attempts to characterize acidic hydroxyl groups on zeolites [10].

2. *Adsorption of* NH_3

The earliest infrared study of a commercial catalyst [2] sought to identify the nature of the acid sites on a silica-alumina cracking catalyst through the spectral bands produced by adsorbed ammonia. The premise was that Brønsted acid sites would interact with adsorbed NH_3 to form NH_4^+, giving rise to bands near 1450 and 3130 cm^{-1}, while NH_3 held, as such, on Lewis acid sites would show bands near 1630 and 3330 cm^{-1}. Both sets of bands were observed. Mapes and Eischens [2] concluded that most of the NH_3 was adsorbed on Lewis sites on silica-alumina predried at 773 K, but the Brønsted acidity, shown by the NH_4^+ bands, increased when water was added to the catalyst. This study was followed by many others wherein NH_3 was used as a probe to investigate the nature of acid sites on silica-alumina, alumina and other oxides [12–14]. These studies generally confirmed the existence of both types of acid sites on catalysts and some supports and also yielded evidence for some dissociative adsorption of NH_3 to form $-NH_2$ groups or NH_2^- ions ($NH_3 + O^= \rightarrow NH_2^- + OH^-$) on both silica [56] and γ-alumina [57, 58] pretreated in vacuum at high temperatures. Residual chloride impurities in some aluminas (*e.g.* Alon C) [3] or in silicas can also cause complications in studies of NH_3 adsorption (*e.g.* $Si{-}Cl + 2\,NH_3 \rightarrow NH_4Cl + Si{-}NH_2$) [56]. Ammonia can be held in various ways on a wide variety of sites and is thus not always an ideal probe for investigation of surface acid sites.

3. Adsorption of Pyridine, Piperidine, and Other Nitrogen Bases

An alternative approach to determining Lewis and Brønsted acidity was taken by Parry [70]. Pyridine was selected as a probe because, being a weaker base than NH_3, it would presumably be held only by the stronger acid sites. An absorption band near 1540 cm^{-1} is characteristic of pyridinium ions, while spectral differences in the regions 1640–1580 cm^{-1} and 1500–1440 cm^{-1} permitted physically adsorbed pyridine to be distinguished from pyridine coordinated to Lewis acid sites. From spectra of adsorbed pyridine, shown in Figure 7, Parry was thus able to confirm that silica-alumina holds both Lewis and Brønsted acid sites. The latter increased in number with increasing addition of H_2O. On silica, pyridine was held by H-bonding interactions with surface hydroxyl groups. Pyridine adsorbed on alumina showed only Lewis acid sites, however. The hydroxyl groups on alumina are, of course, all weak Brønsted acids, but do not form pyridinium ions with pyridine.

Pyridine adsorption has since been widely used to characterize the surface acidity of various oxide supports and catalysts [12–14], zeolites [10], and other high area solids [65]. It has also been used to demonstrate changes in acidity resulting from halide treatment [71] or addition of metals or metal oxides to supports. Pyridine is currently usually regarded as a better probe than NH_3 for studying acidity because it does not adsorb as extensively on weak acid sites and is more stable toward dissociative adsorption on certain sites. Complete stability of pyridine on active catalysts may not always be safely assumed, however. Eberly [72] has reported C—H bands at 2970 and 2880 cm^{-1}, indicating possible formation of $—CH_3$ groups when pyridine was adsorbed on H—Y zeolite at 533 K, and others [73] have suggested that

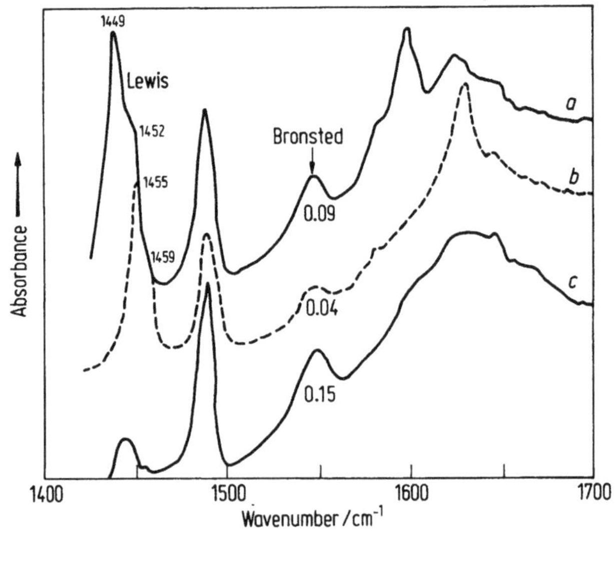

Figure 7. Pyridine adsorbed on a silica-alumina cracking catalyst. *a* Sample equilibrated with pyridine and evacuated at room temperature; *b* after evacuation at 573 K; *c* after adding 0.05 m mole of H_2O. Optical densities (absorbances) are shown for the 1540 cm^{-1} band. (Reproduced with permission from ref. [70])

pyridine forms 2–2′-dipyridyl by cleavage of C—H bonds when adsorbed on Pt/Al_2O_3, giving rise to bands at 1535, 1430, and 1325 cm^{-1} which are not seen for pyridine on alumina.

On zeolites pyridine can adsorb on Brønsted sites, cations, or "true" Lewis acid sites (= tricoordinated aluminum cations as defined by Ward [10]). In any case the number of either Brønsted or Lewis sites on which pyridine can be adsorbed apparently greatly exceeds the number of sites active for hydrocarbon reactions [10].

Piperidine, variously sterically hindered pyridines [74], and other nitrogen bases have all also been used with some success to characterize surface acidity of catalysts, but the possibility of reactions of these probe molecules must be kept in mind.

4. *Adsorption of CO and CO_2*

The choice of probes used for characterization of surface acid sites normally depends on some preconception of the nature of these sites or of the alternative types of sites to be distinguished from one another. Although CO might not be an obvious choice for study of acid sites, it can in fact serve rather well to reveal strong Lewis acid sites and exposed cations. Usually CO is held less strongly than NH_3 or other nitrogen bases and is thus held more selectively by strong Lewis acids. Depending on the cation to which CO is attached, the frequency of the C—O stretching vibration, typically in the 2150 to 2250 cm^{-1} range, can often provide evidence on the nature and abundance of exposed cations on zeolites [10] and other catalysts. The weak binding of CO by most cations means, however, that for quantitative work either lower temperatures or higher pressures must be used than would be required for probes which are more-strongly held.

A band for weakly-adsorbed CO at 2193 cm^{-1} was noticed by Eischens and Pliskin [3] early in infrared studies of CO on NiO but was attributed to the asymmetric vibration of Ni...O$\overset{...}{=}$C$\overset{...}{=}$O. Taylor and Amberg [75] attributed similar bands seen on zinc oxide catalysts to CO adsorbed by dipolar forces. Gardner and Petrucchi [76] have related the vibrational frequencies of adsorbed CO with the number of valence electrons in the molecule. On this basis bands at 2184 and 2205 cm^{-1} were attributed to adsorbed CO^+ and CO^{2+} ions respectively and bands at other frequencies to CO "intermedions" having fractional numbers of valence electrons. Little [12] has discussed the relative merits of some of these explanations.

Further work on supported nickel [66], zeolites [10, 77], and other catalysts [78–80] has established the general occurrence of weak bonding of CO giving rise to bands near 2200 cm^{-1}. Both the nature of the exposed cation and that of neighboring anions [78] can influence the binding of CO, but it appears that cations (Lewis acid sites) play the predominant role in adsorption of CO which gives bands above 2140 cm^{-1}. The exact nature of the binding remains somewhat mysterious, but apparently σ-bonding to the surface cation plus some charge transfer leads to a strengthening of the C—O bond. An empirical relationship between the CO frequency and formal charge and ion radius has been proposed for metal ions of Group I, II, III [81].

Adsorption of CO_2, held as such, without formation of carbonates, bicarbonates, or other identifiable surface compounds also appears to occur selectively on certain sites containing exposed cations and anions (Lewis acid-base sites), giving rise to a band in the 2350 to 2390 cm^{-1} region. The adsorbing sites on dry alumina [78, 82, 83] and silica-alumina [78, 83, 84] which have been called "α-sites", appear to certain unusually reactive oxide ions or other anions, such as fluoride, adjoining exposed (CUS) aluminum ions.

Similar adsorption of CO_2 is also observed on zeolites [10, 85]. As shown in Figure 8 the frequency of the CO_2 band varies with cation electrostatic field. Thus adsorbed CO_2, like adsorbed CO, held on a catalyst in which the same anion neighbors can always be expected to adjoin exposed cations, can provide information about the types of cations exposed and their numbers. In addition, both CO_2 and CO can readily be displaced by other adsorbates (*e.g.* butene) which are held on the same types of sites and can thus provide insert line reading additional information on changes in site availability during polymerization or other reactions of hydrocarbons [84].

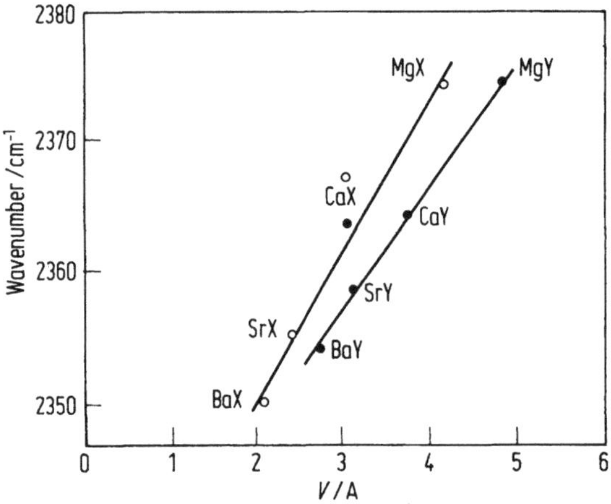

Figure 8. CO_2 frequencies *vs.* cation electrostatic field. (Reproduced with permission from ref. [85])

The inevitable presence of an adjoining Lewis base (anion) is something too easily overlooked in considering acid sites on catalysts, however. The anions can also be coordinatively unsaturated and reactive. They can play a role in catalytic reactions by, for example, temporarily accepting a proton from NH_3 or a hydrocarbon molecule. Unusual reactivity of some surface oxide ions can be shown by the ease with which they exchange with the oxygen atoms in CO_2 [86].

5. *Adsorption of NO*

Adsorption of NO has also often been used to study exposed metal cations on catalyst surfaces. The coordination complexes formed by NO with metal atoms are similar to those formed by carbon monoxide, but the stretching frequency of the NO bond shows greater variation in complexes than that of the CO bond in carbonyls. Discussion of the bonding in nitric oxide complexes can be found elsewhere [12, 87] and will not be repeated here. Basically, however, we are again dealing with σ-bonding from the attached atom (N in this case) plus M $\rightarrow$ N π-bonding. Various degrees of ionicity and "bent" or linear bonding are possible with NO. Bridging (either double or triple bridging) NO groups are also found, although apparently less commonly than bridging CO groups [12, 87]. The spectrum of NO adsorbed on metals, metal oxides and salts was extensively investigated at an early date by Terenin and Roev [88] and NO has been used in various investigations since [12, 87, 89–94]. Generally, the use of NO to study reduced transition metals suffers from the disadvantage that NO reacts more readily to oxidize or reduce [92] metals present on a catalyst surface than CO or CO_2, although the latter probes cannot be regarded as chemically inert either. Regardless of its potential for alteration of the oxidation state of metal atoms or ions on a catalyst surface, nitric oxide is often useful for characterizing transition metal ions on a supported catalyst [87, 89, 93, 94] and can aid in pinpointing active sites present at low concentrations on catalyst surfaces. The use of NO to supplement studies of adsorbed CO and other probes can provide additional insight on the nature of the adsorbing sites.

D. Supported (Reduced) Metal Catalysts

1. General

Infrared study of reduced metal catalysts on high-area oxide supports dates back to the pioneering work of Eischens and collaborators [3] discussed above and the slightly later work of Yang and Garland [95] and of Sheppard and coworkers [96]. Study of such catalysts, typically using probes such as CO but also including hydrocarbons, H_2 and other molecules, has continued actively to the present day and promises to remain an active research area for years to come. In contrast to work described above in which CO and other probes were used to study cations or cation-anion sites on a catalyst surface, the main objective in most work on supported metals has been characterization of fully reduced metals. These were thought to be held as very small crystals on a support (such as silica) which was, it was thought and hoped, fairly inert. From the beginning, however, work with alumina and other supports has paralleled work on silica-supported metals. From the beginning, it was also noted that the support can apparently influence the nature of the supported metal. As techniques have improved and research has become more sophisticated, it has become increasingly clear that support effects can often be extremely important. Much present infrared study of supported metal catalysts is directly concerned with support effects of one sort or another. This section will briefly discuss some of the effects observable and present a few examples to illustrate the extent to which such effects can modify the adsorptive, and probably the catalytic, properties of metals. An excellent summary of current thinking on the vibrational spectra of CO adsorbed on reduced metal catalysts, along with a comprehensive review of work in this area through 1977, is given by Sheppard and Nguyen [11]. An earlier review by Ford [97] is also still useful. The metals which have been the most studied are catalytically important ones, Pt, Pd, Ni, and also Cu.

2. *Carbon Monoxide Adsorption on Pt*

The wide use of Pt catalysts coupled with their ease of reduction and resistance to reoxidation led to intensive study of such catalysts by infrared methods at an early date [3, 98]. Spectra of CO adsorbed on supported Pt characteristically show an intense, moderately broad band between 2020 and 2100 cm^{-1} arising from CO linearly attached to Pt and, at times, a much weaker band, apparently caused by bridged CO, in the 1800–1900 cm^{-1} region.

Sheppard and Nguyen [11] have summarized and discussed frequencies observed for CO on a variety of substrates and also on single crystals, foils, and evaporated Pt films. The exact frequency of the linear CO band changes with surface coverage and other factors. Oxidation of supported Pt samples causes the linear CO band to shift upward in frequency to around 2100 cm^{-1} or even higher. The appearance and behavior of the bridged CO bands on supported Pt is also variable. There is usually reason to be cautious in accepting bands below 2000 cm^{-1} on alumina-supported metals as arising from bridged CO, however, because incomplete reduction or the water gas shift reaction ($CO + H_2O \rightarrow CO_2 + H_2$) can produce CO_2 which can when adsorbed on alumina give strong bands in this region. This problem does not exist with silica-supported catalysts, but problems with background subtraction can sometimes create artificial bands on silica-supported metal catalysts.

There seems to be little doubt that at least four factors can be important in determining the frequencies exhibited by CO adsorbed on supported Pt, and on other metals as well. These are as follows:

a) Surface coverage with CO (effects of dipole-dipole interactions).
b) Inductive effects of ligands — including surface atoms of the support.
c) Crystal face exposure of metal crystallites.
d) Particle size effects — *i.e.* relative abundance of edges, corners, and high index crystal planes.

These effects have all been discussed extensively elsewhere.

In studies of supported Pt or other metals it is usually impossible to define variables closely enough to permit unambiguous description of the dispersed metal. Careful study of a particular catalyst can, however, greatly limit the range of uncertainty and provide information useful in optimization of the performance of commercial catalysts [52, 99]. Obviously, however, the catalyst studied should either be or closely approximate the catalyst of practical interest. Such limitation is not, or need not be, a problem in effectively studying supported Pt catalysts. Most catalysts used in commercial reforming of petroleum naphthas contain 1% or less of Pt dispersed on high area γ- or η-alumina. Apparently for historical reasons, most published infrared studies of supported metals have been made on catalysts holding 5–20 wt.%

of metal. Because high metal concentrations generally produce poorer metal dispersion (*i.e.* larger crystal sizes and lower specific surface areas for the metal), studies of Pt at concentrations more typical of those used commercially, and on supports more like those normally used, are generally of greater interest to the industrial research worker. Studies of such "real" catalysts are actually not much more difficult than those of the more artificial model catalysts studied in most of the earlier work. It is thus surprising that more attention has not been given to studies at lower metal concentration. One reason often advanced for the use of silica supports and high metal loadings is that support effects are thus minimized. Support effects can, however, play an important role in the performance of a catalyst, as they evidently do in the case of Pt reforming catalysts.

Figure 9 illustrates effects observed on the linear CO band when HCl was adsorbed on a 10% Pt/Al_2O_3 catalyst holding previously adsorbed CO [73]. The adsorbed HCl clearly shifts the frequency to higher values. The same shift was seen in experiments where the HCl was chemisorbed first and then CO added.

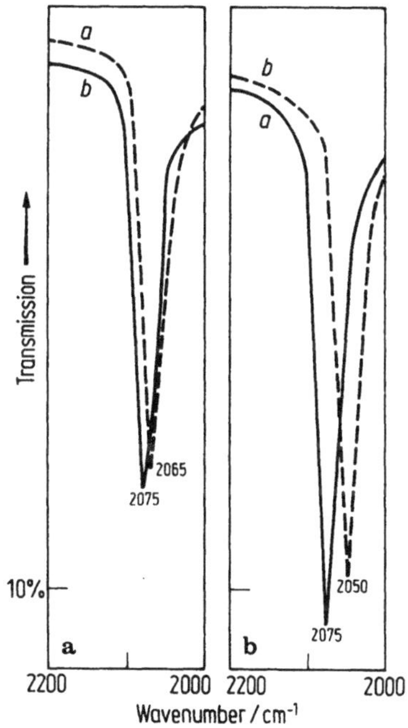

Figure 9. Adsorption of HCl on Pt/Al_2O_3: **a** (*a*) Pt/Al_2O_3 + 0.2 monolayer of CO; (*b*) after introduction of 6.67 kPa HCl at room temperature. **b** (*a*) Pt/Al_2O_3 + 0.2 monolayer of CO after treatment under 6.67 kPa of HCl at 573 K during 30 min; (*b*) after desorption of HCl at 473 K during 4 hr. (Reproduced with permission from ref. [73])

Figure 10 shows effects observed when pyridine was similarly adsorbed on 10% Pt/Al_2O_3 holding previously adsorbed CO [73]. The frequency shift in this instance was to lower frequency. Similar frequency shifts were previously reported by Queau and Poilblanc [100], in studies of CO on evaporated Pt films and were attributed, as in this case, to the effects of Lewis bases in increasing the electronic density on Pt crystallites, thus promoting back-

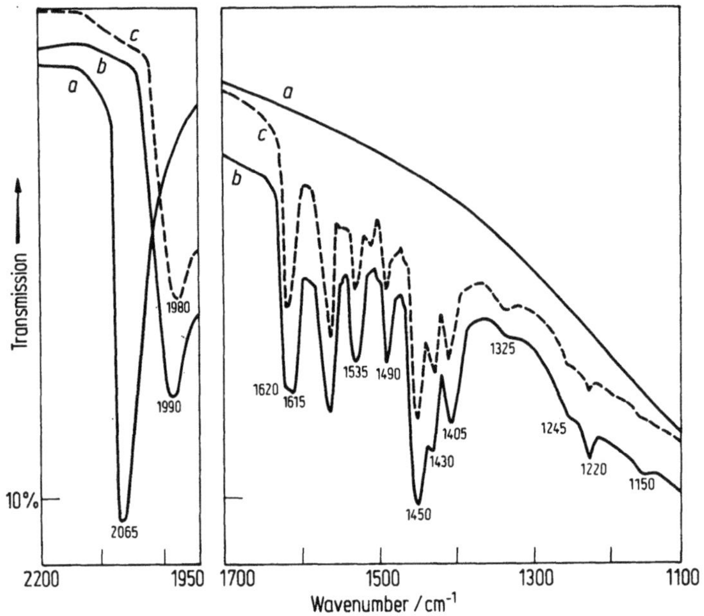

Figure 10. Infrared spectra of pyridine adsorbed on Pt/Al_2O_3; *a* Pt/Al_2O_3 + 0.2 monolayer of CO; *b* introduction of pyridine at room temperature; *c* evacuation of sample at 423 K. (Reproduced with permission from ref. [73])

bonding of *d* electrons to the CO. Shifts of the CO band to higher frequencies caused by HCl and oxygen, were thought by Primet *et al.* [73] to require attachment of oxygen or chlorine (electron acceptor) atoms directly to the same Pt atom holding CO. The highest shift ($\Delta\nu = 75\ cm^{-1}$) apparently corresponded to attachment of two chlorine atoms to the same Pt atom holding CO. Similar shifts of CO/Pt bands with HCl and O_2 were also seen previously by Eischens [3]. Primet *et al.* [73] also noted the apparent formation of adsorbed carbonyl chloride complexes of Pt after treatment of their catalyst with phosgene.

Figure 11 shows similar spectra obtained for CO adsorbed on a used commercial 0.8% Pt/Al_2O_3 catalyst before and after a typical "rejuvenation" treatment in $CCl_4 + O_2$ at 770 K [99]. Evidence for chlorocarbonyl formation is again seen in the CO bands near 2140, 2162 and 2200 cm^{-1}. Adsorption of CO on the rejuvenated catalyst was quite high, approaching 1 molecule per Pt atom. The influence of chloride (and possibly other support effects) is again seen by the high frequency of the CO band on rejuvenated catalysts. Surface area of such Al_2O_3-supported Pt is lost at a fairly rapid rate on heating in O_2 at 870 K or higher, and lost area is restored by rejuvenation treatments similar to that of Figure 11. Evidently Pt halides or chlorocarbonyl complexes can effectively redisperse Pt crystallites on alumina supports.

The extent to which Pt dispersed on an alumina support can be influenced, for example, by Lewis acid or base sites on the alumina surface evidently depends on both the dispersion and the concentration of the supported Pt.

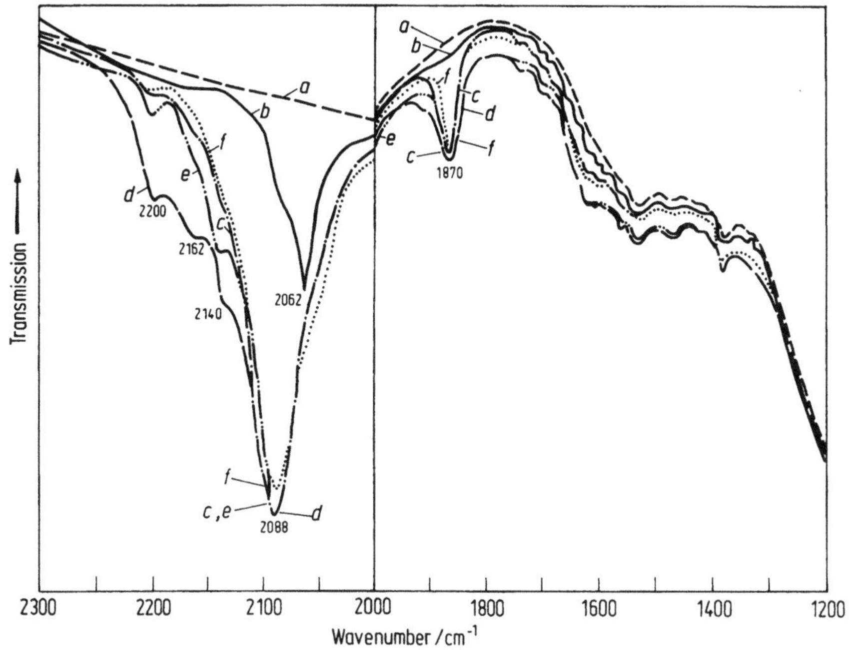

Figure 11. Spectra of CO adsorbed on rejuvenated, used 0.8% Pt/Al_2O_3: *a* after heating catalyst in O_2 at 873 K and reducing at 773 K; *b* a + 1.0 kPa of CO; *c* after rejuvenation of catalyst by heating in $CCl_4 + O_2$ at 773 K, reduction in H_2 at 773 K, and addition of 1.0 kPa of CO; *d* c + 26.7 kPa of CO; *e* after evacuation to 3 Pa; *f* after evacuation to $< 1 \times 10^{-2}$ Pa. (Reproduced with permission from ref. [99])

If factors such as CO coverage effects are to be important on Pt, the Pt crystal surfaces must be sufficiently developed to permit such effects. Isolated Pt atoms, for example, could hardly exhibit coverage effects or, in fact, any of the effects associated with crystal surfaces, edges or corners. Any meaningful discussion of supported Pt catalysts or of other supported metals should, therefore, first attempt to define the actual degree of dispersion of metal in the catalysts of interest. This is one aspect which has often been ignored in extrapolations from single crystal work. At present, evidence exists to indicate that at very high dispersion Pt, Ir, and probably other metals may often exist as small "rafts" or "rugs" on the surface of an alumina support. Any theoretical treatment of the observed band shifts and differences in CO binding energies must inevitably be very approximate at present, but empirical correlations with catalytic performance are clearly in sight. A thorough systematic infrared study of CO adsorbed at various pressures on a series of supported Pt crystallite catalysts made on the same support and differing only a) in metal concentration, or b) in crystallite size has evidently not yet been reported. Such a study might further clarify the present picture. Obviously, however, much work would be entailed in such research.

A striking feature in the adsorption of CO on Pt seems to be the very strong preference for linear, as opposed to bridged bonding. Some bridged bonding occurs, as can be seen for example in Figure 11, but usually only on highly dispersed Pt. In contrast to bridged CO on Pd, for example, the bridged CO on Pt is held more weakly than is the linear CO. When CO is bridge-bonded on Pt, the bridged CO bond evidently appears only at high coverages, accompanied by a shift in the frequency of the band for pre-adsorbed linear CO [99].

Although the support effects normally seen with supported Pt catalysts appear relatively minor when compared with those observed, as will be shown, with more reactive metals, it should not be assumed that they are unimportant or that they could not become much more striking under some circumstances. A recent EELS study of the effect of potassium on adsorption of CO on the Pt (111) crystal face [101] is of particular interest in this respect. The authors observed that when a half monolayer of potassium was deposited on the Pt surface the bridge-bonded CO vibrational frequency decreased to as low as 1400 cm^{-1} while CO preferentially occupied bridge bonded instead of linearly bonded surface sites. The adsorption energy also increased from 113 to 163 kJ $mole^{-1}$. This was attributed to enhanced electron back donation from Pt to the CO. Such effects could be very important in explaining the catalytic effects of potassium and other promoters on metal catalysts.

3. Carbon Monoxide Adsorption on Pd

Extensive study has also been made of CO on supported Pd catalysts, single crystals, and polycrystalline foils and films, and this continues to be another active area for research. As first observed by Eischens and coworkers [98] on silica-supported Pd, the main feature of CO adsorbed on Pd is the predominance of bridge-bonded CO species giving rise to bands in the 1820 to 1990 cm^{-1} region. The frequency of the band maximum in this region shifts to higher frequencies with increasing surface coverage. Several bands are, in fact, observed in this region, each tending to move to higher frequencies at higher CO coverages. These sub-bands were interpreted by Eischens *et al.* as arising from CO held on different homogeneous areas of the Pd surface, possibly representing exposed major crystal planes of well-formed single crystals [98]. At CO coverages above 50% a linear CO band was seen around 2060 cm^{-1}, and this shifted with increasing coverage to about 2090 cm^{-1}. This band corresponded to CO held more weakly than the bridged species and was thought to reflect filling of adjacent double sites, and thus to bear no relation to the type of face exposed.

Based on this and more recent work [102–104], Sheppard and Nguyen [11] have assigned the two strongest bridge-bonded CO bands, 1985 and 1955 cm^{-1}, to CO on Pd (100) and (111) planes, respectively, and the "tail" to lower wavenumbers (between 1930 and 1850 cm^{-1}) as representing (110) or similar planes and/or (100) or (111) planes of small diameter. They consider this the best example where the infrared spectra from adsorbed CO can be convincingly interpreted in terms of surfaces of metal crystallites. The linear bands near 2090 are attributed to linear CO weakly held on "fill-in" sites on (111) planes. The more strongly-held linear CO responsible for the band near 2060 cm^{-1} was assigned to CO on other sites, (110) planes, etc. Eischens' original views thus appear to be essentially substantiated.

The complexity of the band structure revealed in studies of Pd catalysts, and clear evidence from changes in the spectra obtained by Baddour *et al.* [102] as a function of the degree of "break-in" that changes in the relative exposure of different crystal planes occur during reaction, illustrate the major problems facing attempts to relate certain types of surface sites to high catalytic activity. Beyond possible differences in the exposure of different crystal planes, surface coverage with CO, C, O, and other molecules, if also present and adsorbed by a Pd catalyst, will also affect the frequencies of the bands observed. As previously indicated, support effects can also be very important.

The apparent strong preference of Pd catalysts for bridged adsorption of CO contrasts, as previously noted, with the preference of Pt for linear bonding. Explanation for this apparently does not lie in the crystal planes preferentially exposed or in the spacing of atoms in these planes, but rather in the inherent chemistry of metals, *i.e.* in the bonding orbitals available at the surface.

4. Carbon Monoxide Adsorption on Ni

Because CO participates in reactions such as methanation and steam reforming of hydrocarbons over nickel catalysts, adsorption of CO on Ni has received considerable attention. Because Ni is much more easily oxidized than either Pt or Pd, greater interaction with adsorbed molecules, and with supports can be expected. Support effects and C and O overlayers may perhaps be absent or overlooked with Pt or Pd catalysts, but must certainly be taken seriously when dealing with Ni catalysts. Supported Ni may often not be fully reduced or even reducible by usual reduction procedures [66]. Beyond this, dissociative chemisorption of probes such as CO and NO or reaction with residual gases such as O_2 or H_2O cannot often be eliminated in usual experiments on supported Ni. In dealing with the catalytic behavior of supported Ni, where active sites need not be present at high concentration to profoundly affect the selectivity, possible support effects must be very carefully evaluated. Alumina is notorious for strongly interacting with supported nickel, and incorporation of some Ni ions in the surface (or bulk) of alumina supports where they are difficult or impossible to reduce is probably the rule rather than the exception. Even on silica, however, Ni ions which are difficult to reduce can be held in the surface and be important catalytically [66].

Studies of CO on Ni single crystals by EELS and infrared [105] and infrared studies of evaporated Ni films have yielded bands at frequencies generally similar to those seen for Pd, including both linear (2050–2085 cm^{-1}) and bridged (1570–1970 cm^{-1}) CO. Bands are also seen above 2130 cm^{-1}, in some instances, reflecting oxidized Ni sites. Prior adsorption of O_2 or other molecules which give partial surface overlayers of adatoms tends to weaken the binding of the CO adsorbed subsequently.

As with most other metals, infrared studies of supported Ni to date have generally been of catalysts containing 4 to 20% Ni usually on silica [106, 107], although some of the earliest work on supported Ni [108, 109] used an alumina (Alon C) support. The linear CO bands reported on Ni have generally been relatively more intense than the bridged CO bands when compared to Pd catalysts.

Although CO chemisorption is generally regarded as non-dissociative the effects of dissociative adsorption of both CO and CO_2 on the spectra of CO subsequently adsorbed are well illustrated in recent FTIR studies [107]. Figures 12 and 13 show spectra apparently indicating dissociative chemisorption of CO and CO_2 respectively on pure Ni/SiO_2. Although surface reorganization leading to preferential exposure of different crystal planes may also be involved, Martin *et al.* [107] present good evidence indicating dissociation to yield surface carbide and bound atomic oxygen. Although not always explicitly stated, it appears that CO was weakly adsorbed on the nickel surfaces.

Figure 14 shows additional spectra illustrating the effects of carbon deposition (from CO) on alumina-supported Ni catalysts [110]. The CO bands were interpreted on the basis that carbide carbon was formed and held by the surface while CO which formerly adsorbed in two-fold bridged sites

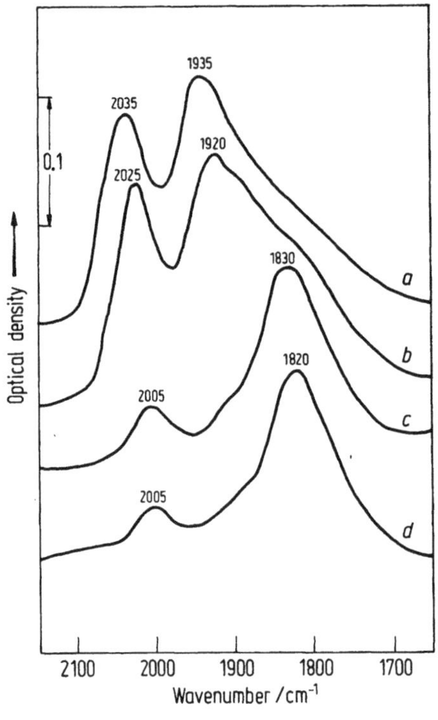

Figure 12. Infrared spectra obtained during heat treatment of chemisorbed CO on Ni/SiO_2 without cryopumping of evolved gases. *a*, 15 ml (NTP) of CO per g of Ni, adsorbed at 300 K. This was followed by heat treatment in the closed system; *b* for 1 hr. at 380 K; *c* for 2 hr. at 480 K; *d* for 1 hr. at 600 K. (Reproduced with permission from ref. [107])

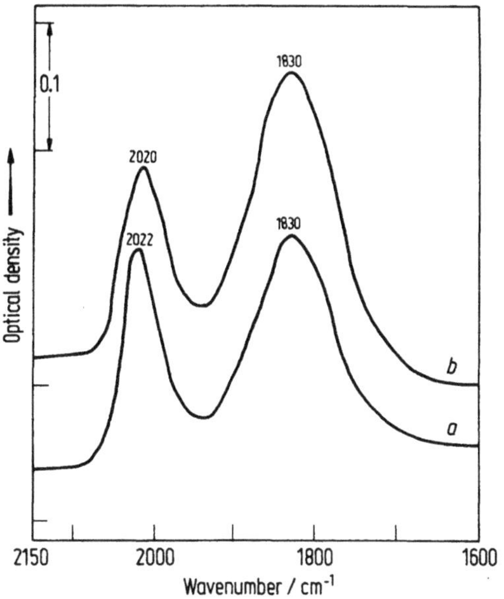

Figure 13. Infrared spectra of CO_2 adsorbed on pure Ni/SiO_2 catalysts. *a* CO_2 adsorbed at 300 K under 0.13 kPa of equilibrium pressure; *b* Sample (a) heated in closed system for 2 hr. at 600 K. (Reproduced with permission from ref. [107])

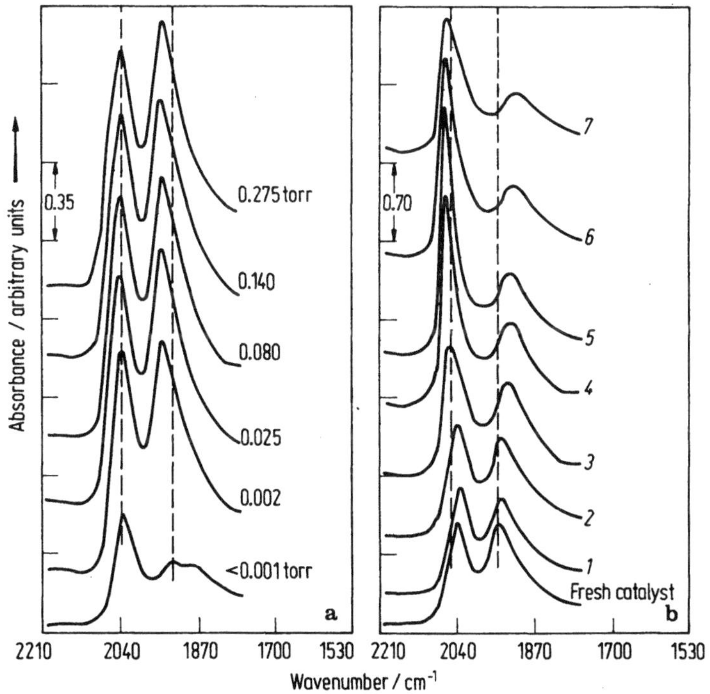

Figure 14. Absorbance spectra of **a** CO adsorbed at increasing pressures on a fresh Ni catalyst. **b** CO adsorbed at 17 Pa on a Ni catalyst after increasing deposits of carbidic carbon. (Reproduced with permission from ref. [110])

($\sim 1950\ cm^{-1}$) was forced into a decreasing number of three-fold bridged sites ($\sim 1870\ cm^{-1}$) or held more weakly than before as linear CO, initially with a somewhat higher frequency. The detailed interpretation of these and other spectra must, in light of the foregoing discussion, be accepted with some reserve, but both band frequencies and CO binding energies can clearly be affected by C or other overlayers.

When dealing with Ni at low concentrations ($<1\%$) and/or at high temperatures in the presence of steam, CO, and CO_2, the possible effects of support interactions and of surface overlayers or even bulk carbide formation can never be ignored. Other factors often involved in practical catalysis also require attention. Nickel (plus vanadium and iron) contaminants on cracking catalysts, for example, presently constitute an important problem in petroleum refining, increasing coke and gas yield undesirably. The Ni on such catalysts is frequently held at concentrations of less than 0.1%. Computerized infrared offers very good prospects for effective study of such catalysts, but further work is needed before reliable information can be routinely obtained on such samples.

5. *CO and NO Adsorption on Supported Mo Catalysts*

The increasing importance of hydrosulfurization used commerically to remove sulfur from petroleum, has inspired much research on the catalysts used in this process. These catalysts usually contain 7 to 10% molybdenum, plus smaller amounts of Co or Ni promoters. Similar Mo-containing catalysts are also used for coal liquefaction and other catalytic processes. None of the transition metals in these catalysts can be regarded as unreactive toward CO or NO probes or toward oxide supports. In addition, under reaction conditions such hydrodesulfurization catalysts also differ chemically from their original oxide forms. Infrared spectroscopy can, however, be used to provide useful insights on the nature and number of Mo (and/or Ni and Co atoms) exposed on these catalysts, either after reduction of the original oxide forms in H_2 or after treatment with sulfides followed by partial reduction.

Figure 15 shows spectra obtained for CO and NO adsorbed on a MoO_3/ /Al_2O_3 catalyst after various prereduction treatments [93]. Absence of CO bands in the 2100–2000 region indicates that after reduction in H_2 at 970 K where reoxidation by H_2O can readily occur few, if any, reduced Mo^0 atoms are exposed on the catalyst surface. The exposure of Mo^0 can be somewhat increased by reduction procedures at 970 K wherein the H_2O generated is more effectively removed as indicated by spectrum b. Only after reduction at 1070 K with effective removal of H_2O, however, can significant exposure of Mo^0 be achieved. Under commercial conditions, of course, such extreme dryness and such high temperatures are not normally achieved and much exposure of Mo^0 would not be anticipated. The spectra of adsorbed NO indicate that although exposed ionic Mo sites remain after 1070 K reduction their nature has been changed so that the relative intensities of the two characteristic NO bands are now different than after reduction at lower temperatures.

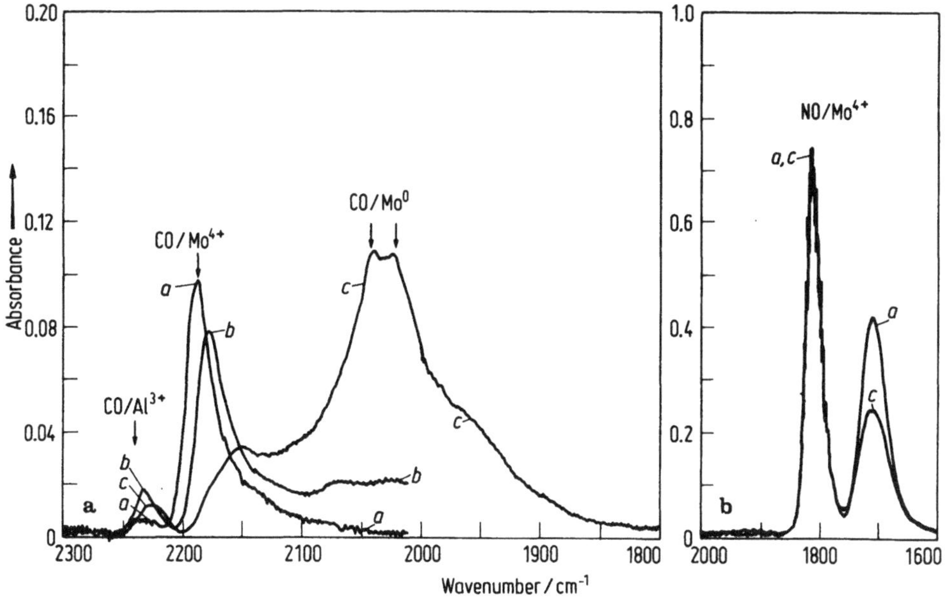

Figure 15. Effects of high temperature prereduction of 12% MoO_3/Al_2O_3: **a** Adsorbed CO; *a*) — After 30 min reduction in H_2 at 973 K (without change of H_2) + addition of 1.2 kPa of CO; *b*) After 1 hour reduction (fresh sample) at 973 K, including four complete changes of H_2 in cell, + CO addition; *c*) After 1 hour reduction at 1073 K with 4 changes of H_2 + CO addition. **b** Adsorbed NO; *a*) and *c*) as for *a*) and *c*) above except NO added after removal of adsorbed CO by evacuation. (Reproduced with permission from ref. [93])

As indicated by the spectra of Figure 16, however, pretreatment of MoO_3/Al_2O_3 with sulfides can apparently lead to significant exposure of reduced Mo after reduction in H_2 at temperatures where normally little or no exposed Mo^0 or $Mo^{\delta+}$ would be seen [93]. The frequency of the CO band on the reduced sulfided catalyst is here seen near 2100 cm^{-1}, apparently reflecting the inductive effects of neighboring Mo ions held in sulfides or of partial sulfide overlayers on reduced Ni crystallites. The CO held on such $Mo^{\delta+}$ sites is weakly bound, presumably again because of the effects of S, C, or O partial overlayers or simply because the slightly electropositive nature of the Mo atoms to which the CO is bound tends to prevent back donation of electrons. Adsorption of CO on fully reduced Mo would be expected to give some dissociative adsorption, giving no detectable infrared bands, but further non-dissociative, linear bonding of CO should then occur on unoccupied sites where dissociative adsorption does not readily occur.

Dissociative adsorption of NO may similarly alter the oxidation state of exposed Mo^0 or of Mo ions originally exposed on the surface. Here again, however, some NO often appears to be molecularly adsorbed on the altered sites and can provide eivdence on the effects of various pretreatments on the relative exposure of Mo on different catalyst preparations. The maximum exposure of Mo, either as reduced Mo^0 or as Mo ions seen in infrared studies to date represents only 2 to 8% of the total Mo content on

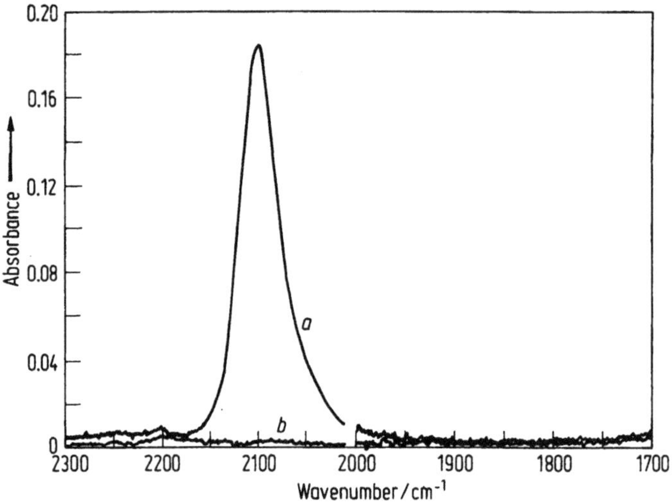

Figure 16. Effects of presulfiding 8% Mo/Al_2O_3: *a* CO adsorbed at 1.0 kPa after presulfiding at 523 K; *b* After 5 min. evacuation. (Reproduced with permission from ref. [93])

Mo/Al_2O_3 catalysts, depending on reduction conditions and on the particular catalyst. Through use of isotopically-labelled NO, the typical site for NO adsorption can be shown to hold two NO molecules either as dinitrosyl or as dimer [52, 93, 94].

The sites which hold CO and NO on Mo/Al_2O_3 catalysts appear to be catalytically active, but more evidence is needed on the nature and numbers of these sites, and on the role of promotors such as Ni and Co before complete understanding of these catalysts is possible.

6. Studies of "Alloy" Catalysts

The possibility that two or more different metals held on the same support may interact to form supported bi- or multi-metallic clusters or "alloys" has been a major topic in several recent infrared studies [111–113, 135]. Revived interest in alloy catalysts stems from the introduction of commercial $Pt{-}Re/Al_2O_3$ reforming catalysts about 10 years ago which outperformed the Pt/Al_2O_3 catalysts then universally used. Although conceptually acceptable, alloying of metals at low concentrations on oxide supports is rather difficult to demonstrate experimentally in most cases. Infrared spectroscopy has been used in attempts to learn, through study of adsorbed CO and NO probes, whether bimetallic clusters are actually formed on "alloy" catalysts. Results to date generally seem inconclusive at best. In principle, interactions between unlike metal atoms which differ in their binding of CO or NO, as shown by the frequencies of bands for these adsorbed probes, might lead to alteration of the band frequencies. Thus an A + B alloy might show a band for linearly adsorbed CO at 2075 cm^{-1} while the individual metals A and B would show bands at 2100 and 2050 cm^{-1}. Such "electronic" effects have been reported [112, 135].

Alternatively geometric effects may be important. If CO characteristically adsorbs in a bridged form on pure A giving a band or bands below 2000 cm^{-1} alloying with B, which normally holds CO only in a linear form, might by decreasing the number of AA sites decrease or eliminate the characteristic bridged bonding of CO shown by pure A. Thus Pd + Ag, for example, would if alloyed be expected to show less bridged bonding of CO than if there were no interaction between the two metals. Such behavior has been observed [113]. Other examples exist, but caution seems warranted in interpretation of the data.

If "alloying" is interpreted to mean any interaction between two dissimilar metal atoms, including interactions of metal cations with reduced metal atoms or small crystallites, alloying often exists on oxide supports and can be demonstrated by infrared and other techniques. In the usual sense, where alloying implies close association of fully reduced metal atoms of different kinds, the problem of showing such interactions at low metal concentrations on oxide supports becomes very difficult. In the case of typical Pt—Re reforming catalysts, infrared study, although not entirely excluding some interaction, provides no evidence for alloying of the Pt and Re atoms on alumina [52, 99]. At higher concentrations, however, alloying may be seen in some cases [114]. Because the simultaneous presence of two metals may, through decreased support interaction or for other reasons, facilitate reduction of one or both of the metals, care must be taken in attributing observed band shifts to alloying. In a recent paper [112], for example, a CO band near 2100 cm^{-1} attributed to CO/Cu^0, might conceivably instead reflect CO held by Cu^{1+} or $Cu^{\delta+}$ sites. Thus changes observed in this band, taken as evidence for alloying with Ni, might alternatively be explained as caused by more-complete reduction of Cu. In most cases, no H "spillover" or similar effects need be invoked to explain more-complete reduction of a metal when the total concentration of metal on the support is increased by addition of a second metal. Support interactions which can stabilize oxidized metal A at a low limited number of surface sites are at least partially saturated by metal B when B is added to A + support. If interactions of the few support sites with metal B occur in strong preference to those with A then A may be more reducible in the presence of B even at the same or lower metal concentration. Changes in the electronic state of a supported metal when a second metal is added thus do not necessarily show direct association of the two metals in a cluster or "alloy".

A better case can be made for alloying based on a geometrical factor in adsorption of CO, but even here existing evidence is hardly unambiguous, given present uncertainty as to why bridging occurs in one case and not another. If surface oxide formation from CO_2 can prevent bridging adsorption of CO on A, and B can produce CO_2 from CO with carbide formation on B, bridging adsorption of CO on A might be prevented by addition of B without direct alloying. Until such time as the individual components and support interactions are more clearly understood, and until the complex band structures often seen can be more surely resolved or deconvoluted, conclusions in this type of investigation will probably remain open to considerable doubt.

7. *Adsorbed Hydrocarbons on Supported Transition Metals*

Much infrared study of adsorbed hydrocarbons on oxide-supported metal catalysts was carried out in the 1950's and 60's [3, 12]. Many of the conclusions are undoubtedly still valid. Research in this field faces major potential problems, however, because of the reactivity of hydrocarbons with clean metal surfaces or on acid sites on the support. Supported nickel catalysts for example, can contain unreduced Ni ions which constitute acid sites capable of forming oligomer or polymer from olefins. Such polymer can give spectral bands which can easily be misinterpreted as arising from a simple adsorbed species produced from the molecule added originally [66]. Platinum and palladium catalysts are apparently safer in this regard, but caution is certainly in order when interpreting spectra obtained for hydrocarbons adsorbed on any active catalyst. The recent work of Haaland [115], using FTIR to study cyclohexane and benzene adsorbed on 10% Pt/Al_2O_3 (Alon C) illustrates both some of the problems entailed in reactions of hydrocarbons with supported metals and also the use of FTIR in such studies.

E. Catalyst Surfaces Under Reaction Conditions — "Active Intermediates"

The search for active intermediates on catalyst surfaces has long been a theme of infrared research, but rather limited real success has been achieved to date. Various adsorbed hydrocarbon species have been observed at room temperature which can be hydrogenated and desorbed at slow rates, but in most cases these appear to be too stable to be important intermediates in fast reactions. It is difficult to prove that any observed adsorbed species is an important reaction intermediate, but the chances for doing this are obviously improved if the species can be observed under reaction conditions and shown to change in concentration with changing conditions in parallel with the production of some reaction product.

The most successful attempt to date to identify an active intermediate on a catalyst appears to be the work of Chang, Conner, and Kokes [116] who used infrared to study isomerization of 1-butene on ZnO. Plots of the intensities of four surface species formed from 1-butene are shown as a function of time in Figure 17. It was concluded that an anti π-allyl adsorbed species is the intermediate when 1-butene isomerizes to 2-butenes on ZnO. This study was only possible, however, because ZnO is relatively transparent to infrared and is also a poor catalyst so that isomerization rates were low. Characterization of adsorbed intermediates on ZnO has been well discussed by Kokes [117].

Attempts to find and identify intermediates in more practical catalytic systems have, on the whole, been unsuccessful to date. In studies of hydrocarbon reactions, particularly at realistic pressures, gas phase and physically

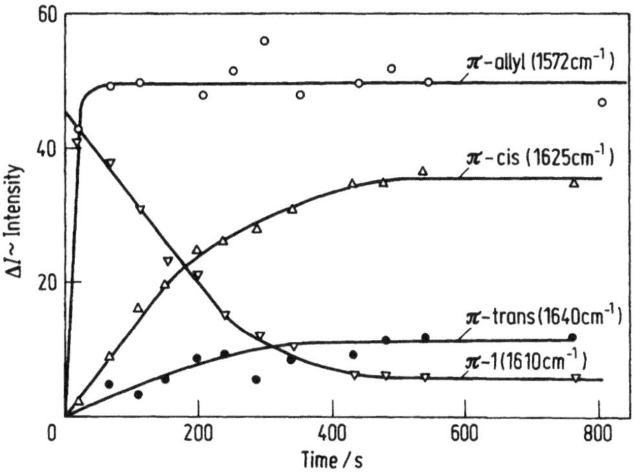

Figure 17. Intensities of surface species formed from 1-butene *vs.* time. (Reproduced with permission from ref. [116])

adsorbed hydrocarbon usually cause strong absorption of infrared in regions where bands from adsorbed intermediates might also be expected to occur. High temperatures may reduce surface coverage with adsorbed intermediates to below detectable levels. Meaningful infrared studies of complex hydrocarbon mixtures adsorbed on catalyst surfaces are obviously not presently feasible.

The best chances for success currently appear to lie in studies of catalyzed reactions of CO. In such reactions the adsorbed CO, possibly adsorbed H_2, most of the products and some postulated intermediates can, in principle, be observed and distinguished by infrared study. Several infrared studies of the synthesis of hydrocarbons from CO and H_2 over oxide-supported Ru catalysts under reaction conditions have been published in recent years [41, 118, 119]. Although various surface species were observed and identified, no direct spectroscopic evidence was obtained on the nature of the active intermediates. Possibly such intermediates are carbidic in nature, however, and thus not detectable by the procedures employed.

The synthesis of methanol from CO + H_2 over CuO/ZnO catalysts would seem to offer the best opportunity to see adsorbed active intermediates. Here both reactants, products, and intermediates might be observed and identified under reaction conditions. At least one study of this process has been carried out under fairly realistic conditions using FTIR [42]. Spectra indicated the presence of carbonate, carboxylate, hydride, hydroxide, and formate on both ZnO and CuO/ZnO catalysts at 470 K and 50 atm pressure, but the mixed oxide had additional bands, attributed to a methoxide surface species. If an active intermediate was seen in this work, it was not clearly identified or shown to be an intermediate.

Also favorable for infrared study of possible intermediates is the reaction of NO + CO over Pt and other noble metal-on-alumina catalysts. This catalytic reaction is important in connection with control of automotive exhaust emissions. Unland [120] studied the reaction of NO + CO over

Pt, Pd, Rh, Ir, and Ru catalysts, all supported on alumina, and observed under reaction conditions at elevated temperatures formation of a strong band in the 2260 to 2270 cm^{-1} region for all catalysts except Ru/Al_2O_3. The band was assigned to an isocyanate surface species. The weakness or absence of this band on Ru/Al_2O_3 appeared to correlate with the unique low ammonia-forming properties of this catalyst during NO_x removal from automobile exhaust. Hydrolysis of the isocyanate species provides a pathway to NH_3 formation, suggesting a possible role of the isocyanate surface complex as an intermediate. Complications exist, however, because NH_3 can be formed and decomposed in other ways as well. Further research on the location and behavior of surface isocyanate on oxide-supported Pt catalysts under reaction conditions [121, 122] has added further complexity to the picture. The isocyanate appears to be largely located on the support and is on the whole a rather stable species [121].

Other similar examples could be cited of fairly stable intermediate species identified by infrared study, but, their true roles in catalytic reactions remain uncertain. It is very unlikely, of course, that activated complexes will be found at significant concentrations on any active catalyst. Fairly stable intermediate compounds, adsorbed reactants or products, are much more likely to be seen. Better spectrometers and computer treatment of spectral data can increase sensitivity, but it will probably be some time before active intermediates can be successfully studied in most instances.

This does not mean that studies of catalysts subjected to realistic reaction conditions are futile. Much important information on catalyst deactivation and changes in catalyst surface chemistry caused by variations in pretreatment can be obtained in this way. Although of considerable theoretical interest, identification of reaction intermediates is certainly not necessary for improved understanding of catalyst surface chemistry.

5. Complementary Vibrational Spectroscopic Techniques

This chapter has been concerned almost exclusively with the use of conventional transmission infrared spectroscopy for studies of typical high-area solid catalysts, although mention has been made of other methods more suitable for studies of low area metal surfaces. Transmission infrared spectroscopy is, by far, the most generally useful vibrational spectroscopic technique for catalysis research, but a number of other techniques do exist for obtaining vibrational spectra of adsorbed molecules and surface groups on solids. These techniques have been reviewed recently [15–17]. These alternative techniques, which should be regarded as complementary in many cases, are most useful in special situations where, for one reason or another, conventional infrared spectroscopy cannot be used to obtain necessary information.

A. Optical Vibrational Spectroscopies

Most closely related to conventional infrared transmission spectroscopy are infrared emission spectroscopy [3, 123] and various techniques using photons to excite vibrational transitions. These include techniques measuring infrared absorption (*e.g.* on reflection, either specular or diffuse and either internal or external), and photoacoustic spectroscopies. Such techniques all offer, at least in principle, the advantage of permitting study of a catalyst surface in contact with a gas or even a liquid. In practice, the difficulties of obtaining sufficiently short gas or liquid path lengths make application of some of these techniques difficult or impossible with gases at appreciable pressures or with liquids.

For studies of high-area catalysts the advantages of the alternative infrared methods are most apparent with very opaque samples, *e.g.* carbon-supported or coked catalysts or those with very high metal loadings, and consist principally of the effort saved in not having to use extraordinary, time consuming, measures to prepare samples thin enough for study by conventional transmission infrared. Such convenience comes at a price, however, since it means losing a constant effective sample thickness and thus hurts prospects for quantitative comparisons of samples. The effective thickness varies in diffuse reflectance, photoacoustic, ATR and other methods with the amount of scattered light (*i.e.* with particle size and refractive index) and with frequency, depending on the presence of absorption bands. Moreover, since the study of surface species requires penetration of an appreciable thickness of sample, these methods usually fail for studies of surfaces on the same samples that are difficult to prepare for transmission studies. The main advantage, as indicated, is the convenience in easily obtaining a qualitative spectrum of an opaque solid. There may, however, be special cases where, for example, emission spectroscopy or ATR (MIR) could be used for *in-situ* studies of high area catalysts under high pressure where conventional transmission methods would be difficult or impossible to use.

The potential advantages of Raman spectroscopy in studies of high area catalysts seem more significant [124–126]. In addition to providing supplementary information theoretically unobtainable by infrared (*i.e.* bands from symmetrical vibrations not involving a change in dipole moment if polarizability is changed by the vibration) Raman spectroscopy permits obtaining data in regions of the spectrum where strong, broad absorption bands normally preclude infrared study on common oxide supports. Raman spectroscopy is normally a much less sensitive technique than infrared, however, and laser irradiation is usually essential to obtaining useful Raman spectra of adsorbed species. For supported metal catalysts, laser irradiation usually means excessive heating of the sample, which can cause desorption or decomposition of adsorbed molecules. Cooling and rotating the sample, in addition to defocusing and moving the laser, are some of the measures employed to minimize heating in a recent study of CO on Ni/SiO_2 [125]. Studies of adsorbed pyridine have been popular, and sensitivity has been greatly improved in recent years [126].

For low area metal samples, as previously discussed, the use of reflection-absorption infrared (IRRAS) offers major potential advantages [127], particularly in permitting study of adsorbates in contact with the gas phase at reasonable pressures. Such methods should permit in the near future fairly routine infrared studies of absorbed molecules on metal single crystal planes at low coverages.

Prospects for effective Raman study of adsorbates on some low area metal surfaces through surface-enhanced Raman scattering (SERS) also appear promising [128, 129]. This new technique may in special cases also permit study of metal surfaces under reaction conditions, but it will not be applicable to typical commercial catalysts. Further development of the method and better understanding of the phenomena involved, appear to be needed before serious applications will be possible, however.

The best optical spectroscopic method for studying oxide single crystal surfaces appears, at present, to be internal reflection spectroscopy, wherein a "totally" internally reflected infrared beam interacts with adsorbed molecules or surface groups on the external surface of a transparent solid to produce an infrared spectrum [130]. The method, as used to date, lacks sensitivity, but major improvement appears possible through computer treatment of data and other refinements. Most crystalline oxides, halides, sulfides, and many other solids transmit infrared over a useful range of frequencies and spectra of surface species on such solids should be obtainable by internal reflection techniques. Unfortunately, the long path length usually needed within the solid for internal reflection studies usually limits study to the region above 2000 cm^{-1}. Although internal reflection methods appear to have the potential to ultimately provide the same sort of information on oxide and other insulator surfaces that reflection-absorption methods can give on metal single crystal surfaces, the published work using this technique has been unimpressive. Serious questions exist as to the significance of some of the results reported.

B. Neutron Scattering Spectroscopy

Neutrons are inelastically scattered fairly strongly by hydrogen and penetrate most solids quite well. Thus neutron scattering by hydrogen or hydrogen-containing molecules held on the surface of a solid provides, in principle, a good method for obtaining vibrational spectra of such adsorbates, particularly on high area solids, such as powdered metals, which are opaque to infrared. The surface location of the hydrogen-containing species responsible for the vibrational spectra obtained must,of course, be established, (just as in conventional IR studies). The technique has been discussed in recent reviews [17, 131].

Although attractive in principle, and invaluable in special cases, the technique requires high intensity neutron sources, not commonly available, and will probably never be a practical technique for routine characterization of catalyst surfaces.

C. Electron Vibrational Spectroscopy

As mentioned earlier in this chapter, electron energy loss spectroscopy (EELS) has proven invaluable in obtaining spectra of surface species on low area metal surfaces [17, 101]. EELS is analogous to Raman spectroscopy except that electrons, rather than photons, interact with a vibrating molecule and lose energy equivalent to one quantum of vibrational energy. The technique is highly sensitive, but the resolution obtainable is usually not high (50 to 80 cm^{-1}) and requirement for ultrahigh vacuum precludes study of surfaces under reaction conditions. The electron bombardment essential in such spectroscopy can also cause alteration in the oxidation state of surface atoms. Nevertheless, the fact that vibrational spectra can be obtained for adsorbates on single crystal metal surfaces over a broad frequency range makes the technique an extremely valuable one.

A second electron spectroscopic technique for surface studies which has received considerable publicity in recent years is inelastic electron tunneling spectroscopy (IETS) [132]. This is also a highly sensitive technique, but resolution can be increased to about 10 cm^{-1} when the sample is at 1 K. Monoenergetic electrons are obtained by quantum mechanical tunneling through a thin insulating gap between two metals near absolute zero. Vibrational transitions occurring in the insulating gap (~20 Å thick) cause excess current at applied voltages resonant with these vibrational transitions. The most commonly used "sandwich" is $Al—Al_2O_3—Pb$, and the adsorbate to be studied is on the oxide layer. Small particles of metal can be evaporated onto the oxide layer to act as chemisorption sites in a simulated alumina supported catalyst [133]. The technique can provide data useful in some cases in interpreting other types of vibrational spectra, but the fact that the adsorbate must be essentially sealed off in metal, and may be affected by the metal overlayer, restricts its utility. It is certainly not applicable to practical catalysts under realistic conditions.

6. Future Prospects for Infrared Spectroscopy in Catalytic Research

The literature on infrared studies of catalyst surfaces presently contains a substantial amount of chaff along with much wheat. Separating the two is not always easy. Given the complexities of the systems studied and poor understanding of many of the variables, this situation is not likely to change greatly in the near future. At present, through selective choice of references, one can support almost any position. Particularly disturbing is an evident trend toward presenting a few spectra of surface species, with minimum experimental detail, in conjunction with other data to support some particular hypothesis.

The euphoria common in the initial development of any new technique must at some point will be replaced by a more realistic view of problems and pitfalls if real progress is to continue. Although a need exists for more careful work and greater caution in interpretation of spectra than often seen, infrared spectroscopy still offers extremely important theoretical and practical advantages and should continue to yield important insights on catalyst surface properties.

Improved tools, particularly FTIR and computerized dispersive transmission IR spectroscopy, greatly enhance the information obtainable on high-area catalysts and should ultimately permit quantitative evaluation and comparison of important features of catalyst surface chemistry on a fairly routine basis. In many instances, even empirical information should prove practically important in catalyst development and evaluation, eliminating the need for much present testing in catalytic reactors. New tools, although often very useful, are just tools and must be properly used if something other than data proliferation is to result. Progress seems at present less limited by poor tools than by the inherent complexity of the subject. As better understanding of solid surface chemistry develops, through the synthesis of reliable information from various surface techniques, the reliability of interpretation of infrared spectra of surface species can also be expected to improve.

The goals of catalytic research are varied. Complete understanding of catalytic reaction mechanisms, including the nature of adsorbed intermediates, is, of course, highly desirable but hardly seems attainable in the present state of chemical knowledge. Catalysis is primarily an applied science, however, and as such should reasonably expect major assistance in reaching the goals of better catalysts and improved catalytic processes from better fundamental understanding of catalyst surface chemistry. This is an area in which infrared spectroscopy will undoubtedly make further major contributions in future years, given persistent effort and patient support.

References

1. Kiselev, A. V., Lygin, V. I.: Russian studies prior to 1956 involving the use of infrared spectroscopy in adsorption and surface chemistry. In: Infrared spectra of adsorbed species, Little, L. H. Academic Press, New York 1966, pp. 213–227
2. Mapes, J. E., Eischens, R. P.: J. Phys. Chem. **58**, 1059 (1954)
3. Eischens, R. P., Pliskin, W. A.: The infrared spectra of adsorbed molecules. In: Adv. Catal. **10**, 1 (1958)
4. Blyholder, G.: Spectra of adsorbed species. In: Experimental methods in catalytic research, (Anderson, R. B., ed.). Academic Press, New York 1968, pp. 323–360
   a. Blyholder, G., Neff, L. D.: J. Phys. Chem. **66**, 1464 (1962). (1962)
5. Pritchard, J., Catterick, T.: Infrared spectroscopy. In: Experimental methods in catalytic research, Vol. III, Characterization of surface and adsorbed species. (Anderson, R. B., and Dawson, P. T., eds.). Academic Press, New York 1976, pp. 281–318
6. Haller, G. L.: Infrared spectroscopy. In: Spectroscopy in heterogeneous catalysis. (Delgass, W. N., Haller, G. L., Kellerman, R., Lunsford, J. H.), Academic Press, New York 1979, pp. 19–57

7. Hair, M. L.: Transmission infrared spectroscopy for high surface area oxides. In: Vibrational spectroscopies for adsorbed species, (Bell, A. T., and Hair, M. L., eds.). Washington, D.C.: American Chemical Society 1980, pp. 1–11
8. Chesters, M. A., Sheppard, N.: Chem. Brit. **17**, 521–532 (1981)
9. Bell, A. T.: Applications of Fourier transform infrared spectroscopy to studies of adsorbed species. In: Vibrational spectroscopies for adsorbed species, (Bell, A. T., and Hair, M. L., eds.). Washington, D.C.: American Chemical Society 1980, pp. 13–35
10. Ward, J. W.: Infrared studies of zeolite surfaces and surface reactions. In: Zeolite chemistry and catalysis. Rabo, J. (ed.). Washington, D.C.: American Chemical Society 1976, pp. 118–284
11. Sheppard, N., Nguyen, T. T.: The vibrational spectra of carbon monoxide chemisorbed on the surfaces of metal catalysts — a suggested scheme of interpretation. In: Adv. infrared Raman spectros. **5**, 67–148 (1978)
12. Little, L. H.: Infrared spectra of adsorbed species. Academic Press, New York 1966
13. Hair, M. L.: Infrared spectroscopy in surface chemistry. New York: Marcel Dekker, Inc. 1967
14. Kiselev, A. V., Lygin, V. I.: Infrared spectra of surface compounds. John Wiley & Sons, New York-Toronto 1975 (Russian edition, 1972)
15. Willis, R. F. (ed.): Vibrational spectroscopy of adsorbates. Springer, Berlin Heidelberg New York, 1980
16. Bell, A. T., Hair, M. L. (eds.): Vibrational spectroscopies for adsorbed species. Washington, D. C.: American Chemical Society 1980
17. Haller, G. T., Catal. Rev.-Sci. Eng., **23**, 477–504 (1981)
18. Herzberg, G.: Molecular spectra and molecular structure. I. Diatomic molecules. II. Infrared and Raman spectra of polyatomic molecules. Van Nostrand, New York 1945 (new edition 1960)
19. Wilson, E. B., Decius, J. C., Cross, P. C.: Molecular vibrations. McGraw-Hill, New York 1955
20. Horák, M., Vitek, A.: Interpretation and processing of vibrational spectra. Wiley-Interscience, Prague 1978
21. Nakamoto, K.: Infrared spectra of inorganic and coordination compounds, 3rd edition. Wiley-Interscience, New York 1978
22. Bellamy, L. J.: Infrared spectra of complex molecules, 3rd edition. Vol. 1. Chapman and Hall, London 1975
23. Colthup, N. B., Daly, L. H., Wiberley, S. E.: Introduction to infrared and Raman spectroscopy, Academic Press, New York 1975
24. Potts, W. J., Jr.: Chemical infrared spectroscopy, Vol. I, techniques. New York: John Wiley & Sons 1963
25. Eischens, R. P.: Acc. Chem. Res. **5**, 74 (1972)
26. Tompkins, H. G.: Infrared reflection-absorption spectroscopy. In: Methods and phenomena: their application in science and technology. Vol. I — Methods of surface analysis, (Czanderna, A. W., ed.). Elsevier Scientific Publ. Co., New York 1975, pp. 447–472
27. Hollins, P., Pritchard, J.: Intermolecular interactions and the infrared reflection adsorption spectra of chemisorbed carbon monoxide on copper. In: vibrational spectroscopies for adsorbed species, (Bell, A. T., and Hair, M. L., eds.). Washington, D.C.: American Chemical Society 1980, pp. 51–74
28. Blyholder, G.: J. Chem. Phys. **36**, 2036 (1962)
29. Blyholder, G., Tanaka, M.: J. Phys. Chem. **76**, 3180 (1972)
30. Smith, A. W., Quets, J. M.: J. Catal. **4**, 163–171 (1965)
31. Harrod, J. F., Roberts, R. W., Rissman, E. F.: J. Phys. Chem. **71**, 343 (1967)
32. Parkyns, N. D., Patrick, J. B.: J. Sci. Instrum. **43**, 695 (1966)
33. Zecchina, A., Versino, C., Coluccia, S., Borello, E.: J. Chim. Phys. **67**, 1237 (1970)
34. Kember, D. R., Sheppard, N.: J. Chem. Soc. Faraday Trans. 2, **77**, 1309 (1981)
35. Amenomiya, Y.: Appl. Spectrosc. **37**, 487 (1978)
36. Peri, J. B.: Actes Cong. Int. Catal., 2nd 1960, Paris: Ed. Technip. 1961, p. 1333
37. Seanor, D. A., Amberg, C. H.: Rev. Sci. Instr. **34**, 917 (1963)
38. Bhasin, M. M., Curran, C., John, G. S.: Spectrochim. Acta **23A**, 455 (1967)
39. King, S. T.: Appl. Spectrosc. **34**, 632 (1980)

40. Moon, S. H., Windawi, H., Katzer, J. R.: Ind. Eng. Chem. Fundam. (1981) 396
41. King, D. L.: J. Catal. **61**, 77 (1980)
42. Edwards, J. F., Schrader, G. L.: Appl. Spectrosc. **35**, 559 (1981)
43. Hecker, W. C., Bell, A. T.: J. Catal. **71**, 216 (1981)
44. Penninger, J. M. L.: J. Catal. **56**, 287 (1979)
45. Tinker, H. B., Morris, D. E.: Rev. Sci. Instrum. **43**, 1024 (1972)
46. Eischens, R. P.: Science **146**, 486 (1964)
47. Delgass, W. N.: presented at spring symposium of the Chicago Catalysis Club, May, 1979
48. Jakobsen, R. J.: Application of FT-IR to surface studies. In: Fourier transform infrared spectroscopy, volume 2, (Ferraro, J. R., and Basile, L. J., eds.). Academic Press, New York, 1979, pp. 165–191
49. Yang, R. T., Low, M. D., Haller, G. L., and Fenn, J.: J. Colloid Interface Sci. **44**, 249 (1973)
50. Griffiths, P. R., Sloane, H. J., Hannah, R. W.: Appl. Spectrosc. **31**, 485 (1977)
51. Chenery, D. H., Sheppard, N.: Appl. Spectrosc. **32**, 79 (1978)
52. Peri, J. B.: Preprints-symposia, Miami Beach, Fla., Sept. 10–15, 1978: Division of Petroleum Chemistry, Inc., American Chemical Society **23**, 1281 (1978)
53. Hirschfeld, T.: Quantitative FT-IR, a detailed look at the problems involved. In: Fourier transform infrared spectroscopy, volume 2, (Ferraro, J. R., and Basile, L. J., eds.). Academic Press, New York 1979, pp. 193–242
54. Grasselli, J. G., Griffiths, P. R., Hannah, R. W.: Appl. Spectrosc. **36**, 87 (1982)
55. McDonald, R. S.: J. Am. Chem. Soc. **79**, 850 (1957)
56. Peri, J. B.: J. Phys. Chem. **70**, 2937 (1966)
57. Peri, J. B., Hannan, R. B.: J. Phys. Chem. **64**, 1526 (1960)
58. Peri, J. B.: J. Phys. Chem. **69**, 211 (1965)
59. Peri, J. B.: J. Phys. Chem. **69**, 220 (1965)
60. Borello, E., Della Gatta, G., Fubini, B., Morterra, C., Venturello, G.: J. Catal. **35**, 1 (1974)
61. Morterra, C., Ghiotti, G., Garrone, E., Boccuzzi, F.: J. Chem. Soc., Faraday Trans. 1, **72**, 2722 (1976)
62. Knözinger, H., Ratnasamy, P.: Catal. Rev.-Sci. Eng. **17**, 31 (1978)
63. Carter, J. L., Lucchesi, P. J., Yates, D. J. C.: J. Phys. Chem. **68**, 1385 (1964)
64. Peri, J. B.: The nature of ultrastable faujasite. In: Catal. Proc. Int. Congress Catalysis, 5th 1972, Vol. 1, (Hightower, ed.), North-Holland Publ. Co., Amsterdam 1973, pp. 329–341
65. Peri, J. B.: Discuss. Faraday Soc. **52**, 55 (1971)
66. Peri, J. B.: Discuss. Faraday Soc. **41**, 121 (1966)
67. Scherzer, J., Bass, J. L.: J. Catal. **46**, 100 (1977)
68. Carter, J. L., Yates, D. J. C., Lucchesi, P. J., Elliott, J. J., Kevorkian, V.: J. Phys. Chem. **70**, 1126 (1966)
69. Chang, C. C., Conner, W. C., Kokes, R. J.: J. Phys. Chem. **77**, 1957 (1973)
70. Parry, E. P.: J. Catal. **2**, 371 (1963)
71. Hughes, T. R., White, H. M., White, R. J.: J. Catal. **13**, 58 (1969)
72. Eberly, P. E.: J. Phys. Chem. **72**, 1042 (1968)
73. Primet, M., Basset, J. M., Mathieu, M. V., Prettre, M.: J. Catal. **29**, 213 (1973)
74. Dewing, J., Monks, G. T., Youll, B.: J. Catal. **44**, 226 (1976)
75. Taylor, J. H., Amberg, C. H.: Can. J. Chem. **39**, 535 (1961)
76. Gardner, R. A., Petrucchi, R. H.: J. Phys. Chem. **67**, 1376 (1963)
77. Angell, C. L.: J. Phys. Chem. **70**, 2420 (1966)
78. Peri, J. B.: J. Phys. Chem. **72**, 2917 (1968)
79. Borello, E., Zecchina, A. Morterra, C., Ghiotti, G.: J. Phys. Chem. **73**, 1286 (1969)
80. Zecchina, A., Garrone, E., Ghiotti, G., Coluccia, S.: J. Phys. Chem. **79**, 972 (1975)
81. Rebenstorf, B., Larson, R.: Acta Chem. Scand. A **34**, 239 (1980)
82. Peri, J. B.: J. Phys. Chem. **70**, 1482 (1966)
83. Peri, J. B.: J. Phys. Chem. **70**, 3168 (1966)
84. Peri, J. B.: Proc. Int. Congr. Catal., 3rd Vol. II, 1964, North-Holland, Amsterdam 1965, pp. 1100–1112
85. Angell, C. L.: J. Phys. Chem. **70**, 2420 (1966)
86. Peri, J. B.: J. Phys. Chem. **79**, 1582 (1975)

87. Zecchina, A., Garrone, E., Morterra, C., Coluccia, S.: J. Phys. Chem. **79**, 978 (1975)
88. Terenin, A. N., Roev, L. M.: Spectrochim. Acta **15**, 274, 946 (1959)
89. Peri, J. B.: J. Phys. Chem **78**, 588 (1974)
90. Chao, C. C., Lunsford, J. H.: J. Am. Chem. Soc. **93**, 6794 (1971)
91. Jermyn, J. W., Johnson, T. J., Vansart, E. T., Lunsford, J. H.: J. Phys. Chem. **77**, 2964 (1973)
92. Che M., Dutel, J. F., Primet, M.: Molecular sieves. Int. Conf. Molecular Sieves, Zeolites 3rd, Zurich 1973, p. 394
93. Peri, J. B.: J. Phys. Chem. **86**, 1615 (1982)
94. Millman, W. S., Hall, W. K.: J. Phys. Chem. **83**, 427 (1979)
95. Yang, A. C., Garland, C. W.: J. Phys. Chem. **61**, 1504 (1957)
96. Little, L. H., Sheppard, N., Yates, D. J. C.: Proc. R. Soc. **A259**, 242 (1960)
97. Ford, R. R.: Carbon monoxide adsorption on the transition metals. In: Adv. Catal. **21**, 51 (1970)
98. Eischens, R. P., Francis, S. A., Pliskin, W. A.: J. Phys. Chem. **60**, 194 (1956)
99. Peri, J. B.: J. Catal. **52**, 144 (1978)
100. Queau, R., Poilblanc, R.: C.R. Hebd. Seances Acad. Sci-C **274**, 337 (1972)
101. Garfunkel, E. L., Crowell, J. E., Somorjai, G. A.: J. Phys. Chem. **86**, 310 (1982)
102. Baddour, R. F., Modell, M., Goldsmith, R. L.: J. Phys. Chem. **74**, 1767 (1970)
103. Palazov, A., Chang, C. C., Kokes, R. J.: J. Catal. **36**, 338 (1975)
104. Bradshaw, A. M., Hoffmann, F.: Surf. Sci. **72**, 513 (1978)
105. Campuzano, J. C., Greenler, R. G.: Surf. Sci. **83**, 301 (1979)
106. Primet, M., Dalmon, J. A., Martin, G. A.: J. Catal. **46**, 25 (1977)
107. Martin, G. A., Primet, M., Dalmon, J. A.: J. Catal. **53**, 321 (1978)
108. Garland, C. W.: J. Phys. Chem. **63**, 1423 (1959)
109. Yates, J. T., Garland, C. W.: J. Phys. Chem. **65**, 617 (1961)
110. Moon, S. H., Onuferko, J. H., Windawi, H., Katzer, J. R.: J. Vac. Sci. Technol. **18**, 467 (1981)
111. Grill, C. M., McLaughlin, M. L., Stevenson, J. M., Gonzalez, R. D.: J. Catal. **69**, 454 (1981)
112. Hierl, R., Knözinger, H., Urbach, H.: J. Catal. **69**, 475 (1981)
113. Soma-Noto, Y., Sachtler, W. M. H.: J. Catal. **32**, 315 (1974)
114. Charcosset, H., Frèty, R., Leclercq, G., Mendès, E., Primet, M., Tournayan, L.: J. Catal. **56**, 468 (1979)
115. Haaland, D. M.: Surf. Sci. **111**, 555 (1981)
116. Chang, C. C., Conner, W. C., Kokes, R. J.: J. Phys. Chem. **77**, 1957 (1973)
117. Kokes, R. J.: Acc. Chem. Res. **6**, 226 (1973)
118. Dalla Betta, R. A., Shelef, M.: J. Catal. **48**, 111 (1977)
119. Ekerdt, J. G., Bell, A. T.: J. Catal. **58**, 170 (1979)
120. Unland, M. L.: J. Catal. **31**, 459 (1973)
121. Solymosi, F., Volgyesi, L., Sarkaw, J.: J. Catal. **54**, 336 (1978)
122. Hegedus, L. L., Chang, C. C., McEwen, D. J., Sloan, E. M.: Ind. Eng. Chem. Fundam. **19**, 367 (1980)
123. Primet, M., Fouilloux, P., Imelik, B.: J. Catal. **61**, 553 (1980)
124. Egerton, T. A., Hardin, A. H.: Catal. Rev.-Sci. Eng. **11**, 71–116 (1975)
125. Krasser, W., Fadini, A., Renouprez, A. J.: J. Catal. **62**, 94 (1980)
126. Heaviside, J., Hendra, P. J., Paul, S. O., Freeman, J. J., Friedman, R. M.: Appl. Spectrosc. **35**, 220 (1981)
127. Golden, W. G., Dunn, D. S., Overend, J.: J. Catal. **71**, 395 (1981)
128. Creighton, J. A.: Raman spectroscopy of adsorbates at metal surfaces. In: Vibrational Spectroscopy of Adsorbates, (Willis, R. F., ed.), Springer, Berlin Heidelberg New York 1980
129. Van Duyne, R. P.: In: Chemical and Biological Applications of Lasers, Vol. 4, (Moore, C. B., ed.). Academic Press, New York 1978, p. 101
130. Haller, G. L., Rice, R. W., Wan, C. Z.: Cat. Rev. — Sci. Eng. **13**, 259 (1976)
131. Wright, C. J.: Neutron scattering studies. In: Vibrational Spectroscopy of Adsorbates, (Willis, R. F., ed.). Springer, Berlin Heidelberg New York 1980, pp. 111–124
132. Wolfram, T. (ed.): Inelastic electron tunneling spectroscopy, Springer, New York 1978

133. Kraeker, R. M., Hansma, P. K.: J. Chem. Phys. **72**, 4845 (1980)
134. Angell, C. L.: Fourier transform infrared spectroscopy in the study of catalysts. In: Fourier transform spectroscopy, Vol. 3, (Ferraro, J. R., and Basile, L. J., eds.), Academic Press, New York 1982, pp. 1–36
135. Dalmon, J., Primet, M., Martin, G., Imelik, B.: Surf. Sci. **50**, 95 (1975)
136. Segawa, K., Hall, W. K.: J. Catal. **76**, 133 (1982)
137. Wieserman, L. F., Hercules, D. M.: Appl. Spectrosc. **36**, 361 (1982)

Chapter 4

X-Ray Techniques in Catalysis

Pierre Gallezot

Centre National de la Recherche Scientifique
Institut de Recherches sur la Catalyse
2, Avenue Albert Einstein
F-69626 Villeurbanne, Cedex
France

Contents

1. Introduction

X-ray diffraction (XRD) was developed during the first half of the century and was soon applied to identify the solid phases involved in heterogeneous catalysts. Identification was based on the Bragg law giving the lattice spacings which are the fingerprints of any crystalline phase. Soon, this powerful analytical tool was complemented by line broadening analysis (LBA) giving the average crystallite size in the directions perpendicular to the reflecting plane, thus providing a direct evaluation of the state of division of the solid phases. The development of small angle X-ray scattering (SAXS) brought another method to determine the particle size and the specific surface area of catalysts. Finally, the theory of X-ray scattering by amorphous matter made possible the determination of the interatomic distances between nearest neighbor atoms in materials which do not even exhibit any diffraction lines.

By 1950, the theories and techniques of X-ray diffraction were well advanced, and their future applications in the characterization of catalysts looked promising. Thus, in 1948, Jellinek and Fankuchen [1] concluded their review article on the application of X-ray diffraction to the study of catalysts: "It has proved a powerful tool over a wide range of application . . . X-ray studies will undoubtedly furnish some of the solution of many as yet unsolved problems". This was unfortunately the last detailed review on the subject in journals dedicated to catalysis. Of course, this apparent lack of interest did not stop routine X-ray diffraction analysis, but the inherent capabilities of the methods and their continuous improvements in theory, instrumentation, and data processing were used minimally during at least two decades. This is probably because much of the attention in the catalysis club was retained by the outburst of spectroscopic methods capable of probing the surface properties of catalysts. However, in the seventies and thereafter, efforts were made to derive more complete and accurate information on the texture and structure of the catalysts by X-ray methods. A number of factors can explain these new trends. First, there is a need for methods like line broadening analysis and small angle X-ray scattering that give a detailed picture of the dispersion of small metal particles on supports, including particle size distribution. Second, the development of zeolite catalysts has prompted many X-ray studies of these crystalline materials, since there is an opportunity to determine the positions of the active sites even in the course of catalytic reactions. Third, progress in the preparative methods of metal or molecular metal cluster supported catalysts have led to solid phases where most of the atoms are on the surface; thus X-ray methods like the radial electron distribution (RED) or the extended X-ray absorption fine structure (EXAFS) can be used to study the structure of surfaces. Fourth, with the growing need for *in situ* characterization of catalysts, X-ray methods are well suited, since measurements can be performed in the broadest range of temperatures and pressures used in catalytic reactions. Finally, instrumentation has been improved to save time in X-ray data recording and processing, thus narrowing the gap between the time scale of reaction rate measurement and that of diffracted intensities measurements.

The story of the application of X-ray absorption spectroscopy to the study of catalysts parallels in many respects that of X-ray diffraction. The discontinuities in the values of the X-ray absorption coefficient within a few electron volts on each side of the absorption edge were explained by Kossel [2] in 1920 in terms of electronic transition from inner levels to bound states. In 1931 Kronig [3] attributed the fine structure extending over several hundreds of electron volts on the high energy side of the edge to the interference between the photoelectrons ejected from the absorbing atom in the solid. X-ray absorption spectroscopy was never routinely applied as X-ray diffraction was, in spite of a review article by Van Nordstrand [4], who showed that valuable information can be obtained by X-ray absorption on the valence state as well as on the short range order of the atoms in a number of compounds. However, except in a few investigations [5, 6], the method was hardly applied to catalysts until the theory of the Kronig structure renamed EXAFS [7] was greatly improved [8], and radiation from electron storage rings became used as an intense source of X-rays. Thus, from the mid-seventies, X-ray absorption spectroscopy was applied as a new tool to study the local arrangement and the electronic structure of atoms in catalysts.

This chapter is intended to give a survey of the application of X-ray methods to the study of catalysts and catalysis. We will not enter deeply into the theory associated with each of these methods, because this would require too lengthy a development in a short space[1]. Only the most used final equations

[1] Although there are scores of books dedicated to X-ray diffraction, most of them deal with crystallography and crystal structure determination, which is of little help in the study of catalysts. Detailed treatments of the diffraction or scattering by imperfect crystals were given by Guinier [9], Warren [10], and Cowley [11]. Books by Klug and Alexander [12], and by Schwartz and Cohen [13] are less theoretical and develop more the experimental methods of X-ray diffraction applied to materials science. Small angle X-ray scattering was treated by Guinier and Fournet [14] and by Glatter and Kratky [15]. Azaroff [16] presented the state of the art in X-ray spectroscopy before the recent development of EXAFS, which is covered in several reviews [17–21].

will be given, stress being laid on their limit of application and on the significance of the parameters thus derived.

The relative advantages and shortcomings of methods will be critically examined. A complete review of the literature covering the study of catalysts by X-ray methods will not be given. Rather, a few investigations will be chosen to exemplify how far it is possible to go in the determination of texture and structure of catalysts. As far as EXAFS is concerned, it is premature to make definitive statements about the usefulness of the technique for catalyst characterization because the technique is in rapid evolution, and there are still few reports on its use. Therefore, stress will be laid on the comparison of EXAFS with the radial electron distribution methods, since both can be used to study the atomic structure of catalysts.

X-ray absorption edge spectroscopy is discussed minimally, and high resolution X-ray emission spectroscopy is not examined. They deserve to be discussed at full length in relation to the theories of electronic structure of solids and to be compared with the other spectroscopic methods used to study the electronic structure of matter. The X-ray emission spectroscopy for chemical analysis induced by electron bombardment (electron microprobe) or by X-rays (fluorescence) is not discussed either. These analytical tools are promising for the characterization of catalysts, since the composition of minute amounts of materials can be determined, especially when these tools are associated with electron microscopy.

2. Phase Identification of Heterogeneous Catalysts

Heterogeneous catalysts are composed of finely divided solid phases which can be studied only by the powder diffraction method. The use of this method for qualitative and quantitative analysis of crystalline phases is well known. A comprehensive account is given in the book by Klug and Alexander [12]. Therefore, the purpose of this section is to stress some of the difficulties which are encountered in phase identification of catalytic materials, because the X-ray reflections are often broad, weak, and blurred by the scattering of the crystal defects. We also want to advocate *in situ* phase identification, taking advantage of the progress in instrumentation.

A. Advances in X-ray Instrumentation

The production of X-rays still relies mostly on sealed X-ray tubes which have been progressively improved in brightness (power per unit area of the target). However, more powerful sources are available now. X-ray generators with a rotating anode provide a higher brightness of the focal spot resulting in a gain in intensity of about one order of magnitude with respect to the sealed tubes, and five additional orders of magnitude can be gained by using the synchrotron radiation emitted by electron storage rings. The availability of synchrotron radiation has been one of the factors in the development of X-ray absorption spectroscopy. One can expect a widespread use of these X-ray sources, especially for the study of poorly crystalline materials and crystal defects which require very long times for accumulation of data with conventional sources.

The other major improvement in instrumentation is in the field of X-ray detection. The scintillation counter is still the most used detector; however, it is progressively replaced either by solid state detectors (Si—Li, Ge—Li) or by position sensitive detectors. Solid state detectors have a much better photon energy resolution than the scintillation counter (100–200 eV compared to 4000 eV); the continuous X-ray spectrum and the K_{β} radiations can be completely eliminated with pulse height analysis but without a metal filter, with a resulting net gain in intensity. With the help of a multichannel analyser, it is possible to analyse the radiation emitted under X-ray excitation and therefore to determine the chemical composition. It is also possible to carry out energy dispersive diffraction: the counter set at a given angle detects the reflections of wavelengths in the continuous spectrum satisfying the Bragg law. This method of detection could be useful to perform *in situ* measurements under high pressure (*e.g.*, during a Fischer-Tropsch reaction) because only small entrance and exit windows are needed in the high pressure reactor. However, the resolution of lines and the measurement of intensities are not so accurate as with conventional, angular dispersive diffraction. It should be used for quick identification, *e.g.*, during *in situ* study of phase transformation.

The position sensitive detectors are gas filled counters with a linear or curved wire collector. An electronic circuit measures the relative time for a pulse to reach both ends of the wire, and thus the position of the photon impact along the wire. The counter at a fixed position can record a typical angular 2θ range of 10°, so that several lines are recorded simultaneously, just as on a photographic plate but at a much higher speed and with the digital output of the line profile. It can also be continuously moved with high angular velocity along the focalisation circle of the diffractometer. A computer is then needed to care for the correct addressing of the diffracted angles and intensities in a multi-channel memory. Any deviation from linearity at a given point of the wire is eliminated by averaging due to continuous scanning. Gain in recording time can be as high as 50-fold relative to conventional scanning with a scintillation counter.

More powerful X-ray sources and faster detecting systems are valuable improvements as far as study of the heterogeneous catalyst is concerned. They reduce the gap between the period of time needed for recording the X-ray pattern and the period of time during which catalyst modifications occur. This opens new prospects for *in situ* characterization. Unfortunately, little progress has been made in commercial attachments for temperature and atmosphere control. The performance of these devices is more oriented toward high temperatures than towards well-defined atmosphere control needed to obtain clean surfaces and unpolluted reagents. The Guinier-Lenne focussing camera (Nonius) with film displacement coupled with temperature control is convenient for *in situ* study of catalytic materials. It has been used in the study of palladium location and dispersion in Y-zeolite during *in situ* treatment and catalytic hydrogenation reaction [22, 23].

B. Sensitivity in Phase Detection

A limitation of X-ray diffraction methods is the lack of sensitivity in detecting a low amount of a given phase in a mixture. This limitation is even more severe in the case of heterogeneous catalysts. The X-ray reflections on highly divided and poorly crystalline phases are difficult to detect because they are weak, broad, and blurred by the scattering due to the various defects present in these phases. Therefore, a given phase might be greatly underestimated or even overlooked: then it is said to be "X-ray amorphous". This terminology bears no important consequence if it is meant that the phase is actually not detected, but it can be misleading in many instances. Thus, the absence of metal diffraction lines on the X-ray pattern of a supported metal catalyst does not mean that the metal is amorphous or atomically dispersed. The limit of detection of a metal phase depends upon several factors, the weight fraction, the square of the atomic number, the size of the particles, and the intensity of the continuous scattering from the

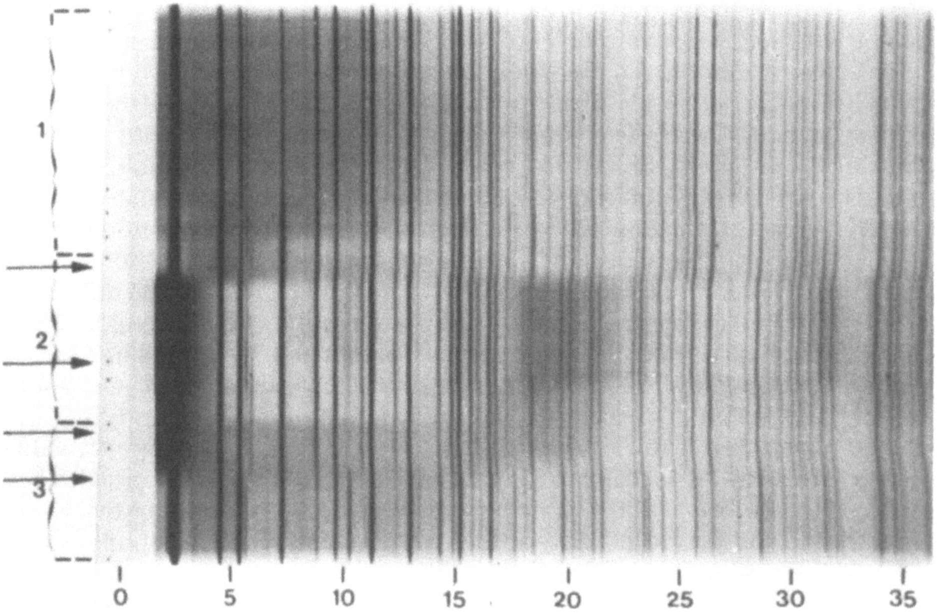

Figure 1. X-ray pattern of a PdHY zeolite taken with a Guinier-Lenné camera (CuKα) under controlled temperature and atmosphere while the film cassette is moving upwards. Step 1, heating from 290 to 400 K under O_2; step 2, heating from 290 to 400 K under H_2; step 3, heating from 290 to 780 K under O_2. (Reproduced with permission from ref. [22])

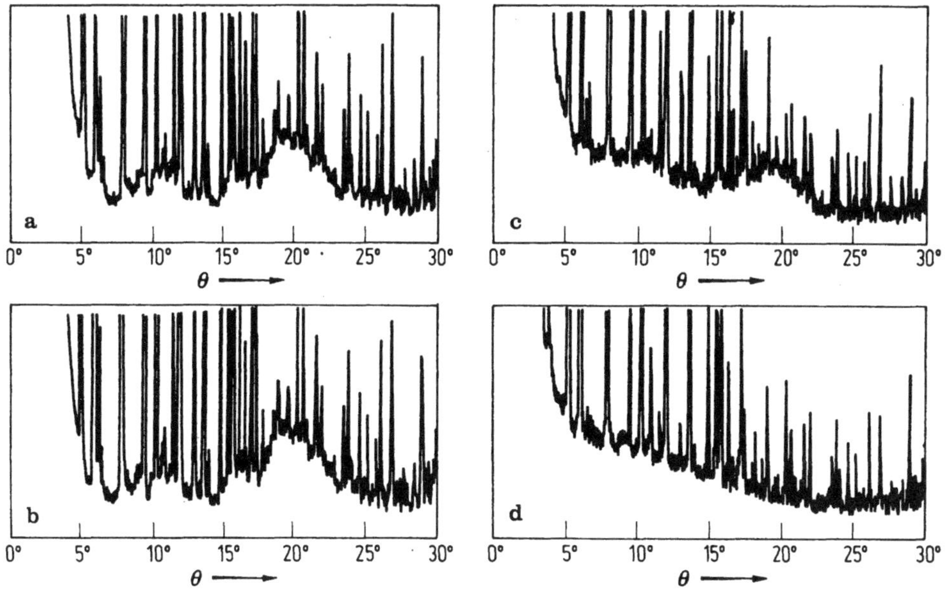

Figure 2. Densitometer recording of the X-ray pattern of Figure 1: **a**, trace at 290 K under H_2; **b** trace at 400 K under H_2; **c** trace at 290 K under O_2; **d** trace at 470 K under O_2. (Reproduced with permission from ref. [22])

support. It is therefore difficult to define in a general way a limit of concentration or a limit of size under which the phase can no longer be detected. Furthermore, these limits depend upon the experimental arrangement; thus, diffraction lines undetected on a conventional pattern taken with a filtered beam could well be detected with a focalisation camera or with a diffractometer equipped with a monochromator which gives a far better peak-to-background ratio. The following examples show that very small metal particles can be detected. In 1–2 wt.% Pt/SiO_2 catalysts studied by Sashital *et al.* [24], platinum particles have not only been detected but their size distribution has been established by line broadening analysis (Section 3.A.3). Figures 1 and 2 show the X-ray pattern of a Pd/NaHY zeolite (10 wt.% Pd) taken with a Guinier-Lenne camera during 290–400 K heating under O_2, 290–400 K reduction under H_2, and 290–780 K reoxidation under O_2 [Bergeret *et al.*, [22]]. The reduction leads to a broad diffraction band extending from $\theta = 15°$ to $\theta = 25°$. It corresponds to the unresolved 111 and 200 reflections of PdH. At the same time the small angle scattering is spreading as far as $\theta = 5°$. It was shown by electron microscopy that the particles are smaller than 1.5 nm; thus, particles that small can still be detected by X-ray diffraction provided the metal concentration is sufficient. The reoxidation of the particles into cations at 470 K was evidenced by the disappearance of the X-ray diffraction band and by the appearance of Pd^{2+} cations in SI′ sites (*c.f.* section 4.A.2b).

C. Identification of Solid Solution

Solid solution of two components in oxides or metals can be detected by the modification of lattice parameters, provided the atoms have a different atomic radius. Since the parameter change is usually linear (Vegard's law), the composition of the solid solution can be established. Figure 3, extracted from the work of Sinfelt and Via [25], shows that the formation of Pt—Ir solid solution is clearly seen in Pt—Ir/SiO_2 catalysts containing 2.5 wt.% of each component; however, it was not possible to characterize Pr—Ir alloys with 1 wt.% of each component because of the lower concentration and larger line width. Inhomogeneities in the alloy compositions lead to a line broadening which is superimposed on the particle size broadening; however, in many cases, the broadening is then asymmetric so that inhomogeneities can be readily detected.

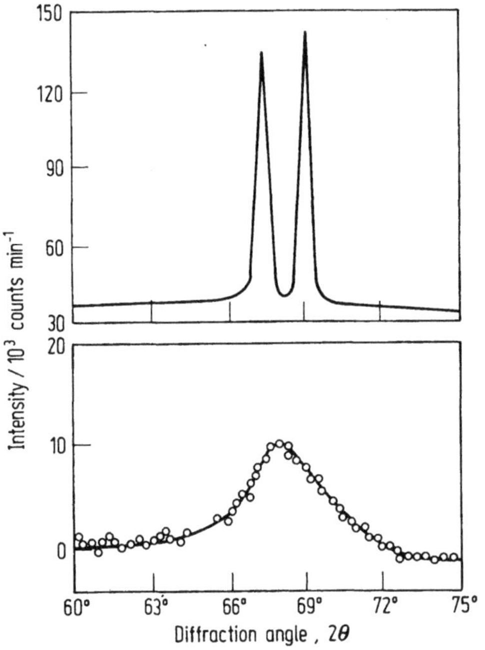

Figure 3. X-ray diffraction pattern (220 reflection) of Pt—Ir bimetallic catalyst supported on SiO_2. Upper field is a reference pattern for a physical mixture of platinum and iridium on SiO_2. Lower field is the X-ray pattern of the catalyst containing 2.5 wt.% each of the component (particles size 2.7 nm measured from line width). (Reproduced with permission from ref. [25])

D. Identification of Disordered Structures

The analysis of X-ray diffraction by various kinds of disordered structures can be found in the books of Guinier [9] or Warren [10]. Substitution and displacement disorders are difficult to analyse in complex systems such as catalysts. Strains and twin faulting in close packed metals can be assessed by line profile analysis. Stacking faults in layer structures are readily recognized from their X-ray patterns. Let c be the axis perpendicular to the layer. Two kinds of defects can be observed:

(1) The spacing d_{001} increases as a result of the intercalation of atoms or molecules between the layers (swelling of clays, intercalation compounds of graphitic carbons), and the d_{001} reflections can be broadened because of a distribution of the d_{001} spacings.
(2) The layers undergo random displacements in their plane (clays, carbons, MoS_2). Then the *hk*0 reflections are asymmetric, the reflection intensity decreasing very slowly with increasing Bragg angle, and no *hkl* lines appear.

From the theory developed by Warren [26], the mean dimension *D* in the plane of the layers can be determined from the displacement of the maximum or from the width at half height of the *hk*0 reflections. Application can be found in the case of MoS_2 catalysts [27].

E. *In Situ* Phase Identification

There are still very few *in situ* X-ray studies of catalysts, although this type of investigation is very rewarding compared to the study of fresh or used catalyst in air. Rashkin and Pierron [28] have studied bismuth molybdate in the course of propene oxidation and have shown that the composition of the crystalline phase is dependent upon temperature and atmosphere

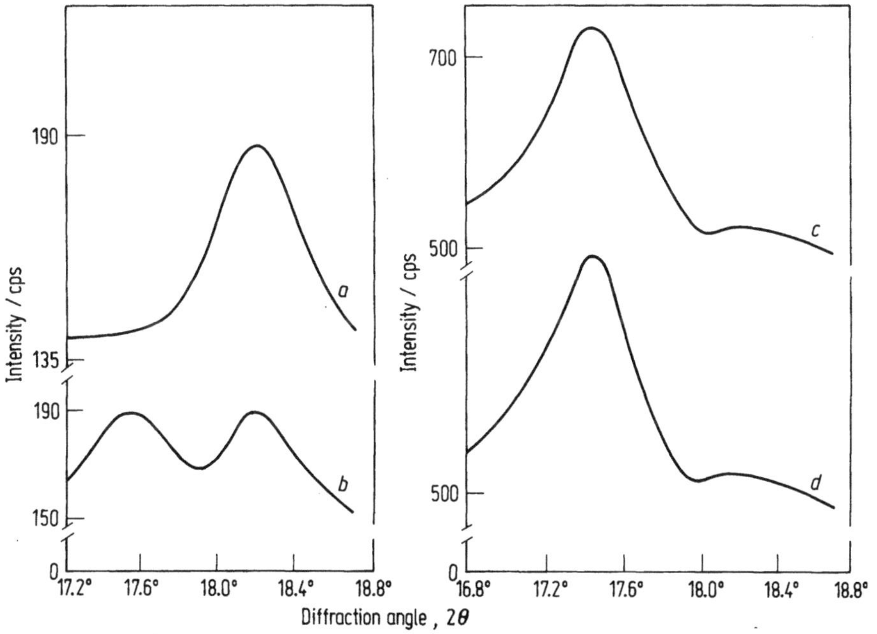

Figure 4. X-ray diffraction pattern (MoKα, 111 reflection) of 2 wt.% Pd/SiO_2 catalyst taken under controlled temperature and atmosphere in the course of methylcyclopropane hydrogenation. Left field: *a* 111 Pd reflection after cooling from 723 to 300 K under He; *b* 111 PdH and Pd reflection during hydrogenation at 300 K. Right field: *c* 111 PdH reflection after cooling from 723 to 300 K under H_2; *d* 111 PdH reflection in the course of reaction. (Reproduced with permission from ref. [29])

conditions. Thus, under reaction conditions, a phase appeared which cannot be detected in air. The study of Pd/SiO_2 in the course of methylcyclopropane hydrogenolysis is a good example of phase identification during catalytic reaction [Nandi *et al.* [29]]. Figure 4 shows the 111 reflection of Pd and PdH under various conditions. When the reaction is carried out on Pd (reduced at 723 K and cooled under He), half of the Pd is converted into PdH (Figures 4a and 4b). On the other hand, if the reaction is carried out on PdH (cooling under H_2), the composition is not altered by the reaction mixture (Figures 4c and 4d). The reaction rate was found slower in this latter case.

3. Determination of the Texture of Heterogeneous Catalysts

Determination of the texture of catalysts, *i.e.*, of the size, shape, porosity, and specific surface area of particles, is a requisite step in understanding the catalytic properties of solids. In this section, two very general X-ray methods to study the texture of catalysts are presented: LBA and SAXS. Information on particle size and shape can also be obtained from the RED method and from the EXAFS. However, these last two methods are primarily used to determine the structure of catalysts. Their applications to texture will be mentioned in sections 4.B and 4.C.

A. Line Broadening Analysis

The width of an X-ray reflection on the lattice planes of crystallites (small single crystals) depends essentially upon instrumental factors when the crystallites are large enough (*e.g.*, >500 nm) and perfect. If the crystal is small and not perfect, the reflection undergoes broadening due to the limited number of reflecting planes stacked together (crystallite size broadening), or to the perturbation of the interplanar spacings (strain or deformation broadening). The purpose of LBA is to derive crystallite sizes after correction for instrumental broadening and separation of the crystallite size broadening from that due to crystal defects. Line broadening analysis can be carried out, either with an elementary analysis of the line width from which a mean crystallite size is derived (assuming that the broadening is due to crystallite size only), or with a complete line profile analysis involving a rigorous correction for instrumental broadening, the separation of crystallite size broadening, and the calculation of particle size distribution.

1. Elementary LBA

Let us consider an X-ray reflection on N lattice planes of spacing d. The thickness of the crystallite $L = Nd$ in the direction perpendicular to the reflecting planes can be obtained from the width $\Delta(2\theta)$ of the reflection corrected for instrumental broadening with the Scherrer equation [30]

$$L = K\lambda/\Delta(2\theta)\cos\theta_0 \qquad (1)$$

where λ is the wavelength, and θ_0 is the Bragg angle at line maximum. The constant K depends upon several factors, including the way $\Delta(2\theta)$ is measured and the shape of crystallites. Instead of measuring the width of the diffraction line at half maximum intensity, it is highly recommended to take the integral line width (total area under the line profile divided by the line intensity at maximum). Indeed, in the general case of non-identical crystallites (*i.e.*, in all practical cases), the size L obtained from the integral line width is a volume weighted average size [9]

$$L = \bar{D}_V = 1/V \int D \, \mathrm{d}v_D$$

where $\mathrm{d}v_D/V$ is the fraction of the volume of the powder for which the diameter normal to the reflecting plane lies between D and $D + \mathrm{d}D$. On the other hand, L measured from the width at half maximum has no clear physical meaning. Moreover, the smaller crystallites contribute mainly to the wings of the line profile; they are therefore almost neglected in $\Delta(2\theta)$ measured at half maximum, and L could be greatly overestimated.

The value of constant K (also called shape factor) could effectively be adjusted for different *hkl* reflections and crystallite shapes in the case of crystallites identical in shape and dimension [12]. However, this is of no practical use as far as catalysts are concerned, and K should be taken as unity in view of all the approximations involved in elementary LBA analysis.

The $\Delta(2\theta)$ width in equation [1] has been corrected for instrumental broadening. This is achieved by measuring the line width b of a standard sample giving a line near θ_0 and containing crystallites large enough and not subject to lattice distortion. If B is the uncorrected line width, a good approximate corection is that given by Warren [31]

$$[\Delta(2\theta)]^2 = B^2 - b^2$$

The measure of B and b should be carried out with a monochromatic radiation or the separation of the $K\alpha_1 - K\alpha_2$ doublet should be corrected, *e.g.*, by the Rachinger correction [32].

The $\Delta(2\theta)$ width is supposed to be due to crystallite size only. To check that crystal defects do not contribute to line broadening, a simple test can be performed. Suppose that a cubic lattice is compressed or stretched by internal stresses. Then, the lattice constant a deviates by Δa from its equilibrium position value. By differentiating the Bragg law, the broadening due to these defects is

$$\Delta(2\theta) = 2 \tan \theta_0 (\Delta a/a)$$

On the other hand, the broadening due to particle size is

$$\Delta(2\theta) = \lambda/L \cos \theta$$

Since the dependencies of $\Delta(2\theta)$ on the Bragg angle or on the wavelength are not the same, measurements of line width at different angles or with different wavelengths can in principle be used to distinguish the cause of line broadening. These criteria are, however, of little help in practice, because the errors in $\Delta(2\theta)$ are large with respect to the difference expected from the above equations. The complete line profile analysis is required to detect the effect of distortion (strain) broadening and to extract its contribution.

To sum up, the Scherrer equation applied to properly corrected integral line width can give a good estimate of the mean crystallite sizes in the direction perpendicular to the reflecting planes when the particles have similar habit and a narrow size distribution. It is well suited to compare the relative size of crystallites in a series of samples. In the case of a narrow size distribution, information on particle shape can be obtained by measuring the size perpendicular to the different reflecting planes.

2. *Line Profile Analysis*

a) Correction for instrumental broadening

The function $h(x)$, describing the experimental line profile, is given by the convolution of the function $f(y)$, representing the pure line profile, and of the function $g(z)$, representing the broadening due to the experimental conditions. To obtain $f(y)$ from the measurements of $h(x)$, the profile of the actual sample, and $g(z)$, the profile of a standard sample containing strain-free, large crystallites, the function, $h(x) = \int_{-\infty}^{+\infty} f(x - z)\, g(z)\, \mathrm{d}z$, must be deconvoluted. This can be done by the Fourier transform method (Stokes method) or by numerical analysis methods.

The Stokes method [33] is based on the Fourier transform of $h(x)$ and $g(z)$ which gives the Fourier coefficients of these line profiles. Then the Fourier coefficients A_L (sine and cosine) of the corrected line profile are given by the ratio of the Fourier coefficients of the $h(x)$ and $g(z)$ profiles. The Stokes correction is calculated from the diffracted intensities measured by step scanning over the entire line profiles of the sample and of a standard. The wavelength does not need to be monochromatic since peak asymmetry due to $K_{\alpha_1} - K_{\alpha_2}$ splitting is corrected. However, the correction method is reliable only if the profile of the standard sample is narrow compared to the profile under investigation.

The corrected line profile is in some instances perturbed by spurious oscillations which are due to end-of-series effects because of the truncation of the data at both ends of the experimental profile. These oscillations occur mainly in the wings of the corrected profile. This is a severe drawback because the wings carry information on the smallest crystallites and the size distribution can be deeply perturbed. These oscillations can be minimized by smoothing the data and by extending the tail of the experimental profile with an asymptotic law [34, 35].

Independently from the Stokes method, the convolution equation can be solved by analytical methods such as those described by Louër *et al.* [36] and Moraweck *et al.* [37]. These methods require some parameter adjustments and more computing time, but are very effective to deconvolute the diffraction data without getting oscillations in the final line profile.

b) Separation between size and distortion broadening

Let A_L be the Fourier cosine coefficients of the corrected line profile. They can be decomposed in two terms, A_L^S a size coefficient and A_L^D a distortion coefficient such that $A_L = A_L^S A_L^D$ [Warren [10]]. The size and distortion component can be separated by the Warren and Averbach method [38], using at least two lines corresponding to successive reflection orders. This method allows the determination of A_L^s and of the mean square strain. However, before applying this method, the absence of distortion can be readily checked by plotting the normalized A_L coefficient for two reflection orders. There is no significant strain if they are equal [24]. In many cases for which it is difficult to measure accurately more than one order of reflection, Ganesan *et al.* [39] have described a method based on the work of Gangulee [40] and Mignot and Rondot [41] for determining the coefficients A_L^S and the mean square strain from the analysis of a single line profile.

c) Derivation of size parameters

Crystallite size parameters can be determined from the Fourier coefficients of the line profile corrected for instrumental broadening and for distortion broadening [Bertaut [42], Warren and Averbach [38]].

A mean surface-weighted diameter $\bar{D}_S$ is obtained from the first derivative of A_L taken at the origin

$$\bar{D}_S = -1/[\mathrm{d}A_L/\mathrm{d}L]_{L\to 0} \tag{2}$$

In practice $\bar{D}_S$ is obtained from the plot of A_L by extrapolating the initial slope on the L axis (*c.f.* Figure 5). The significance of $\bar{D}_S$ has been discussed by Smith [43].

A volume-weighted size distribution $P_V(D)$ can be obtained from the second derivative of the Fourier coefficients

$$P_V(D) = D\left[\frac{\mathrm{d}^2 A_L}{\mathrm{d}L^2}\right]_{L=D} \tag{3}$$

By definition, $P_V(D)\,\mathrm{d}D$ represents the volume fraction of the crystallites with diameter perpendicular to the reflecting planes between D and $D + \mathrm{d}D$. Note that $P_V(D)$ applies to all the diameters whether they are in the same crystallite or not.

On the basis of the crystallite size distribution, a volume-averaged mean diameter $\bar{D}_V$ can be calculated

$$\bar{D}_V = \int D P_V(D)\,\mathrm{d}D \tag{4}$$

This mean diameter is equivalent to that obtained from the Scherrer equation (equation 1) using the integral width of the line.

The size parameters given by equations (2)–(4) are obtained from the Fourier coefficients of the line profile corrected by the Stokes method. A quite different approach consists in calculating a size distribution by solving the integral describing the line profile [Bertaut [42]]. This has been performed with an original analytical method [Moraweck *et al.* [44]].

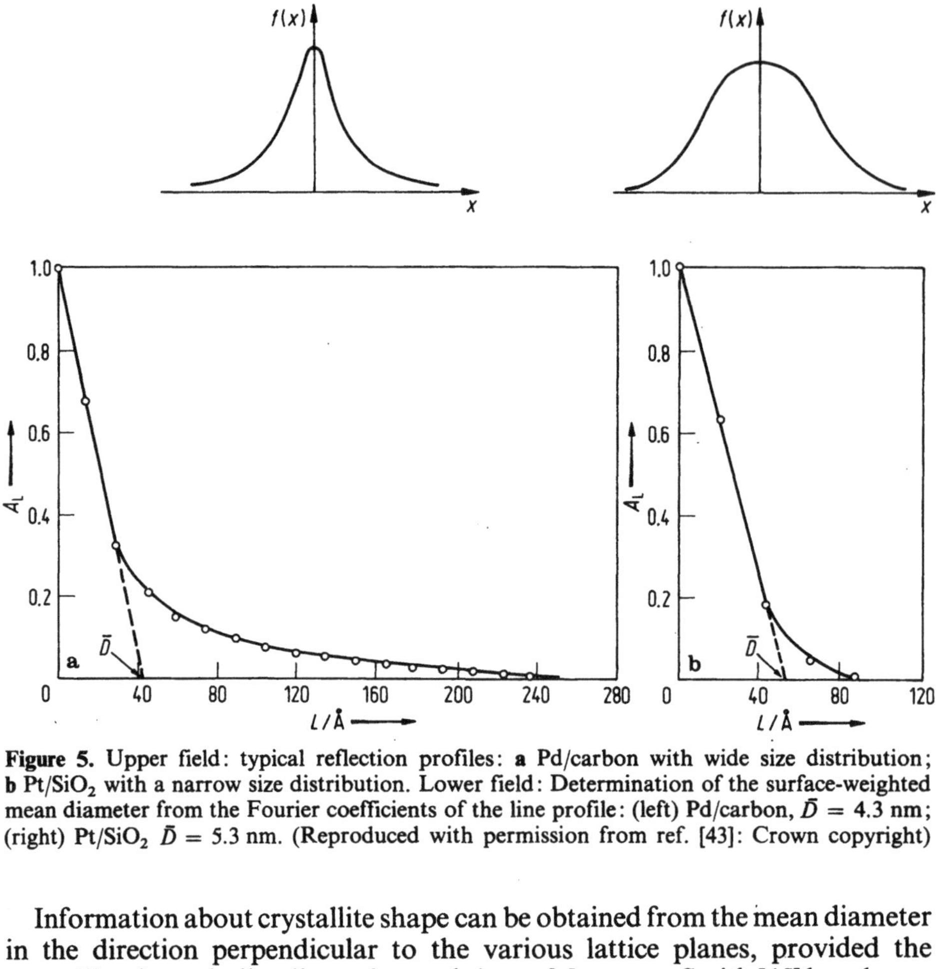

Figure 5. Upper field: typical reflection profiles: **a** Pd/carbon with wide size distribution; **b** Pt/SiO_2 with a narrow size distribution. Lower field: Determination of the surface-weighted mean diameter from the Fourier coefficients of the line profile: (left) Pd/carbon, $\bar{D}$ = 4.3 nm; (right) Pt/SiO_2 $\bar{D}$ = 5.3 nm. (Reproduced with permission from ref. [43]: Crown copyright)

Information about crystallite shape can be obtained from the mean diameter in the direction perpendicular to the various lattice planes, provided the crystallites have similar dimension and shape. Moreover, Smith [45] has shown that information on crystallite shapes can be obtained from the size distributions calculated for different reflections, provided the crystallites have the same shape.

3. Application of LBA to catalysts

The applicability of LBA to catalysts is often restricted for a number of reasons. The reflections from the different solid phases might be superimposed, for instance, in Pt/Al_2O_3 catalysts. Neighboring reflections such as the 111 and 200 lines in f.c.c. metals are overlapping when the crystallites are very small. Then the LBA should be carried out with data recorded on one half of the profile. The other severe limitation comes from the lack of contrast between the intensity of the reflection and that of the overall scattering as discussed in section 2.B. Since several factors are involved, it is not possible to state a definite limit of application of the method. Furthermore, as the amount of the phase under study decreases, the wings of the line profile become too weak to be measured and the smaller particles are not taken into account. So before the method ceases to be applicable, there is a range of concentrations where LBA is less and less accurate. In this range, the mean size is overestimated and the size distribution is displaced to larger size. However, the following examples show that the limit of size which can be measured is much smaller than that usually assumed in many papers (*e.g.* 4 nm). There is also an upper limit of size beyond which crystallite size can no longer be measured because the broadening due to crystallite size becomes small with respect to the instrumental broadening. The limit depends upon the particular experimental setup and upon the error tolerated. As a rule crystallites larger than 100 nm should not be measured by LBA.

A very complete study of 1.1–1.9 wt.% Pt/SiO_2 catalysts has been achieved by Sashital *et al.* [24]. In spite of the low metal concentration, they have derived a complete set of parameters from the wide angle pattern, *viz.*, crystallite size distributions in various directions, mean diameters, mean square strains, root mean square amplitudes of Pt atom vibrations, and changes in lattice parameters. Table 1 shows that there is an excellent agreement between the dispersion determined by chemisorption and from the surface-weighted mean diameters indicating that crystallites and particles are the same entities. In the last three samples, the crystallites are near spherical, whereas in the first, they are bounded by specific planes. This was deduced from the fact that the pattern of sizes in the different directions is similar to that calculated from model crystallites with specific boundary planes. There are two other interesting results in this investigation. There are considerable strains in the non-spherical crystallites of sample 7.1 SiO_2—PtCl—S. It was suggested that these distortions could be due to the differential contractions of the Pt crystallites and of the SiO_2 gel because the crystallites and the pores have similar dimensions. Besides, it was shown that the root mean square amplitudes of vibration of the platinum atoms increased with particle dispersion, *i.e.*, with the number of surface atoms. This is in agreement with the EXAFS results reported by Via *et al.* [46] (*c.f.* section 4.C.). A correlation between these atom displacements and the catalytic activity of Pt/SiO_2 in hydrogenation of cyclopropane, methylcyclopropane, and propene has been reported [47].

Table 1. Pt Crystallite Sizes and dispersion in Pt/SiO_2 Catalysts. (Ref. [24])

Catalysts (Pt wt.%)		Size[a]/nm, perpendicular to (hkl) planes				Dispersion/% atom exposed	
		(111)	(100)	(110)	(311)	LBA	Chemisorption
7.1-SiO_2—PtCl—S	(1.9)	13.1	8.2	10.8	8.1	10	7.1
21.5-SiO_2—IonX—L	(1.5)	4.9	4.3	4.5	4.7	25	21.5
27-SiO_2—IonX—S	(1.5)	4.3	3.9	4.2	3.1	27	27.3
40-SiO_2—PtCl—S	(1.9)	3.0	2.6	2.2	2.0	44	39.8

[a] Surface weighted mean diameter obtained from the first derivative of the Fourier coefficients.

Table 2. Mean Crystallite size/nm from LBA. (Ref. [43])

Catalyst	$\bar{D}_S$ [a]	D_I [b]	$D_{1/2}$ [c]
Pd/Carbon	4.3	7.2	13.0
Pt/Silica	5.3	6.0	5.9

[a] Surface weighted mean diameter derived from line profile analysis.
[b] Mean diameter calculated with the Scherrer equation from integral line width.
[c] As b, from line width at half height.

The study of Pt/SiO_2 and Pd/C catalysts by Smith [43] is a good illustration of LBA on typical line profiles. The line profile of Pt/SiO_2 is broad and flat (Figure 5b); it corresponds to a narrow size distribution, while that of Pd/C is tall with extended wings and corresponds to a large size distribution. Figure 5 shows how the surface-weighted mean diameter $\bar{D}_S$ is determined from the Fourier coefficient of the line profile. Table 2 gives the value of $\bar{D}_S$ and its comparison to D_I and $D_{1/2}$ determined with the Scherrer equation from the integral line width and from the line width at half maximum respectively. It is conspicuous that the three mean diameters compare well when there is a

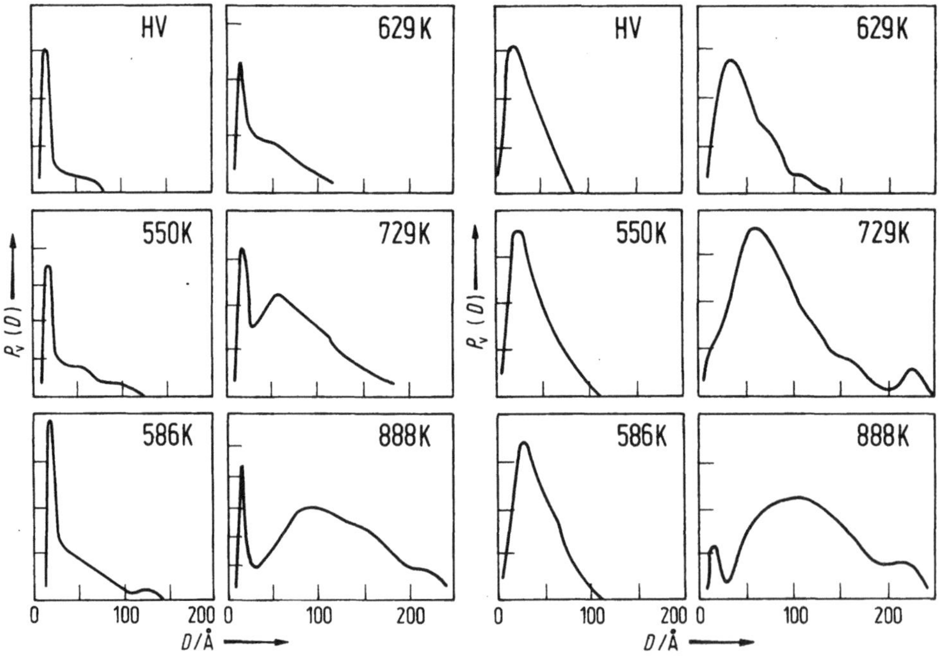

Figure 6. Crystallite diameters distribution, $P_v(D)$ from line broadening analysis obtained from the second derivative of the Fourier coefficients. Left field: distribution in the direction perpendicular to the (111) planes of NiO; right field: distribution perpendicular to the (200) planes. Temperatures of sintering in air are given on curves. (Reproduced with permission from ref. [48])

narrow size distribution, but that the Scherrer equation, especially with the line width at half maximum applied to profile of type (a), greatly overestimates the mean size because the smaller crystallites contributing to the wings of the profile are not taken into account.

An interesting LBA study of NiO sintering has been carried out by Moraweck *et al.* [48]. The NiO obtained by thermal decomposition of $Ni(OH)_2$ under vacuum is formed in thin sheets about 3 nm thick and 100 nm long with the (111) plane parallel to the sheet as shown by electron microscopy and diffraction. Values of size distribution along the [111] and [200] directions and at different temperatures of sintering are given in Figure 6. In the [111] direction, the maximum of $P_v(D)$ at 2 nm corresponds to the thickness of NiO sheets. It is noteworthy that the sintering is very difficult in this direction; the maximum is not displaced, and at 888 K a number of sheets have not coalesced. The distribution in the [200] direction (Figure 6) or in the [220] direction indicates that the large sheets detected by electron microscopy are formed by much smaller crystallites. The maximum of the distribution is shifted towards large size, indicating that crystallite sintering occurs within a given sheet much more readily than the coalescence of sheets in the [111] direction. Another detailed study of NiO sintering has been reported by Ganesan *et al.* [39]. Various mean size parameters, including those determined by an original single profile analysis technique, were compared. The mechanism of NiO sintering was inferred from the change in particle size distribution.

Hosemann and Bagchi [49] have developed a theory of diffraction by crystallites perturbed by liquid-like distortions which were named paracrystals. The paracrystalline distortions of a family of lattice planes hkl are evaluated with the distortion parameter g_{hkl} given by

$$g_{hkl} = (\overline{d^2_{hkl}}/\bar{d}^2_{hkl} - 1)^{1/2} ,$$

where $\bar{d}_{hkl}$ is the average of the interplanar spacings. The fundamental postulate of the theory [49, 50] is that g_{hkl} is connected with the average size $\bar{L}_{hkl}$ of the paracrystal by the relation

$$\alpha^*_{hkl} = g_{hkl}(N_{max})^{1/2} = g_{hkl}(\bar{L}_{hkl}/\bar{d}_{hkl})^{1/2}$$

where N_{max} is the maximum number of planes stacked, and α^*_{hkl} is a constant with value in the range 0.1–0.2. This relation means that paracrystals cannot grow beyond a limit determined by the distortion parameter g which depends upon the concentration of defects in the lattice.

The line broadening due to the size of paracrystal and to the distortions has been analysed theoretically, and a graphical method has been used to extract the L_{hkl} and g_{hkl} values from the integral widths of at least three reflection orders.

The theory was applied to Al_2O_3-promoted iron catalysts [50, 52]. It was concluded that the promotor induces paracrystallinity in the α-Fe lattice. The paracrystal distortions are due to Al_2FeO_4 groups replacing seven Fe atoms in the lattice. These endotactic groups are distributed at random. They do not change the average lattice spacings, but produce a lattice distortion inducing paracrystallinity. The high stability of the surface of the catalyst was attributed to the paracrystallinity. According to the law given before, the distorted lattice cannot grow beyond a size limit L. However, when the iron promoted catalysts are treated in H_2 above 770 K, the distortion parameter g decreases and becomes zero at 1220 K. The disappearance of paracrystallinity was attributed to a decrease in the number of endotactic groups, either because of their clustering inside the iron lattice or because of their migration to the surface.

B. Small Angle X-ray Scattering

1. Derivation of Parameters from SAXS

Just as a beam of visible light passing through mist is scattered by droplets, a beam of X-rays can be scattered by small particles. In both cases the maximum angle of scattering is given by the ratio λ/D, where λ is the radiation wavelength and D is the particle diameter. It is clear that the central peak of scattered intensity gets broader as the particle size decreases. With wavelengths of the order of 0.1 nm, the usual range of sizes which can be measured is 1–100 nm. A complete treatment of the SAXS theory can be found in the book of Guinier and Fournet [14]. Most of the later improvements in theory and applications are given in the book of Glatter and Kratky [15]. Only the final expressions giving the parameters of interest are given along with their significance and domain of applicability. The form of the equation depends upon the particular geometry of SAXS recording, either by point collimation (pinhole collimator, double curved crystal monochromator) or by linear collimation (slits, single curved crystal monochromator). The equations are given for point and linear collimation as a function of the scattering vector

$$s = 2 \sin \theta/\lambda \simeq 2\theta/\lambda$$

a) Mean size parameters and specific surface area

SAXS parameters are always derived from an analysis of the profile of the SAXS curve. In the simplest case where the particles are identical, in random orientation, and far from each other, Guinier [9, 14] has shown that the SAXS curve $I(s)$ can be represented by an exponential approximation

$$I(s) = (\Delta\varrho)^2 \, V^2 \exp(-4\pi^2 s^2 R_G^2/3) \tag{5}$$

where $\Delta\varrho$ is the difference in electron density between the particles and the surrounding medium, V is the volume of the particle, and R_G is the radius of gyration (or Guinier radius) of the particle. According to equation (5), R_G can be obtained from the slope of the linear plot of $\ln I(s)$ *vs.* s^2. There are definite relations between R_G and the dimension of regular solids. Thus, for a sphere of radius R, $R_G = R(3/5)^{1/2}$. If the particles are not identical, R_G is an average radius where the large particles are weighted much more heavily than the small ones. The Guinier plot should be made in the central part of the

SAXS curve where the exponential approximation is valid. The Guinier plot provides a simple way to derive a mean size parameter with a precise physical meaning, without going into absolute intensity measurements.

The integral intensity scattered at small angles is

$$\int_0^\infty 4\pi s^2 I(s)\, ds = (\Delta\varrho)^2 V \quad \text{(point collimation)} \tag{6}$$

$$\int_0^\infty 2\pi s I(s)\, ds = (\Delta\varrho)^2 V \quad \text{(linear collimation)} \tag{6'}$$

Absolute measurements of $I(s)$ allow the determination of the volume of the particle if $(\Delta\varrho)^2$ is known, or the reverse.

Porod [53] has shown that the intensity of the SAXS in the wing of the curve is such that the product $I(s) \times s^4$ (or $I(s) \times s^3$ in linear collimation) tends asymptotically to a constant value

$$\lim\, (I(s) \times s^4) = [(\Delta\varrho)^2/8\pi^3] \times S \tag{7}$$

or

$$\lim\, (I(s) \times s^3) = [(\Delta\varrho)^2/16\pi^2] \times S \tag{7'}$$

where S is the total area of the interface between two phases. In the case of a solid-air system, S is the specific area of the solid phase. The Porod law is verified when the product $I(s) \times s^4$, or $I(s) \times s^3$ in linear collimation, tends to a constant asymptotic value at large s. The specific surface area is then deduced from the asymptotic value. The Porod law can be applied whatever the dimension, shape, and porosity of particles, and it is still valid in concentrated systems because the interferences between particles have little effect on the wing of the SAXS curve. If the solid phase is in the form of non-porous particles, an equivalent mean diameter can be calculated assuming spherical shape, $\bar{D} = 6/\varrho S$, where ϱ is the density of the particle.

A combination of equation (6′) and (7′) can be used to calculate a mean radius R_P (Porod radius) without going into absolute intensity measurements [54, 55]

$$R_P = 3 \int_0^\infty 2\pi s I(s)\, ds / 16\,\pi^2 \lim\, (I(s) \times s^3) \tag{8}$$

b) Calculation of particle size distribution

When the particles are assumed to scatter indenpendently, the scattered intensity is proportional to the integral

$$I(s) \simeq \int_0^\infty P(D)\, I(D, s)\, dD$$

where $I(D, s)$ is the intensity scattered by a particle of diameter D, and $P(D)$ is a distribution function such that $P(D)\, dD$ is equal to the probability that the particle has a diameter between D and $D + dD$. There are three methods to evaluate $P(D)$. First, the distribution $P(D)$ is supposed to have a given form described by a simple analytical expression containing one or more adjustable parameters which are fitted so that the calculated curve matches the experimental one. This approach has been used by Whyte *et al.* [55]. A log-normal distribution was assumed, and the moments of distribution were calculated from the values of R_G and R_P. Second, the integral can be solved by numerical analysis methods without any assumption as to the form of the distribution [56, 57]. Third, the most used methods are based on the Fourier transform of the scattered intensities [58–61].

The Fourier transform of $I(s)$ gives the function $\gamma(r)$ (characteristic function of the particle) or $P(r)$, the Patterson function ($\gamma(r) = P(r)/\varrho^2 V$) where ϱ is the density of the particles. The function $\gamma(r)$ has been used to calculate chord length distributions [59]. A chord is a segment in a scattering medium with both ends at the boundary with another scattering medium. Chords are suitable to characterize porous solids. $\gamma(r)$ has also been used to calculate various kinds of diameter distribution assuming a given particle shape [58]. Thus, for spherical particles, the volume-weighted distribution of particle diameters $f_v(D)$ is given by [60, 61]

$$f_V(D) = -(D^3/3)\,\frac{\mathrm{d}}{\mathrm{d}D}\left(\frac{\gamma''(D)}{D}\right) \tag{9}$$

A volume-weighted mean diameter can be calculated from the distribution

$$\bar{D}_V = \int_0^{D_{\max}} D f_V(D)\,\mathrm{d}D \tag{10}$$

Surface-weighted distribution and mean diameter can also be obtained from $\gamma(r)$.

The various expressions of SAXS parameters given so far are valid for a two-phase system. They can be applied to three-phase systems with some assumptions [Renouprez and Imelik [62]], or the scattering due to one of the phases can be suppressed experimentally by the pore masking technique. These points are discussed in section B.2.

c) SAXS instrumentation

The measurement of the intensities scattered near the direct beam cannot be carried out with conventional diffractometer or camera. The sample is examined by transmission with a collimation system, giving a very narrow beam without parasitic scattering. This is achieved either with a system of slits or with a curved crystal monochromator focussing the beam on the counter slit. Recent advances in instrumentation (position sensitive and solid state detectors, synchrotron radiation, small angle neutron scattering) have been reviewed by Schelten and Hendricks [63].

2. *Application of SAXS to catalysts*

a) Determination of the specific surface area

One of the most useful applications of SAXS to catalysts is the possibility of determining the specific surface area, S, of the solid phase. In a two-phase system (solid-void), S encompasses all the interface area including intra-

Table 3. Comparison Between Specific Surface Area Measured by BET and SAXS (Ref. [64])

Samples	$S/m^2\ g^{-1}$ (SAXS)	$S/m^2\ g^{-1}$ (BET)	Remarks
SiC	0.34	0.35	non-porous
SiO_2 (aerosil)	42	41	non-porous
NiO	107	118	non-porous
SiO_2 (gel)	640	680	porous
Petroleum coke	40	0.9	closed porosity

particle pores, even if they are closed. This is a possible cause of discrepancy between the BET area, which does not take into account closed porosity. The agreement between the BET and the Porod area is usually good, especially on materials of high surface and of low density. The limit of application is reached when the domain of angles, over which the Porod law is followed, is too near the direct beam. Specific areas as low as 0.5 $m^2\ g^{-1}$ can be measured on low density supports with a well collimated beam. Table 3 gives a comparison between the BET and the Porod areas for a number of porous and non-porous solids [Renouprez *et al.* [64]]. The low-density porous materials should be dehydrated prior to SAXS measurements, because water molecules condensed in the pores decrease the electron density contrast between solid and voids and thus lead to an underestimated S value. Indeed, $(\Delta\varrho)^2$, the square of the difference between the electron density of the two phases, is involved in all the SAXS equations. This is the basis of the pore masking technique, which consists of filling the pores with a liquid of electron density similar to that of the solid phase, thus suppressing the SAXS from this phase (*vide infra*).

The case of three-phase systems (two solid phases + voids) is of great interest since it includes the supported metal catalysts. Designating the support, void, and metal by the subscripts 1, 2, and 3 respectively, there are three interfaces — S_{1-2}, S_{1-3}, and S_{2-3} — which contribute to the total Porod area. Assuming that S_{1-3}, the metal support contact, is negligibly small and that the support texture is not modified by the metal, Renouprez *et al.* [65] have subtracted the S_{1-2} area measured on the metal-free support from the total S to obtain the metal area S_{2-3}. Table 4 shows that there is a good agreement between the metal area determined by SAXS and by H_2 or CO

Table 4. Comparison Between SAXS and Chemisorption Measurements of Metal Area in Pt/Al_2O_3 Catalysts. (Ref. [65])

Pt Content/ wt.%	$S/m^2\ g^{-1}$ (SAXS)	$S/m^2\ g^{-1}$ (H_2 chemisorption)	$S/m^2\ g^{-1}$ (CO chemisorption)
0.6	225	243	218
2.0	125	171	156
3.7	105	105	98

chemisorption in Pt/Al_2O_3 catalysts even when the concentration in platinum is as low as 0.6 wt.%. Goodisman *et al.* [66] have studied Pt/Al_2O_3 catalysts with the aim of determining not only the metal area but also the metal support interface area, S_{1-3}. The procedure requires the knowledge of the volume fraction of the three phases. The results are given in Table 5. The uncertainties in S_{1-3} and S_{2-3} are large due to the inaccurate determination of the volume fraction of the metal, which is small. Nevertheless, this method is valuable to estimate the metal-support contact area, which could be large in the case of raft-like metal particles. In the present case S_{1-3} is about one-fifth of S_{2-3}, the accessible metal area, as it would be for cubes attached with one face to the support surface.

Table 5. Area of Three Interfaces in Pt/Al_2O_3 Catalysts. (Ref. [66])

Pt Content/ wt.%	Volume fraction[a]			Interface area/$m^2\,g^{-1}$ of cat.[a]		
	V_1	V_2	V_3	S_{1-2}	S_{1-3}	S_{2-3}
0	0.214	0.786		377		
4.3	0.208	0.791	0.0014	364	0.64	2.79
7.6	0.204	0.794	0.0025	353	1.11	5.37
11.3	0.199	0.797	0.0038	341	1.63	7.96

[a] Subscripts 1, 2, and 3 refer to support, void, and metal, respectively

The measurement of the metal surface area can also be carried out with the pore masking technique [54, 55, 67], which eliminates most of the scattering from the support, *i.e.*, the S_{1-2} interface area. The pore masking liquids for low density support are usually alkyl iodides. A comparison of their efficiency was discussed by Whyte *et al.* [55]. Care should be taken not to absorb an excess of pore masking liquid, which would increase the overall scattering.

To sum up, the Porod law provides a general method to determine the specific surface area of catalysts. It should be stressed again that the method is valid whatever the dimension, shape, and porosity of particles. It can be applied to any kind of solid with one major exception: the zeolites where the Porod area is very small compared to the BET area. This is because the micropores in these crystalline materials are perfectly ordered according to the lattice symmetry, and there is therefore negligible scattering outside the Bragg reflections.

b) Determination of mean particle size

From the Porod surface a mean surface-weighted particle diameter can be easily calculated assuming spherical shape of the particles. This is a reliable size parameter, provided the particles have no internal pores. The radius of gyration R_G is easily determined from the SAXS curve, but it does not give a reliable mean size in the case of heterodispersed catalysts, because the larger particles are weighted too heavily. In the worst situation exemplified by a study of the porosity of a charcoal carrier [68], R_G was found five times larger than other pore size parameters. Clearly, in this application, the conditions where R_G can be determined were not fulfilled. On the other hand, in a study of 0.35–0.62 wt.% Pt/Al_2O_3 catalyst where the scattering from the support was masked and where the size distribution was comparatively narrow, R_G was in better agreement with other size parameters although it still gave an overestimated value of the particles size [Whyte *et al.* [55]]. The Porod radius, also determined in this study, was in better agreement with the particle diameter determined from chemisorption.

c) Determination of particle size distribution

Particle distribution can be calculated by assuming a given shape for the distribution as described by Whyte *et al.* [55]. However, these distributions are little more informative than the mean parameters R_G and R_P from which they are calculated, and they can lead to misleading conclusions if the actual distribution is bimodal.

Intersting examples of particle size distribution calculated without any assumption other than spherical particle shape have been reported [62, 65, 68–70]. Figure 7 shows the distribution of pores in a porous glass support before and after pore masking [Cocco *et al.* [69]]. Distributions of particle diameters of platinum supported on SiO_2 have been obtained by Renouprez *et al.* [62] without using pore-masking fluid. The SiO_2 was an Aerosil with 50–80 nm spherical particles that gave weak scattering. Figures 8a and 8b show the $f_V(D)$ distributions (equation 9) for 0.5 and 1 wt.% Pt catalysts respectively, which are compared to a volume-weighted diameter distribution obtained by electron microscopy. The oscillations in the SAXS distribution match well those of the TEM distribution. It was shown that for a wide distribution of sizes, the equivalent diameter derived from the Porod surface gives a better representation of the mean particle size than the mean diameter derived from the distribution (equation 10). Figure 9 shows a distribution of Pt particles in 0.3 wt.% Pt/Si—Al glass catalysts determined by SAXS with the pore masking technique, before and after treatment with CO and phosphine. This indicates that the fraction of particles smaller than 2 nm is eliminated by the reagents [Carturan *et al.* [70]].

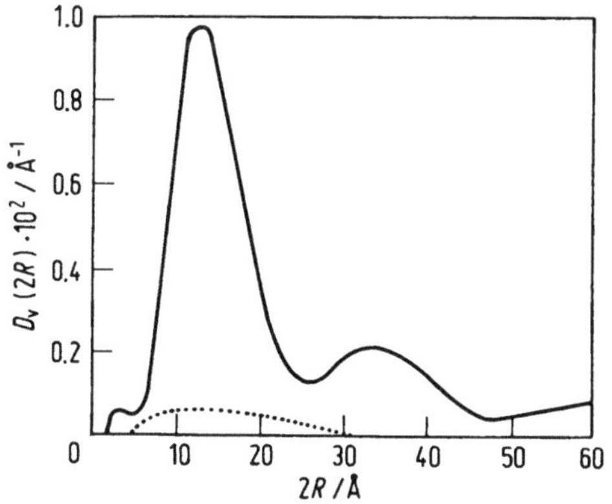

Figure 7. SAXS volume distribution of particle diameters corresponding to a porous glass support. Continuous line, distribution before pore masking; dashed line, after pore masking which eliminates most of the SAXS. (Reproduced with permission from ref. [69])

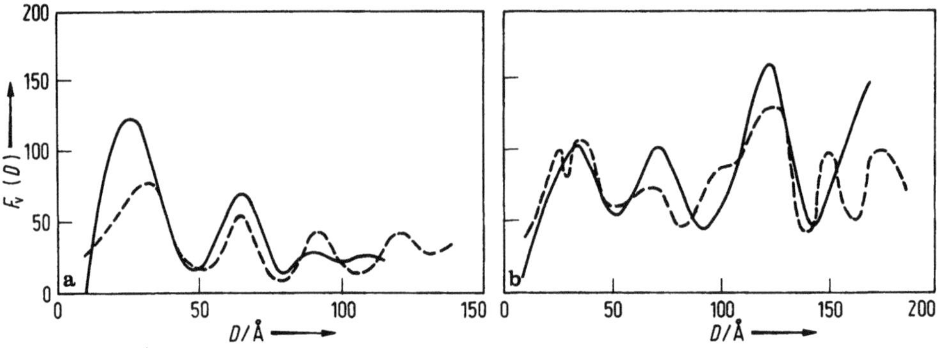

Figure 8. Particle diameters, volume-weighted distribution $f_v(D)$ from SAXS (dashed line) and electron microscopy (continuous line). **a** 0.5 wt.% Pt/SiO_2; **b** 1 wt.% Pt/SiO_2. (Reproduced with permission from ref. [62])

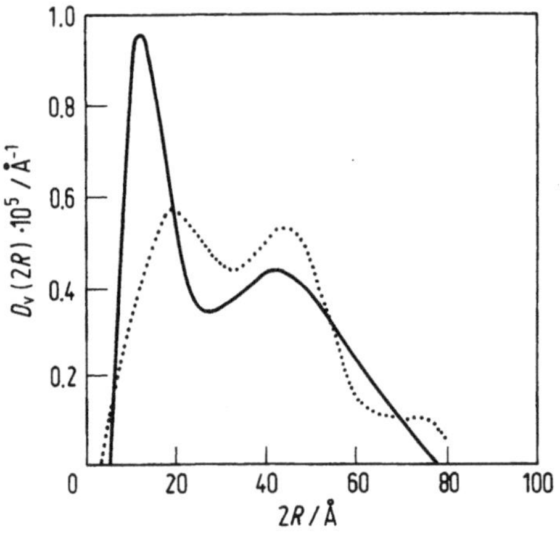

Figure 9. Particle diameters distribution from SAXS on 0.3 wt.% Pt/porous glass catalyst with pore masking (*c.f.* Figure 7). Continuous line, inital distribution; dashed line, distribution after treatment with CO and PPh_3 which eliminates the smallest particles. (Reproduced with permission from ref. [70])

Distribution of diameters of platinum particles supported on Y-zeolite was also obtained without pore masking, since the scattering from the zeolite support was very small with respect to that of platinum [Gallezot *et al.* [71]]. When the zeolite was activated and reduced at 573 K, the maximum of the distribution (1.1 nm) was in good agreement with the Porod equivalent diameter (1.2 nm) because all the particles were located in the zeolite supercages within a narrow size range. However, when the zeolite was activated at 873 K and reduced at 573 K, the largest fraction of platinum was atomically dispersed in the sodalite cages as shown by crystal structure analysis (*c.f.* section 4.A.). These isolated atoms were not taken into account in the Porod area so that the equivalent diameter (2 nm) was much larger than the maximum of the size distribution (0.9 nm) which corresponds to platinum particles in the zeolite supercages. This example shows that isolated atoms or clusters of a few atoms might not be taken into account by SAXS because their scattering is spread over large angles, in spite of the fact that the SAXS is measured with short wavelengths (*e.g.* MoK_α).

C. Comparison between LBA and SAXS

SAXS has been much less applied than LBA to the study of catalyst texture. This is partly due to the fact that SAXS requires a special instrumentation to measure the intensity scattered in the vicinity of the direct beam, whereas LBA measurements are carried out with conventional equipment. The other reason is that derivation of the size parameter is viewed as more cumbersome with SAXS than with LBA. It is true that with SAXS an analysis of the scattering curve is required, whereas a mean size can be obtained from simple line width measurement from LBA. However, this mean size is often not very informative, and a better description of the catalyst texture requires a complete line profile analysis which is at least as time consuming as the complete SAXS analysis. Moreover, the determination of the specific area by SAXS can be much easier in a series of closely related samples provided S is known from the absolute intensity measurement on a given sample. The areas of the other samples can be determined by measuring the scattering at one angle in the region where the Porod law is followed.

SAXS possesses five advantages over LBA as far as the determination of the catalyst texture is concerned. First, it is much more sensitive. The examples given previously show that particle size distribution can be obtained for supported catalysts containing 0.5 wt.% of platinum, whereas LBA requires at least three times that amount in favorable cases (*e.g.*, 5 nm particles), and still higher metal concentrations are needed if smaller particles are present (2–3 nm). Second, whatever the concentration, particles smaller than 1.5 nm cannot be measured in practice by LBA because the overlap of diffraction lines prevents line profile analysis, whereas particles as small as 0.5 nm can be measured by SAXS. Third, LBA is hampered very often by the superposition of diffraction lines from different phases, whereas SAXS analysis is possible in two solid phase systems, provided the electron density contrast between the solid phases is sufficient. Fourth, LBA gives a crystallite size which might be smaller than particle size, thus bearing no relation to the specific surface. In contrast, a complete description of the catalyst texture, including particle size, pore size, and specific area, is obtained from SAXS and can be used to calculate intrinsic catalytic activity, to interpret an activity pattern as a function of particle size, or to account for a particular mass transfer problem in the catalyst pores. Fifth, the particle sizes of poorly crystalline or even amorphous solids can still be measured by SAXS, whereas LBA is no longer applicable.

With respect to other methods used in the study of catalyst texture, both methods can be carried out *in situ* during treatments or catalytic reactions. And, in contrast with TEM, they give an overall averaged picture of the catalyst texture.

Although both methods can give information on particle shape, there are very few convincing results, because the particles in catalysts are always to some extent heterodispersed which prevents the determination of shape. They are also seldom uniform in shape. However, LBA can give a good estimate when the crystallites exhibit marked shape anisotropy.

4. Determination of the Atomic Structure of Catalysts

Heterogeneous catalysts are generally composed of several solid phases which cannot be studied separately because they are intimately connected, and any attempt to separate them would result in a modification of their texture, structure, and catalytic properties. Moreover, these solids are highly divided with a large fraction of atoms on the surface which do not have the same coordination as in the bulk. These solids might also exhibit a number of crystal defects (*e.g.*, substitution and displacement disorder due to non-stoichiometry) or even amorphous regions which complicate the structure analysis to an extreme. It is therefore generally not possible to determine the structure of heterogeneous catalysts, *i.e.*, a complete set of distances and angles between all the atoms in these solids, and especially on their surface. However, in many cases it is possible to determine a local atomic arrangement, *i.e.*, the interatomic distances and coordination number around a given atom. Two methods can be used for that purpose: the radial electron distribution method (RED), which is based on wide angle X-ray diffraction, and the extended X-ray absorption fine structure (EXAFS), which is based on X-ray absorption spectroscopy. Before discussing these methods, we shall first examine the case of a group of crystalline catalysts of growing importance, the zeolites, whose structure can be determined by the well-known methods of crystal structure analysis.

A. X-ray Diffraction Studies of Zeolite Structure

Zeolite structures are built on a three-dimensional arrangement of AlO_4 and SiO_4 tetrahedra sharing corner oxygen atoms. The porous alumino-silicate frameworks thus formed have a well-defined structure from which the large internal surface can be described accurately. There are several points to be made about the structure of zeolites in relation to their catalytic properties. First, the structure of the zeolite framework has to be established, since the molecular sieve properties and the shape selectivity depend on the structure of the microporous lattice. Second, the framework structure has to be studied after modification by synthesis or chemical treatments (*e.g.*, of ratio and order of Si and Al atoms in tetrahedral sites which modify the acidic properties of the zeolite). Third, the positions of the exchangeable cations have to be determined to account for the catalytic properties in relation to the cation position and coordination in the zeolite cages. Fourth, the porous lattice can accommodate various kinds of catalytically active species such as metal coordination compounds, molecular metal clusters, and metal clusters. Locations, dispersions, and interactions with the framework have to be established.

These questions can be answered by the methods of determination of crystal structure described in many textbooks. However, some adaptations are required. In most cases, the zeolite microcrystals are not large enough to be mounted on a four-circle diffractometer or to be subjected to modification treatments. Then the crystal structure should be determined from the X-ray powder pattern. This is possible when the lattice symmetry is such that only a few parameters need to be determined. Thus, in faujasite (X, Y) type zeolites, there is only one independent tetrahedron per unit cell, and the oxygen atoms are in special positions, so that only 10 atomic coordinates are needed to describe completely the zeolite framework. These unknown parameters can be determined from about 100 independent reflections, corresponding to about 200 structure factors. Note that when a reflection corresponds to several structure factors, its intensity is apportioned among the components in proportion to the square of the structure factors calculated from a trial structure.

1. Structure and Modification of Zeolite Framework

Most of the zeolite framework structures have been determined from single crystal study and are well established [72, 73]. However, closely related structures can be hard to distinguish from each other. Thus, Shermann and Bennett [74] have shown that several natural and synthetic zeolites have almost identical X-ray patterns, although their framework topology is different. Dislocation and twinning might add serious complications to the identification of framework topology. Furthermore, variations in lattice parameters and framework symmetry have been observed for many zeolite systems [75–77].

The second role of the X-ray method is to study the modification of zeolite structure as a result of thermal or chemical treatments. A typical example is the dealumination of faujasite type zeolite, which occurs under hydrothermal treatment or chemical complexation. Maher *et al.* [78] have shown by crystal structure analysis from powder data that the removal of Al atoms during the ultrastabilization leaves tetrahedral vacancies which are subsequently filled by silicon atoms. Gallezot *et al.* [79] have shown similarly that the occupancy factor of the tetrahedron sites does not change upon EDTA dealumination, which was also explained by a replacement of Al by Si atoms.

Since the properties of zeolites depend upon their crystalline character, it is essential to determine the crystalline fraction. The usual procedure is to take the ratio between the sum of the intensities of a set of lines from the actual sample over that of a reference sample. There is an important limitation which should be kept in mind, *viz.*, when the crystalline domains become smaller than a few tens of nanometers, there are too few unit cells stacked to produce X-ray reflections (there are only about 4 unit cells along the edge of a 10 nm cube). Although these domains are crystalline, they are not taken into account in the crystalline fraction, since they do not contribute to the reflections. This is a possible cause of discrepancy between the X-ray method and the other methods of evaluation of the crystalline fraction (infrared spectroscopy, adsorption capacity measurements). Finally, the crystalline fraction (percentage of crystalline phase) should not be confused with the degree of crystal perfection in that crystalline fraction. Thus, the crystalline fraction decreases upon dealumination because part of the zeolite is destroyed, but the order of the atoms in this fraction is improved when Al atoms are replaced by Si atoms.

2. *Localization of Active Species in Zeolites*

The zeolite frameworks can host various prosthetic species which might act as active sites or promoters of catalytic activity. These species (cations, protons, metal atoms, coordination compounds) can all be located by crystal structure analysis.

a) Position of cations

In dehydrated states, the exchangeable cations are bonded to framework oxygen atoms but remain coordinatively unsaturated. There is always a larger number of sites than of cations; therefore both the atomic coordinates and the occupancy factors of the cation sites have to be determined. When the sites are outside symmetry elements, the number of sites equivalent by symmetry is large, and site occupancy is low. The best accuracy is obtained when the high atomic number elements are in special positions. When there are cations of two or more different elements, there is no exact solution, because the electron density on a given site is a function of the atomic number and of the occupancy factor. Then site attribution is based on crystal chemistry data (bond lengths between the cation and framework oxygen) and on the comparison between the chemical and crystal composition. These difficulties hold whether the structure is determined from powder or from single crystal diffraction.

A complete review of cation position in zeolites has been published [80]. A few examples of relations between the cation positions determined by X-ray diffraction and the catalytic properties of Y-zeolites are given below.

Gallezot *et al.* [81] have shown that 11.1 Cu^{2+} cations were located in SI′ sites (sodalite cages) in a dehydrated CuY zeolite (*c.f.* Figure 10 for site nomenclature). In subsequent structure determination from single crystal data, 11.4 Cu^{2+} were found in SI′ [Maxwell and De Boer [82]]. The excellent agreement between these investigations shows that reliable cation population can be obtained from powder data. The Cu(I′) cations occupy inaccessible cages. However, it was shown that they can migrate towards the supercages

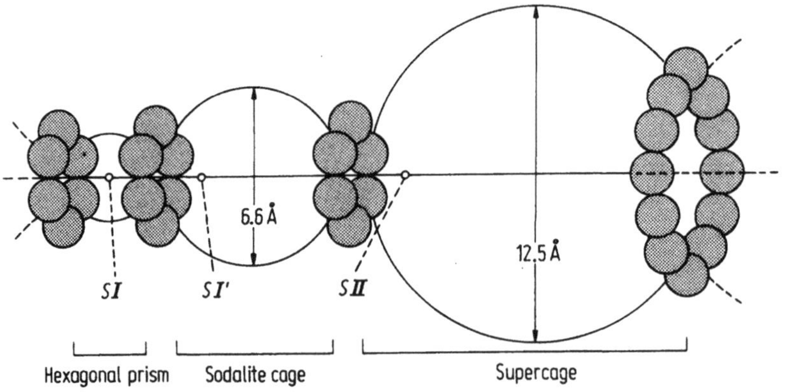

Figure 10. Position of cation sites in faujasite type zeolites.

upon adsorption of reagent molecules (olefins, aromatics, pyridine). Therefore, in spite of their initial inaccessibility, they can be available after migration to catalyse a reaction.

The X-ray study of Ni^{2+} ion positions in $Ni_{19}Y$ and $Ni_{14}Y$ zeolites has shown that 11.3 and 11.7 Ni^{2+}/unit cell (u.c.) occupy the SI sites in the hexagonal prisms [Gallezot *et al.* [83, 84]]. The agreement is also excellent with the 10.6 Ni^{2+}/u.c. population of SI sites found in a single crystal of Ni faujasite [85]. These cations do not migrate out of their hidden positions except if traces of water are present and if the reagents entering the supercages can displace the H_2O molecules coordinated to Ni^{2+} cations. Accordingly, it was shown [86] that in dehydrated NiY zeolite containing 19, 14, and 10 Ni^{2+}/u.c., there is a good correlation between the catalytic activity in the reaction of cyclotrimerization of acetylene into benzene ($Ni_{19} > Ni_{14} > Ni_{10} = 0$) and the number of Ni^{2+} ions outside hexagonal prisms (5.8; 2.3; 0.0). The protons cannot be located directly because of their very small scattering factor. However, their interaction with the framework oxygen produces a lengthening of the corresponding T—O bonds (T = Si or Al). The protons are therefore detected from the set of T—O distances [87, 88]. This does not mean that they are permanently fixed on a given oxygen atom, but there is a time averaged probability of presence depending on the extent of the T—O bond elongation.

b) Metal location in zeolite

Starting with transition metal cations bonded to the zeolite framework on well-defined cation sites, four different situations can be expected upon reduction:

(I) The atoms remain on similar sites but at a longer distance from the oxygen atom.
(II) They remain in the same type of cage but on random sites.
(III) They migrate out of the cage or along pores and agglomerate to form particles encaged or occluded in the zeolite.
(VI) They migrate out of the zeolite crystals.

Cases (I) and (II) are favorable as far as crystal structure determination is concerned, because the atoms still obey the periodicity of the zeolite lattice. Thus, the crystal structure of partially reduced AgA [89, 90] and AgF [90, 91] (F = faujasite) zeolites has been determined from single crystal and powder X-ray diffraction data. Positively charged Ag_6 and Ag_3 silver clusters have been located on well-defined sites in the sodalite cages of these zeolites. The crystal structure studies of $Pd_{12.5}Y$ [92] and $Pd_{13.7}Y$ [22] zeolites have shown that after activation in O_2 the Pd cations occupy the SI′ sites at $x = y = z = 0.044$ strongly interacting with the 0(3) framework oxygen (Table 6). Upon reduction at 423 K, the Pd atoms in $Pd_{13.7}Y$ [23] occupy SI′ sites at $x = y = z = 0.082$ at a longer distance from the 0(3) atoms (0.274 nm) than before reduction (0.207 nm). The AgY and $Pd_{13.7}Y$ samples are typical of case (I), since the reduced atoms are in well-defined sites and still interact with the framework oxygen atoms. On the other hand, the reduction

Table 6. Distribution of Palladium and Platinum atoms in the Sodalite Cages of PdY and PtY Zeolites

Samples and treatments (K, atmosphere)	Population of zeolite sites[a]				Interatomic distances/nm[b]		Ref.
	SI′ (0.04)	*SI′ (0.08)	SII′ (0.17)	U (0.125)	SI′-0(3)	*SI′-0(3)	
$Pd_{12.5}Na_{19.5}H_{11.5}Y$ (873, O_2)	10.6 Pd				0.201		[92]
$Pd_{12.5}Na_{19.5}H_{11.5}Y$ (873, O_2:473, H_2)		4 Pd		9 Pd		0.272	[92]
$Pd_{13.7}Na_{9.3}H_{19.2}Y$ (823, O_2)	12.0 Pd				0.207		[22]
$Pd_{13.7}Na_{9.3}H_{19.2}Y$ (823, O_2: 423, H_2)		8.8 Pd	1.8 Pd			0.274	[23]
$Pd_{13.7}Na_{9.3}H_{19.2}Y$ (823, O_2: 300, C_6H_6)	4.1 Pd	5.7 Pd			0.219	0.270	[23]
$Pt_{10}Na_{17}H_{19}Y$ (873, O_2)	9.8 Pt				0.220		[71]
$Pt_{10}Na_{17}H_{19}Y$ (873, O_2: 573, H_2)		3(0.065) Pt		8 Pt		0.254	[71]

[a] Coordinates of the sites ($x = y = z$) are given in parentheses. (Sites U correspond to Pd or Pt atoms uniformly distributed in a 0.25 nm radius sphere at the center of sodalite cages).
[b] Between species in SI′ sites and framework oxygen atoms (3).

of $Pd_{12.5}Y$ at 473 K is an example of case (II) because after reduction most of the atoms are in random positions in the same type of cage as before reduction. They have been localized using a liquid scattering function corresponding to a model where the Pd atoms are distributed at any place within a 0.25 nm radius sphere in the sodalite cage. The fact that there are more Pd atoms occupying SI′ sites in $Pd_{13.7}Y$ than in $Pd_{12.5}Y$ zeolite is probably due to a stronger interaction of the reduced species with the zeolite framework, because the former zeolite is more acidic than the latter. The adsorption of C_6H_6 on the oxidized form of the $Pd_{13.7}Y$ zeolite [23] also produces a limited reduction of type (I), since almost 6 atoms move from SI′ (0.044) to SI′ (0.082) (Table 6). The structure of a $Pt_{10}Y$ zeolite [71] activated at 873 K and reduced at 573 K is another example of case (II), since the Pt atoms were found mostly in random sites in the sodalite cages (Table 6). In the PdY and PtY zeolites there are statistically only one or two atoms per sodalite cage, so that the metal is truly atomically dispersed. In case (III), where the encaged particles are formed at the expense of migrating atoms, there is usually a very irregular occupancy of cages. In this case and whenever the particles are on the external surface (case IV), the scattering by the metal atoms is independent from that of the zeolite lattice, so the atoms cannot be located by crystal structure analysis. However, the metal dispersion can then be characterized by LBA or SAXS.

c) Location of coordination complexes

Organometallic complexes can be introduced in the porous zeolite lattice by adsorption, provided their molecular dimensions do not prevent them from entering the pores. Molecules of the metal carbonyls $Ru_3(CO)_{12}$, $Re_2(CO)_{10}$, and $Mo(CO)_6$ adsorbed in Y-zeolite have been located from powder data using a liquid scattering function [Gallezot *et al.* [93]]. Coordination complexes can also be formed by adsorption of organic molecules which complete the

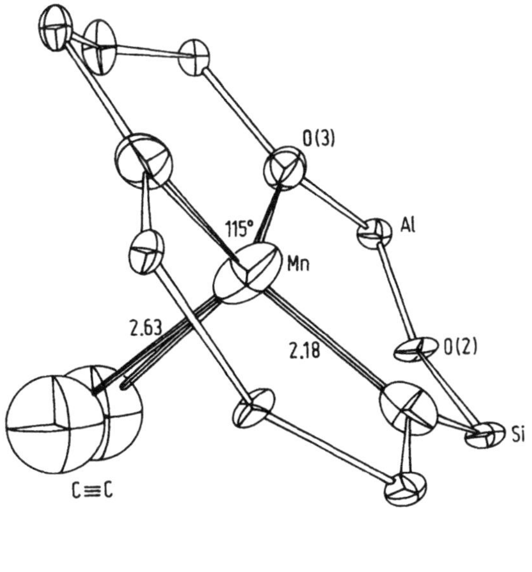

Figure 11. Coordination of Mn^{II} cation in $Mn_{4.5}Na_3$ — A.4.5 C_2H_2. Atoms are represented by their ellipsoids of probability. (Reproduced with permission from ref. [94])

coordination of the cations. The precise stereochemistry of transition metal complexes in zeolite A has been determined by Riley and Seff from single crystal data [94]. Figure 11 shows the stereochemistry of Mn^{II} coordinated to three framework atoms at 0.218 nm and by a C_2H_2 molecule with the C atoms equally distant from Mn^{II} at 0.263 nm.

3. Conclusion

Zeolites are unique among heterogeneous catalysts insofar as the nature, the number, and the position of the active sites can be determined by X-ray methods, even during *in situ* catalytic reaction.

The X-ray methods give a time and space averaged picture of the large internal surface, and there is no hope of measuring the local and transient states involved in catalysis. However, the location of the framework and extraframework atoms during the steady state of the reaction can at least eliminate a number of false hypotheses, *e.g.*, on the state of reduction, or on the accessibility of the active or presumably active species.

B. Local Atomic Arrangement Studied by RED

1. Scattering by Atoms without Long Range Order

A clear-cut distinction is often made between crystalline solids giving strong and sharp diffraction lines and amorphous solids giving weak and broad scattering bands. In fact, all the intermediate structures are possible between perfectly ordered and totally disordered matter. Except for monoatomic gases, atoms always exhibit a short range order. Thus, in vitreous silica the tetrahedra are almost perfect, but the Si—O—Si angles vary between 120 and 180°, which results in a wide distribution of Si—Si distances around 0.312 nm [95]. The distinction between amorphous and crystalline forms of matter on the basis of the appearance of lines on diffraction patterns is completely misleading in the case of very small particles, where the atoms can be as well-ordered as in large particles but their X-ray patterns exhibit very broad bands. This is readily understood from the Debye equation (96) that expresses the intensity scattered by randomly oriented matter.

Let us consider N particles composed of atoms of types m and n separated by the distance r_{mn}, whose scattering factors are f_m and f_n. The scattered intensity I(s), a function of $s = 2 \sin \theta/\lambda$, reads

$$I(s) = N \sum_{m} \sum_{n} f_m f_n \sin (2\pi s r_{mn})/2\pi s r_{mn} \qquad (11)$$

Figure 12 shows the interference function $I(\theta)/Nf_{Pt}^2$ calculated for N identical f.c.c. clusters of platinum containing 14 atoms. It is clear that the pattern is quite different from that of a powder composed of larger crystallites with the same structure, since it exhibits only two bands and resembles hardly the usual set of reflections from the f.c.c. structure. Therefore, particles smaller than 1–1.5 nm constitute a particular category of solids which have no long range order because of their small size. Their structure can be studied by the methods applied to non-crystalline materials.

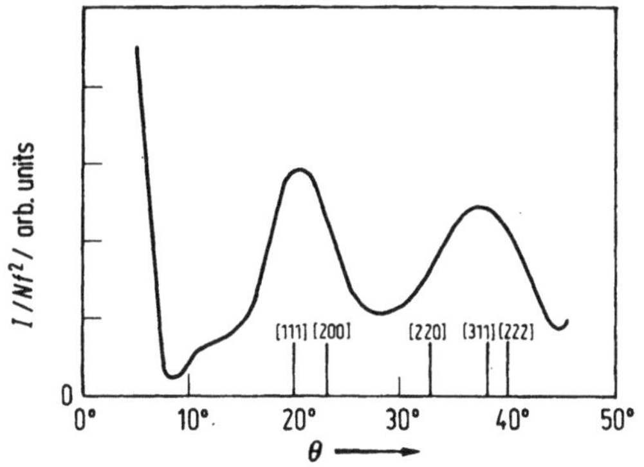

Figure 12. Interference function calculated with the Debye equation for a 14 atom cubic Pt cluster. Positions of the diffraction lines for bulk metal are given for comparison

It is possible, theoretically, to determine the short range order of atoms by calculating a scattering pattern with the Debye formula and by comparing it to the experimental pattern. However, calculation of models is plagued by the multiplicity of parameters involved (number of atoms, structure, shape). In practice, structure information on non-crystalline materials is obtained by the radial electron distribution (RED) method.

2. *The RED Method*

The RED method and its applications to the structure of catalysts have been discussed by Ratnasamy and Leonard [97]. Complete theoretical treatments can be found in the textbook of Warren [10]. Direct information on the structure can be obtained by the Fourier transform of the intensity scattered by matter. Starting from the Debye equation and considering a continuous distribution of atomic density around a given atom, Zernicke and Prinz [98] have derived the following radial distribution function (RDF) in the case of N identical atoms

$$4\pi r^2 \varrho(r) = 4\pi r^2 \varrho_0 + 8\pi r \int_0^\infty s i(s) \sin(2\pi r s)\, ds \tag{12}$$

$4\pi r^2 \varrho(r)\, dr$ is the average number of atoms situated between two spheres of radius r and $r + dr$, and ϱ_0 is the average atom density. Furthermore,

$$i(s) = [I(s)/Nf^2] - 1$$

where $I(s)$ is the experimental intensity in absolute units (electron^2) corrected for polarization and for the Compton (inelastic) scattering. To obtain $I(s)$ in absolute units, the measured intensities are matched at high Bragg angles with the sum of the calculated independent scattering (Nf^2) and the Compton scattering, both available in the International Tables for X-ray Crystallography.

Equation (12) is valid only for identical atoms; however, distribution functions can be calculated for a solid composed of non-identical atoms by making the approximation that all the scattering factors decrease *homothetically* with s. Then, $f_m = K_m f_e$, where f_e is an average scattering factor per electron and K_m is the effective electron number in the atom m. The radial distribution function reads

$$\sum_{uc} K_j 4\pi r^2 \varrho_j(r) = 4\pi r^2 \varrho_e \sum_{uc} K_j + 8\pi r \int si(s) \sin(2\pi rs)\, ds \qquad (13)$$

where ϱ_e is the average electron density in the sample. The summation is over all the atoms in the stoichiometric formula or unit of composition (uc).

$$i(s) = \left[I(s)/N - \sum_{uc} f_j^2\right] \Big/ f_e^2$$

where N is the number of units of composition in the sample.

Equation (13) represents the superposition of the radial distribution function for each kind of atom. Each peak on the radial distribution function corresponds to a pair of atoms separated by a distance r, and its surface is proportional to the number of these pairs and to the product of the atomic number of these atoms.

More rigorous methods have been used to calculate the RDF of solids containing non-identical atoms. They are based on a fitting process between the RDF calculated by Fourier transform of the intensity data and the RDF calculated from models of pair distribution [10, 95, 100]. Significant improvements in the calculation of the RDF were designed to get rid of the oscillations which affect the distribution at small r values [101–103]. These oscillations are due to the truncation effects in the calculation of Fourier series and to the errors made in measuring weak intensities at large Bragg angles. Whatever the mathematical treatments, the derivation of RDF relies on the integral in equations (12) or (13). The final quality of the structure information obtained from the RED method depends upon how far in reciprocal space the intensities are measured and upon the accuracy of these measurements. Indeed, the RDF is calculated from the Fourier transform of the product $si(s)$, which may have a sizable value at large s, even if $i(s)$ is small. In practice, the X-ray pattern is recorded with monochromatized MoKα radiation (0.071 nm) by step scanning from $2\theta = 1°$ to $2\theta = 140$–$150°$. The counting statistics should be good throughout the scattering pattern. With conventional equipment, a week might be required for complete recording.

3. Application to Non-crystalline Catalysts

The RED method has been mainly applied to determine the short range order in vitreous solids, polymers, and carbons; few studies have been reported on the active phases of catalysts.

In a early work [104], the structures of SiO_2 gels were studied by RED as a function of the preparation conditions. Figure 13 shows the RED corresponding to an aged gel (curve A), a dialyzed gel (B), and a freshly precipitated gel (C). The extent of organization of the structure is much greater in the aged gel. The area of the first peak corresponding to Si—O (O—Si) interatomic vectors is in good agreement with the theoretical area S calculated in the unit of composition SiO_2: $S = (n_{Si-O} + 2n_{O-Si})\, K_{Si}K_O = (4 + 2\times 2) \times 14 \times 8 = 896$ electron^2, where n_{Si-O} is the number of O atoms surrounding

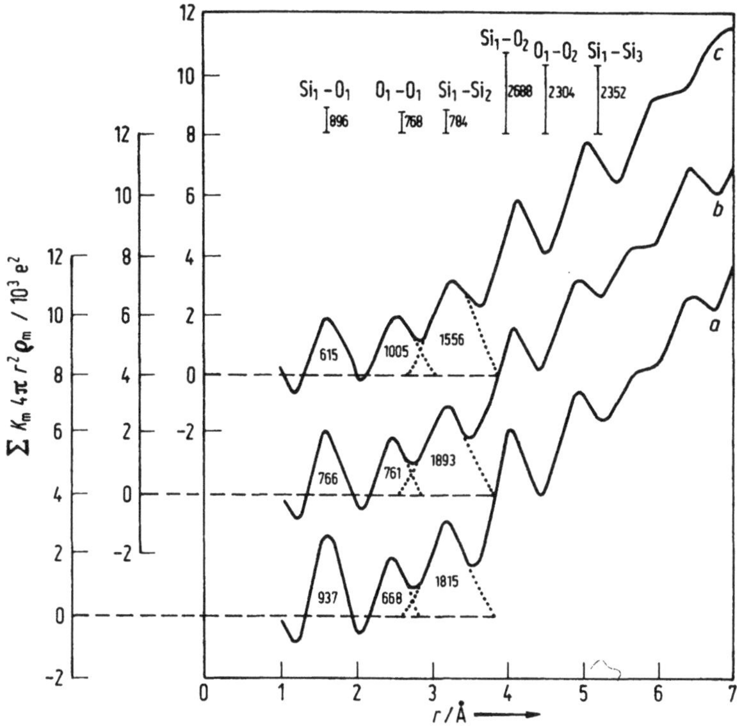

Figure 13. Radial electron distribution of SiO_2 gels. A, gel aged 2 months at *pH* 5.5; B, gel dialysed 8 days after precipitation at *pH* 3.5; C, fresh gel precipitated at *pH* 2. (Reproduced with permission from ref. [104])

Si, n_{O-Si} is the number of Si atoms surrounding O, and K_{Si} and K_O are the effective numbers of electrons in Si and O. Figure 14 gives the average Si—O distances as a function of the amount of water (hydroxyl groups) in different gels. It is clear that the lengthening of the Si—O bonds is proportional to the number of Si—OH groups. The upper curve corresponds to Na-rich gels; the presence of sodium increases the number of non-bridging oxygen atoms. The calcination of these gels results in a decrease of Si—O bonds and in an increase of the Si—O—Si angles which become comparable to those observed in vitreous silica.

Léonard *et al.* [105, 106] have studied in detail the structure of alumina and of silico-alumina by RED. The relative fractions of Al atoms in octahedral and tetrahedral sites were determined. The structures of molybdenum sulfide [107] or Co—Mo sulfide [108] were studied by RED to determine the structure of the active phase at a composition $r = Co/(Co + Mo) = 0.3$ corresponding to a maximum of catalytic activity in thiophene hydrodesulfurization. The RED data calculated for the definite compound $CoMo_2S_4$ and for a mixture of Co_9S_8 and MoS_2 were compared to the experimental RED data for the catalyst of the same composition ($r = 0.3$). It turned out that the structure of the catalyst was not compatible with that of $CoMo_2S_4$, but

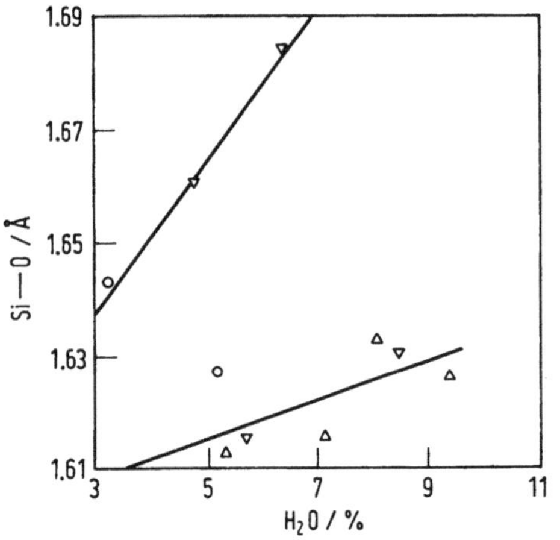

Figure 14. Si—O bond lengths (Å) determined by RED. Modification as a function of constitution water content. Upper curve corresponds to Na-rich gel (>0.3%). (Reproduced with permission from ref. [104])

could be accounted for either by a mixture of MoS_2 and Co_9S_8 or by a solid solution of Co in MoS_2.

4. Application to Metal Clusters

Although the RED method has been widely applied to study metal alloys with the aim of determining the short range order of components, there was no apparent interest in studying pure metal until it was realized that very small metal particles could have different structure than the corresponding bulk metal because of intrinsic size effects or because of environmental effects (support, adsorbates). Indeed, the fraction of surface atoms is no longer negligible. The RED can give three types of information, *viz.*, particle size, particle structure, and particle morphology.

a) Particle size estimation

The RED method is not primarily designed to provide particle size measurement; rather, size information can be obtained as a by-product of structure study. Since the interatomic vectors cannot be longer than the particle diameter, the longest interatomic vector still observable gives the upper limit of the particle diameter. Thus, Figure 15 shows the RED of 1 nm and 3 nm Pt particles supported in Y-zeolites [Gallezot *et al.* [109]]. It turns out that while large Pt—Pt peaks are still observed at large *r* values in the case of the 3 nm particles, the RED of the 1 nm Pt particles exhibits vanishing peaks at 0.8–1 nm.

b) Atomic structure of clusters

In the last decade, the RED method was applied extensively at the Institute of Catalysis in Novosibirsk (USSR) to study monometallic or bimetallic supported catalysts [110–113]. The study of (W + Pt)/SiO_2 catalysts [111] showed that the Pt and W atoms can be associated in a bimetallic cluster, part of the tungsten being in the metal phase, and the other part, undetected, probably in an ionized state bonded to the support. Another interesting

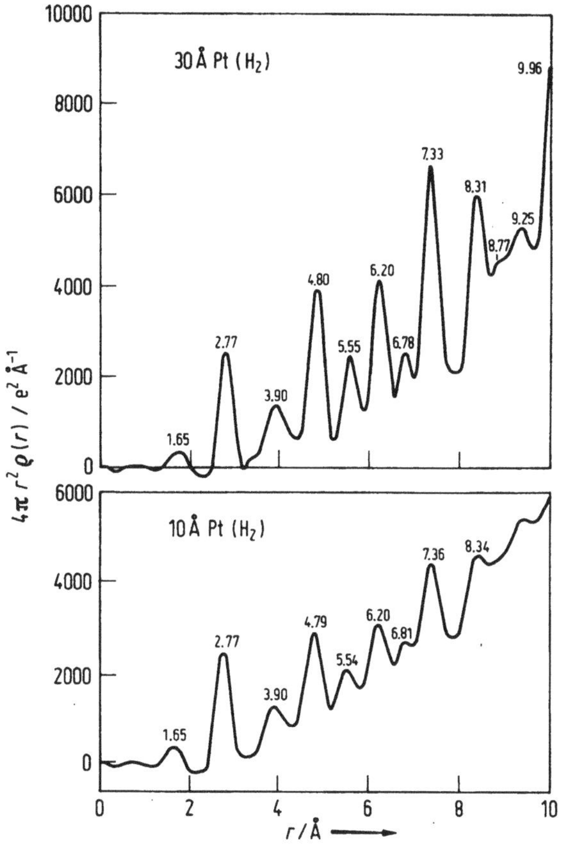

Figure 15. Radial electron distribution of $Pt_{8.7}Y$ zeolite containing 3 nm or 1 nm Pt particles covered with H_2. (Reproduced with permission from ref. [109])

study of $Pt—Re/Al_2O_3$ catalyst showed that the promoter (Re) increases the dispersion of platinum and that it fills the octahedral sites in the spinel structure of the support, since the peak heights from the support atoms are increased upon Re addition [113].

In the previous investigations, the effect of chemisorption on the structure of the particles was not mentioned. Ratnasamy *et al.* [114] were the first to point out that the RED method can probe the structure modifications induced by adsorbates. The RDF of 1.5–3 nm Pt particles supported on alumina or silica-alumina indicated that the structure was f.c.c. upon H_2 reduction, but that O_2 adosption induced a complete disorder in the external layer of the particles. These results were obtained on catalysts containing only 0.5–1 wt.% of platinum; therefore, the peaks corresponding to the Pt—Pt interatomic vectors were very small compared to those of the support atoms.

This type of study was extended by Gallezot *et al.* [109] on 1 nm Pt particles in Y type zeolites, which are well suited materials, since a 100% metal dispersion can still be obtained with metal concentration as high as 15 wt.%. In Figure 16 the RED of a NaHY zeolite (curve 1) is given in com-

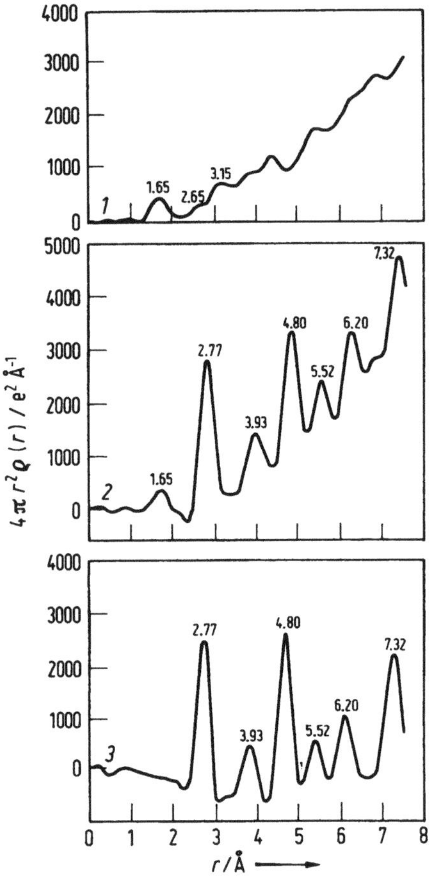

Figure 16. Radial electron distribution: *1* Pt-free NaHY support; *2* $Pt_{11}Y$ zeolite containing 1 nm Pt particles covered with H_2; *3* RED obtained by subtracting curve 1 from curve 2. (Reproduced with permission from ref. [109])

parison with that of a PtNaHY zeolite containing 1 nm Pt particles covered with H_2 (curve 2). The first peaks on curve 1 correspond to T—O, O—O, and T—T distances in the support (T = Si, Al). Then the distribution merges progressively to a continuum of interatomic distances, because about 600 atoms are present in the unit cell of the zeolite. The large peaks on curve 2 can be attributed to the interatomic vectors Pt—Pt, since they correspond to the normal set of distances in f.c.c. bulk platinum. Curve 3 corresponds to the distribution obtained by subtracting the RED of NaHY from the RED of PtNaHY. The peaks from the support atoms are suppressed, and the positions of the Pt—Pt peaks are not appreciably modified. This procedure does not eliminate the interatomic vectors between the platinum and the framework atoms, but these vectors are expected to be quite random.

Figure 17 shows the RED of the 1 nm Pt particles after adsorption of various molecules at 298 K. The particles covered by H_2 (curve 2) exhibit the normal f.c.c. structure [109] and interatomic distances of bulk metal, whereas the structure of the particles outgassed at 700 K is contracted and distorted with respect to the f.c.c. structure [115, 116] (curve 1). The adsorption of H_2S

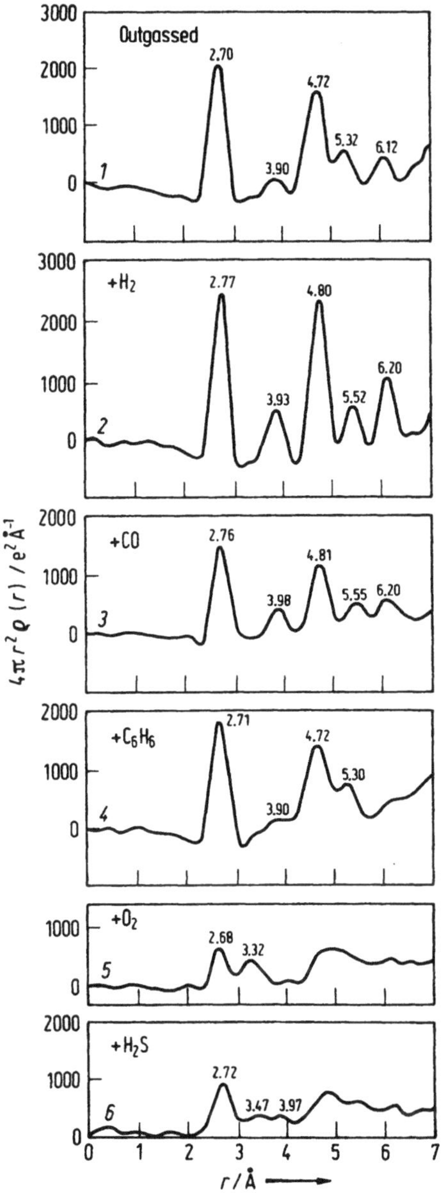

Figure 17. Radial electron distribution for $Pt_{11}Y$ zeolite, (the RED of the support has been subtracted): *1* outgassed 1 mm Pt particles kept under Ar; *2* covered with H_2; *3* covered with CO; *4* covered with C_6H_6; *5* covered with O_2; *6* covered with S. The adsorptions were carried at 300 K. (Reproduced with permission from Gallezot, P. Surface Science **106**, 459 (1981): North Holland Publishing Co.)

(curve 6) or O_2 (curve 5) on the bare or on the H_2-covered particles leads to a nearly complete disorder, since the initial Pt—Pt peaks vanish [109, 117]. New weaks peaks may correspond to distances between Pt atoms bridged by an adsorbed O or S atom. Adsorption of CO on the bare particles (curve 3) produces a complete relaxation of the distances, but the Pt—Pt peaks are 50% smaller [116]. This has been explained by a displacement of the Pt atoms bonded to CO in such a way that some distances are elongated and others contracted

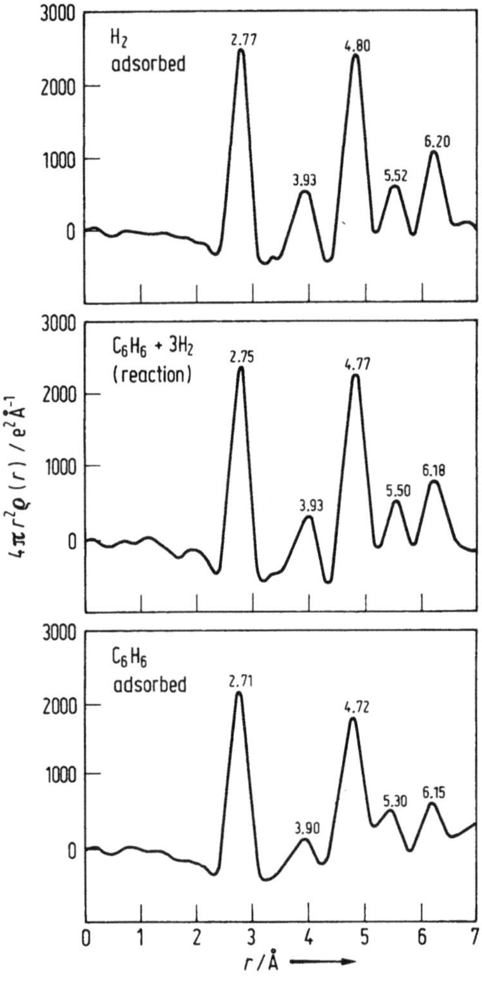

Figure 18. Radial electron distribution for $Pt_{11}Y$ zeolite (the RED from the support has been subtracted): *1* 1 nm Pt particles covered with H_2; *2* during catalytic hydrogenation of benzene with the stoichiometric mixture (the same distribution is obtained in excess of C_6H_6); *3* covered with C_6H_6. (Reproduced with permission from ref. [115])

around a mean value as in the molecular cluster of known structure $(Pt_{19}CO_{22})^{4-}$ [118]. The weakest perturbation of the structure is brought by C_6H_6 adsorption (curve 4) which produces a very small relaxation of the Pt distances [115]. Figure 18 gives the RED of the same catalysts in the course of benzene hydrogenation with the stoichiometric reagent mixture (C_6H_6 + 3 H_2) (curve 2). It is compared to the RED corresponding to H_2-covered particles (curve 1) and to C_6H_6-covered particles (curve 3) [Gallezot and Bergeret [115]]. The RDF (2) is intermediate between RDF (1) and RDF (3), indicating that during C_6H_6 hydrogenation the Pt particles are covered by both adsorbed hydrogen and benzene. Another *in situ* experiment carried out in excess of C_6H_6 gave the same distribution, showing that a large fraction of the Pt surface remains covered by irreversibly adsorbed hydrogen which is not displaced by benzene.

The adsorption of C_4H_{10} (400 K), C_2H_4 (300 K), and the cracking of C_6H_6 at 600 K on the 1 nm Pt particles [119] produced the same effect as CO adsorption, but with a lesser extent of distance relaxation and structural disorder. It was concluded that Pt—C bonds are formed by the dissociative chemisorption of C_4H_{10} and C_2H_4 and by the cracking of C_6H_6.

c) Morphology of small aggregates

A given atom packing in bulk metal gives a characteristic set of peak positions (interatomic distances) and peak heights (coordination number). However, in a small aggregate of the same structure the peak heights change according to the nature of the boundary planes which limit the aggregate in different directions. Conversely, information on the modification of particle morphology can be obtained from an analysis of the peak heights in related samples, provided they have the same structure. Table 7 gives the positions and heights of the Pt—Pt peaks in PtCeY zeolites containing 1 nm Pt particles [119]. After C_6H_6 cracking and H_2 regeneration at 600 K, the structure is similar to that of the starting samples PtCeY (H_2, 300 K). However, it can be noticed that the peaks corresponding to the 1st, 3rd, and 7th coordination shells are higher after the treatment than before, while the other peaks are slightly modified or are smaller. Although complete model calculations are required to account for these changes in coordination number, these effects are consistent with a shape modification resulting in more extended (111) planes.

Table 7. Modification of Pt Aggregate Morphology in PtCeY Catalysts. (Ref. [119])

Sample	Distances of nearest neighbors/nm; and peak heights/e^2 Å^{-1} [b]					
	1	2	3	4	5	7
PtCeY + H_2, 300 K	0.277	0.393	0.481	0.548	0.618	0.730
	2180	760	2215	860	1170	1680
PtCeY + H_2, 300 K[a]	0.276	0.395	0.480	0.550	0.618	0.729
	2300	690	2570	780	1120	1980

[a] After cracking of C_6H_6 at 600 K and regeneration under flowing H_2 at 600 K.
[b] Peak 6 is too weak to be observed in the wing of peak 7.

C. Local Atomic Arrangement Studied by EXAFS

1. Near Edge Absorption Spectroscopy

When the energy E of an X-ray photon is sufficient to promote an electron transfer from a core atomic level (K, L) to unfilled bound states, the X-ray absorption coefficient $\mu = \ln(I_0/I)$ (where I_0 and I are the incident and transmitted intensities) undergoes a steep increase which gives an edge on the μ *vs.* E plot. The position of the edge and the near edge fine structure (extending *ca.* 30 eV above the edge) can give information on the chemical environment of the absorbing atom [4, 16]. This method has a long history; its application was revived in recent years through several studies on the *L* absorption edges of supported metal catalysts [120–126]. Theoretical papers have also appeared on the subject [127, 128].

Near edge spectroscopy appeared in recent years as a by-product of EXAFS because both sets of data are taken in the same experiment. However, edge spectroscopy requires a better energy resolution than that needed for EXAFS and, therefore, more stringent monochromatization conditions.

The effects of the chemical environment on the electronic properties of the metal atoms have been monitored by the displacement of the edge position and by the change in area of the edge peak appearing just above the edge inflection point. The edge peak area, also called the white line because it appears as a clear line on film recording, depends upon the probability of the electron transfer and, therefore, is proportional to the number of unfilled valence states. Thus, the white line area of the PtL_{III} edge is proportional to the number of unfilled states in predominantly the 5*d* band. The analysis and interpretation of the X-ray absorption edge data in relation to band structure are still not well established. Theoretical work is in progress [129, 130]. In the meantime, edge displacement and white line area can be used to compare on a relative basis the modification of charge density on the metal atom in a series of related samples due, for instance, to electron transfer in different chemical environments. Table 8 gives the edge shifts and the normalized edge peak areas in PtY and Pt/Al_2O_3 catalysts. Positive edge shift and increase of edge peak area indicate that the electron deficiency of platinum aggregates increases with the support acidity, the presence of multivalent cations (Ce, Ca), and the adsorption of electron acceptor species (O, S, CO).

Table 8. Edge Shift and Peak Area of L_{III} Absorption Edges in Pt Catalysts

Catalysts[a]		Edge Shift/eV	Normalized peak area	Ref.
Pt		0	1	
PtNaY	(3 nm)	0 ± 0.3	1	[122]
PtNaY	(1 nm)	0 ± 0.3	1.2	[122, 123]
PtNaHY	(1 nm)	0.6 ± 0.3	1.3	[122, 124]
PtHY	(1 nm)	0.6 ± 0.3	1.6	[123]
PtCaY	(1 nm)	0.3 ± 0.3	1.4	[123]
PtCeY	(1 nm)	0.5 ± 0.3	1.5	[122, 123]
PtNaHY	(1 nm) + O_2	0.8 ± 0.3	1.7	[122]
PtNaHY	(1 nm) + NH_3		0.9	[124]
PtNaHY	(1 nm) + H_2S		1.6	[124]
PtNaHY	(isol. atoms)	1.2 ± 0.3	1.3	[122]
Pt/Al_2O_3	(2.6 nm)	0 ± 0.2	1.1	[125]
Pt/Al_2O_3	(1 nm)	0.2 ± 0.2	1.4	[125]
Pt/Al_2O_3	(1 nm) + O_2		2.1	[125]
Pt/Al_2O_3	(1 nm) + CO	0.8 ± 0.3	2.1	[125]

[a] Pt particles are covered with H_2 unless otherwise specified.

2. *EXAFS Theory and Analysis*

When the energy E of the X-ray photon is larger than the threshold energy E_0 of the electron, a photoelectron is ejected from the absorbing atoms with an associated wave vector $k = (2m(E - E_0)/(h/2\pi)^2)^{1/2}$. The emitted photoelectron is backscattered by the surrounding atoms, giving rise to interference between ingoing and outgoing waves and producing oscillations in the absorption coefficient. The fine structure extending several hundred electron volts past the edge is the EXAFS represented by the function $\chi(k) = |(\mu - \mu_0)|/\pi_0$, where μ_0 and μ are the absorption coefficients of the atom in the free state and in the actual solid, respectively.

The modern EXAFS theory has been developed in the seventies [131–134]. Several review articles have appeared on the subject [17–21]. The basic expressions of the EXAFS theory considering atoms as point scatterers are given hereafter [131, 132]. The amplitude of the oscillations $\chi(k)$ are given by

$$\chi(k) = \sum_j A_j(k) \sin [2kr_j + 2\delta_j(k)] \tag{14}$$

where the summation extends over j coordination shells, r_j is the distance from the absorbing atom to atoms in the jth shell, and $2\delta_j(k)$ is the phase shift which is the sum of the phase shifts of the absorbing atom and of the backscattering atom.

The amplitude function $A_j(k)$ for the jth shell ist

$$A_j(k) = (N_j/kr_j^2)\, F_j(k) \exp(-2k^2\sigma_j^2) \tag{15}$$

where N_j is the number of atoms in the jth shell (coordination number), σ_j is the root mean square relative displacement between the emitter and scatterer atoms, and $F_j(k) = f(k) \exp(-2r_j/\lambda(k))$, where $f(k)$ is the backscattering factor and $\lambda(k)$ the mean free path for the electron.

Examination of equation (14) and (15) shows that the local structure around the absorbing atom is described by two parameters, the interatomic distance r_j and the coordination number N_j. The method most used to extract these parameters is based on the Fourier transform of $\chi(k)$

$$\Phi_n(r) = 1/(2\pi)^{1/2} \int_{k_{min}}^{k_{max}} k^n \chi(k) \exp(-2ikr)\, dk \tag{16}$$

Often $n = 3$ for the Fourier transform to give a lower weight to the low energy portion of $\chi(k)$ and be less sensitive to the value of k_{min} and to the threshold energy E_0.

The radial structure function given by equation (16) exhibits peaks at distances $r < r_j$, the differences $r_j - r$ arising from the phase shift. The simplest way of correcting the distances between atoms A and B for the phase shift is to assume that the phase shift is the same for any similar pair A–B. The correction is determined from an EXAFS study of a reference compound of known structure.

Subsequent improvements of the EXAFS theory [see review article of Lee *et al.* [20]] have led to the calculation of phase shift and amplitude function [135] from which $\chi(k)$ is calculated. The distance r_j can be treated as a variable parameter which is adjusted to obtain the best agreement between calculated and experimental EXAFS.

The coordination number N_j and the root mean square displacement of atoms σ_j can be obtained by comparison with a reference compound of known structure, assuming the transferability of phase shift, backscattering amplitude, and electron mean free path from the reference sample to the unknown sample. Pettifer [18] has discussed the problems arising in EXAFS interpretation and the uncertainties in the derivation of the parameters r_j and N_j in relation to the assumptions of phase and backscattering amplitude transferability. The validity of these transfers (or the use of tabulated values) could be dubious if the ionicities of the absorber and scatterer atoms are different in the unknown and reference samples, *e.g.*, because of electron transfer caused by the chemical environment.

3. *Application of EXAFS to Catalysts*

So far, most of the EXAFS studies on catalysts have been carried out on small metal particles on supports with the aim of studying the effect of the carrier or adsorbate on the structure of the metal aggregates. The following monometallic systems have been studied: Au and Pt supported on Al_2O_3, MgO and SiO_2 [136], Ru/SiO_2 [137], Pt/SiO_2 [138], Cu/Al_2O_3 [139], Pt/Y-zeolite [140], Pt/SiO_2 or Al_2O_3 [141], Pt, Os, and Ir supported on SiO_2 and Al_2O_3 [46], Pt/Al_2O_3 [125]. Bimetallic systems, Ru—Cu/SiO_2 [142], Pt—Re/Al_2O_3 [143], Os—Cu/SiO_2 [144], and Co—Mo hydrodesulfurization catalysts [145] have also been investigated.

The study of Ir, Os, and Pt supported on SiO_2 and Al_2O_3 by Via *et al.* [46] provides a good illustration of the capabilities of EXAFS for the study of the structure of metal aggregates. Table 9 gives the average coordination numbers, interatomic distances, and root mean square deviations of distance for the first coordination shell in the different catalysts. The average coordination numbers are obviously smaller than 12, the number of nearest neighbors

Table 9. Structure Parameters Derived from EXAFS Study of Supported Metal Catalysts[e]. (Ref. [46])

Metal[a]	Carrier	N_1[b]	r_1/nm[c]	$\Delta\sigma^2/10^{-6}$ nm^2[d]
Os	SiO_2	8.3	0.2702 (0.2705)	22
Ir	SiO_2	9.9	0.2712 (0.2714)	27
Ir	Al_2O_3	9.9	0.2704 (0.2714)	29
Pt	SiO_2	8.0	0.2774 (0.2775)	18
Pt	Al_2O_3	7.2	0.2758 (0.2775)	37

[a] 1 wt %, dispersion 70–100 %.
[b] Average coordination number, uncertainty ±20 %.
[c] Nearest neighbors distances (uncertainty ±0.001 nm); distances in bulk metal are given in parentheses.
ε $\Delta\sigma^2 = \sigma_1^2$ (cluster — σ_1^2 (bulk).
[e] First coordination shell data.

in f.c.c. platinum, because of the large number of surface atoms. The interatomic distances are equal to those of bulk metals except in the Pt/Al_2O_3 catalyst, which might indicate a carrier effect. Indeed, a contraction of the Pt—Pt distances in 1 nm Pt aggregates supported on acidic PtCaY and PtHY zeolite have also been observed by EXAFS [123] (—0.005 and —0.01 nm respectively with respect to the Pt—Pt distances in PtNaY).

Via *et al.* [46] have also shown that the lower coordination number of atoms is accompanied by an increase in the root mean square of the relative displacement of atoms in agreement with a LEED study [146] on surface Pt atoms and with the LBA study of Pt crystallites in Pt/SiO_2 catalysts [24].

The influence of adsorbed species on the structure of metal aggregates has also been investigated. After outgassing Pt aggregates encaged in Y-zeolite, it was found that the nearest-neighbor distances are contracted (0.266 nm) with respect to Pt bulk value (0.277 nm) [Moraweck *et al.* [140]]. Some of these bare aggregates could exhibit an icosahedral structure. The adsorption of H_2 at 300 K produces a relaxation of the distances which are then comparable to those of f.c.c. platinum. These results are similar to those found by RED [109, 115, 116], except that the Pt—Pt distances in the bare aggregates were slightly longer (0.27 nm, *c.f.* section 4.B.).

Fukushima *et al.* [125] have studied the structure of 1 nm Pt particles supported on Al_2O_3 under H_2, O_2, and CO. The O_2 adsorption at 100 K produces a decrease of the Pt—Pt distance (—0.025 nm) and of the magnitude of the peak in the Fourier transform, but at 300 K the structure appears highly disordered in agreement with earlier RED results [109]. However, there is a discrepancy with RED as far as CO adsorption is concerned, since a 0.012 nm decrease of the Pt—Pt distance was reported, whereas no change in the average Pt—Pt distance was observed by RED [116].

There are few reports on the detection of bonds between the metal atoms and the adsorbed species on heterogeneous catalysts except in the case of oxygen adsorbed on platinum, which gives peak at about 0.167 nm [125, 141] before phase shift correction. On the other hand, distances between metal atoms and ligands have been determined in the case of homogeneous catalysts [147].

The study of Ru—Cu bimetallic clusters [142] exemplifies the unique capability of EXAFS to probe the local environment around each of the components. Thus, the structure around Ru atoms appears very similar to that observed in Ru/SiO_2, *i.e.*, the Ru atoms are coordinated predominantly by other Ru atoms, whereas the Cu atoms have both Ru and Cu neighbors. This is consistent with a model where the copper is present at the particle surface.

It is in principle possible to draw information on particle morphology, provided coordination numbers are available for several coordination shells. Gregor and Lytle [148] have reported an analysis based on the comparison of geometrical shape models and dispersion models which can give clues to the presence of specific particle shape. However, it seems difficult to get reliable information on particle shape when a size distribution is present and from the coordination numbers of the second and third shells which are not accurately known.

D. Comparison Between the RED and EXAFS Methods

1. Features Common to the Methods

RED and EXAFS methods can probe the local atomic structure in amorphous and crystalline solids. They are both based on Fourier transform of scattering data, photoelectron scattering in the case of EXAFS and X-ray wide angle scattering in the case of RED. These Fourier transforms do not give a complete resolution of individual interatomic distances and angles, but rather spherically averaged information on correlation in atomic positions.

Both methods give a radial distribution averaged over all the atoms of a given type; therefore, they predominantly reflect the structure around bulk atoms, except in small particles where the surface atoms are predominant. This is a very important point which should be kept in mind when interpreting RED or EXAFS results. The interatomic distances and coordination numbers are of little use for metallic catalysts if the sample contains large particles or exhibits a wide size distribution. Ideally, investigations should be carried out on particles smaller than 1.0–1.5 nm to study the effect of the support and of the adsorbates with a good sensitivity. Since it has been shown that the structure greatly depends upon the nature and coverage of the adsorbed species on the particle surface, RED or EXAFS investigations on metal catalysts must be performed with a careful control of the atmosphere. More specifically, data taken in air are of little significance, since it was shown that oxygen adsorption produces almost complete disorder among surface atoms [109, 114, 125]. However, both methods remain useful, whatever the particle size, to study the short range order in catalysts containing two or more combined elements as in alloys, mixed oxides, or sulfides.

The last common feature is that these methods are based on X-ray diffraction or absorption data which can be taken *in situ*. Samples can be treated and kept in cells equipped with transparent windows and studied under a wide range of temperature and pressure conditions. They are also well suited to study homogeneous catalysts and, more specifically, homogeneous supported catalysts.

2. Advantages of RED over EXAFS

The RED method is based on wide angle diffraction data which can be recorded with a conventional X-ray source and diffractometer equipped with a high-temperature controlled atmosphere attachment. In favorable cases, the entire pattern can be recorded in three or four days. On the other hand, EXAFS requires either the use of synchrotron radiation, which is sparsely distributed and not convenient for *in situ* experiments needing a bulky side setup (*e.g.*, a catalytic reaction bench) or the use of a rotating anode generator plus specific equipment (analyser crystal, ionization chamber, computerized operation).

The RED processing can be readily carried with a single program correcting and scaling the intensities and calculating the Fourier transform (equation 11). The radial distribution is then obtained without any assumptions and without intermediate data handling. In contrast, EXAFS analysis involves several steps: normalization, background removal, adjustment of E_0, Fourier filtering to isolate the EXAFS of a coordination shell, and adjustment of phase and backscattering amplitude with respect to a reference compound. Therefore, EXAFS data processing needs personal experience and intervention.

The main advantage of RED is that it gives directly the real interatomic distances. Although there are a number of EXAFS studies showing that one can rely on reference compound comparison or on calculated phase shifts and get a precision of 0.001 nm on the first nearest neighbor distances [20], there are still few data on special materials such as heterogeneous catalysts and specifically on metal-adsorbate binding. Thus, metal aggregates smaller than 1 nm depart from the structure of bulk metals taken as reference compounds insofar as their band structure is intrinsically different, or might be modified as a result of electron transfer with the supports or with the adsorbates. Furthermore, the atoms in aggregates are subject to displacement disorder which can be very large in the case of O or S adsorption [109, 117] and still important in CO or hydrocarbon adsorption [116, 119]. Thus, there is a question about the 0.012 nm contraction of the average Pt—Pt distances observed by EXAFS on 1 nm Pt aggregates supported on Al_2O_3 upon CO adsorption [125], which induces both an electron deficiency and a displacement disorder of Pt atoms [116]. Such a contraction was not observed on 1 nm Pt aggregates supported on zeolites [116]. One may wonder whether the contractions of the Pt—Pt distances in the aggregates observed by EXAFS are real or apparent because of a change of phase shift or because of disorder producing a spread of distances. Indeed, it has been shown that the radial distribution obtained by the Fourier transform of EXAFS changes with disorder, peaks get smaller and appear at shorter distances. The limitation of EXAFS in the study of disordered systems has been discussed by Lee *et al.* [20]. Thus, the Nb—Ge distances in Nb_3Ge measured by RED are contracted by 0.012 nm in the amorphous phase compared to the crystalline phase [149], but the contraction measured by EXAFS is 0.024 nm [150, 151].

Another important advantage of RED over EXAFS is that the former method can probe the structure of the successive coordination shells, because X-ray photons are weakly absorbed, and scattering data are available almost from the origin of the reciprocal space. In contrast, the Debye-Waller term $\exp(-2k^2\sigma_j^2)$ and the mean free path factor $\lambda(k)$, damping the value of $F_j(k)$, reduce rapidly the contribution of the second and third coordination shells to the EXAFS. Furthermore, because of the difficulty in analyzing the near edge region, the EXAFS data at $k < 0.3\ nm^{-1}$ which carry information on large r are not available. Indeed, Fourier transforms of EXAFS data exhibit in most cases one or two significant peaks corresponding to the first nearest neighbors of the absorbing atoms. The other peaks when present are greatly displaced from their real value, so that the uncertainty in these bond lengths can be as high as 0.02 nm, and the coordination numbers can be even less accurate. The radial distribution obtained from the RED method provides a more complete description of the atom packing, and eventually of particle morphology, since the complete distribution of interatomic vectors is available. This is conspicuous from the comparison between the distribution obtained by RED for 1 nm Pt aggregates (Figure 16) with the distribution obtained by EXAFS for comparable Pt particles [152].

3. Advantages of EXAFS over RED

EXAFS experiments require a lesser amount of catalyst (*e.g.*, 0.05–0.1 g) than RED (0.5–1 g). Moreover, the concentration of the component under study (*e.g.*, metal supported catalysts) can be much lower than for RED experiments. Thus, EXAFS study can be performed currently with 0.5–1 wt.% of platinum. Similar concentrations have been used in RED study, but the Pt—Pt peaks appeared very faintly with respect to those due to the support atoms. In practice, the RED method needs at least 2–3 wt.% of metal of the third transition row.

The second advantage of EXAFS is that it probes the local environment of the atoms separately, *viz.*, the Fourier transform of EXAFS gives only peaks corresponding to the distances between the absorbing atom and its neighbors. By contrast, the radial distribution given by RED encompasses all the interatomic vectors present in the scattering matter, including necessarily those of the support atoms. In bimetallic catalysts, this advantage of EXAFS is still more valuable, since the environment of each component can be determined and compared. Note, however, that the discrimination of atoms adjacent in the periodic table is not possible in view of the similarity of the amplitudes and phases.

Perhaps the potentially most important advantage of EXAFS over RED is its sensitivity to low Z atoms, *viz.*, the possibility of measuring bond lengths between metal atoms and ligands. This capability has been illustrated by several studies on homogeneous catalysts and on organometallic compounds of biological interest. One can expect that EXAFS could probe the distances between adsorbates and adsorbants in heterogeneous systems, whereas RED gives only the effect of the adsorbates on the arrangement of high Z atoms. However, there are few examples thus far. Indeed, there may be more difficulties in working with heterogeneous catalysts than with organometallic complexes. Thus, in the coordination sphere of a metal surface atom, there are up to 9 other metal atoms (*e.g.*, for the 111 face of an f.c.c. metal) whereas there is, at most, one adsorbed atom which contributes so little to the EXAFS that it can hardly be detected.

5. Concluding Remarks

X-ray diffraction will remain for years a basic tool for the identification of solid phases in heterogeneous catalysts. However, X-ray techniques should not be limited merely to the purpose of phase analysis, since they are powerful tools for evaluating quantitatively parameters describing the intimate texture and structure of heterogeneous catalysts.

In the coming years, one may expect that the full capability of line broadening analysis, *i.e.*, the derivation of crystallite size distribution and crystal defect parameters from line profile analysis, will be used in place of the Scherrer analysis of line width that gives poorer or even misleading information. For a number of reasons developed in Section 3, the characterization of the texture of catalysts can be better achieved with small angle X-ray scattering. The possibility of measuring in a few hours the specific surface areas and the particle size distributions, even for low loaded metal supported catalysts, is worth the investment in SAXS equipment. Provided the X-ray diffractometer or SAXS camera is equipped with heating and controlled atmosphere attachments, the modification of catalyst texture can be studied *in situ* during treatments and catalytic reactions. This cannot be achieved with the other methods used for catalyst texture characterization.

The capabilities of X-ray diffraction and X-ray absorption techniques to determine the atomic structure of catalysts open a large field of investigation hardly explored to date. Determination of the catalyst structure does not mean only the identification of bulk solid phases present in the catalyst, but also the determintion of the local atomic arrangement around the surface atoms acting as active sites in catalytic reactions. This is obvious in the case of zeolite catalysts where the positions of all the atoms building the large inner surface can be pinpointed with an error smaller than 0.001 nm with crystal structure determination methods. The local arrangement of atoms can also be determined, whatever the particle size, by radial electron distribution and by extended X-ray absorption fine structure spectroscopy. Interatomic distances can thus be obtained within 0.001–0.002 nm. When most of the atoms are located on the surface, these techniques can probe the modification of surface structure induced by the particle environment, including the support and the adsorbates. From these modifications it is possible, for instance, to determine whether hydrocarbons or CO molecules chemisorbed on a metal surface in competition with hydrogen are associatively or dissociatively chemisorbed, and what the extent of surface coverage by each of the reagents is. Taking advantage of the greater sensitivity of EXAFS to nearest neighbor atoms and of the greater sensitivity of RED to long range atom arrangement, the combined use of both methods should give a still better picture of the structure of surfaces, including the surface complexes formed between adsorbed reagents and surface atoms. Therefore, these X-ray techniques can probe catalysis on the atomic scale, on real catalysts, and under real conditions of operation.

References

1. Jellinek, M. H., Fankuchen, I.: Advan. Catal. **1**, 257 (1948)
2. Kossel, W. Z.: Z. Physik. **1**, 119 (1920); **2**, 470 (1920)
3. Kronig, R.: Z. Physik. **70**, 317 (1931); **75**, 191 (1932)
4. Van Nordstrand, R. A.: Advan. Catal. **12**, 149 (1960)
5. Lewis, P. H.: J. Phys. Chem. **67**, 2151 (1963)

6. Lewis, P. H.: J. Catal. **11**, 162 (1968)
7. Lytle, F. W., in: Physics of Non-Crystalline Solids. Prins, J. A. (Ed.), Amsterdam: North-Holland, 1965, pp. 12–30
8. Sayers, D. E., Lytle, F. W., Stern, E. A., in: Advances in X-ray Analysis. Henke, B. L., Newkirk, J. B., and Mallet, G. (Eds.), New York: Plenum Press, 1970, pp. 248–271
9. Guinier, A.: X-ray Diffraction. San Francisco—London: Freeman, 1963
10. Warren, B. F.: X-ray Diffraction. Reading, Mass: Addison-Wesley, 1969
11. Cowley, J. M.: Diffraction Physics. Amsterdam—New York: North-Holland, 1981
12. Klug, H. P., Alexander, L. E.: X-ray Diffraction Procedures. New York: Wiley, 1974
13. Schwartz, L. H., Cohen, J. B.: Diffraction from Materials. New York: Academic Press, 1977.
14. Guinier, A., Fournet, G.: Small Angle Scattering of X-rays. New Xork: Wiley, 1959.
15. Glatter, O., Kratky, O. (Eds.): Small Angle X-ray Scattering. London: Academic Press, 1982
16. Azaroff, L. V.: X-ray Spectroscopy. New York: McGraw-Hill, 1974
17. Sandstrom, D. R., Lytle, F. W.: Ann. Rev. Phys. Chem. **30**, 215 (1979)
18. Pettifer, R. F., in: Characterization of Catalysts. Thomas, J. M. and Lambert, R. M. (Eds.). New York: Wiley, 1980, pp. 264–273
19. Lytle, F. W., Via, H., Sinfelt, J. H., in: Synchrotron Radiation Research, Winick, H. and Doniach, S. (Eds.). New York: Plenum, 1980
20. Lee, P. A., Citrin, P. H., Eisenberger, P., Kincaid, B. M.: Rev. Mod. Phys. **53**, 769 (1981)
21. Teo, B. K., Joy, D. C. (Eds.): EXAFS Spectroscopy Techniques and Application. New York: Plenum Press, 1981
22. Bergeret, G., Gallezot, P., Imelik, B.: J. Phys. Chem. **85**, 411 (1981)
23. Bergeret, G., Gallezot, P.: J. Phys. Chem., **87**, 1160 (1983)
24. Sashital, S. R., Cohen, J. B., Burwell, R. L., Butt, J. B.: J. Catal. **50**, 479 (1977)
25. Sinfelt, J. H., Via, J. H.: J. Catal. **56**, 1 (1979)
26. Warren, B. E.: Phys. Rev. **59**, 693 (1941)
27. Pollack, S. S., Makovsky, L. E., Brown, F. R.: J. Catal. **59**, 452 (1979)
28. Rashkin, J. A., Pierron, E. D.: J. Catal. **6**, 322 (1966)
29. Nandi, R. K., Pitchai, R., Wong, S. S., Cohen, J. B., Burwell, R. L., Butt, J. B.: J. Catal. **70**, 298 (1981)
30. Scherrer, P.: Göttinger Nachrichten **2**, 98 (1918)
31. Warren, B. E.: J. Appl. Phys. **12**, 375 (1941)
32. Rachinger, W. A.: J. Sci. Instr. **25**, 254 (1948)
33. Stokes, A. R.: Proc. Phys. Soc. London **61**, 382 (1948)
34. Bergevin, F., De Germi, P.: J. Appl. Cryst. **5**, 416 (1972)
35. Paesescu, P., Manailă, R., Popescu, M., Jijovici, E.: J. Appl. Cryst. **7**, 281 (1974)
36. Louër, D., Weigel, D., Louboutin, R.: Acta Cryst. **A25**, 335 (1969)
37. Moraweck, B., Montgolfier, Ph. De., Renouprez, A. J.: J. Appl. Cryst. **10**, 184 (1977)
38. Warren, B. E., Averbach, B. L.: J. Appl. Phys. **21**, 595 (1950)
39. Ganesan, P., Kuo, H. K., Saavedra, A., De Angelis, R. J.: J. Catal. **52**, 310 (1978)
40. Gangulee, A.: J. Appl. Cryst. **7**, 434 (1974)
41. Mignot, J., Rondot, D.: Acta Met. **23**, 1321 (1975)
42. Bertaut, E. F.: Acta Cryst. **3**, 14 (1950)
43. Smith, W. L.: J. Appl. Cryst. **5**, 127 (1972)
44. Moraweck, B., Montgolfier, Ph. De., Renouprez, A. J.: J. Appl. Cryst. **10**, 191 (1977)
45. Smith, W. L.: J. Appl. Cryst. **9**, 187 (1976)
46. Via, G. H., Sinfelt, J. H., Lytle, F. W.: J. Chem. Phys. **71**, 690 (1979)
47. Otero-Schiper, P. H., Wachter, W. A., Butt, J. B., Burwell, R. L., Cohen, J. B.: J. Catal. **50**, 494 (1977)
48. Moraweck, B., Renouprez, A. J., Imelik, B.: J. Chim. Phys. **70**, 594 (1973)
49. Hosemann, R., Bagchi, S. N.: Direct Analysis of Diffraction by Matter. Amsterdam: North Holland, 1962
50. Hosemann, R., Preisinger, A., Vogel, W.: Ber. Bunsen-Ges. Phys. Chem. **70**, 797 (1966)
51. Ludwiczek, H., Preisinger, A., Fischer, A., Hosemann, R., Schönfeld, A., Vogel, W.: J. Catal. **51**, 326 (1978)

52. Fischer, A., Hosemann, R., Vogel, W., in: Proc. 7th Intl. Congr. on Catalysis, 1980, Seiyama, T., and Tanabe, K. (Eds.). Amsterdam: Elsevier, 1981, Vol. 1, pp. 341
53. Porod, G.: Z. Kolloid. **124**, 83 (1951); **125**, 51 (1952); **125**, 109 (1952)
54. Harkness, S. D., Gould, R. W., Hren, J. J.: Phil. Mag. **19**, 115 (1969)
55. Whyte, T. E., Kirklin, P. W., Gould, R. W., Heinemann, H.: J. Catal. **25**, 407 (1972)
56. Vonk, G. G.: J. Appl. Cryst. **9**, 419 (1976)
57. Glatter, O.: J. Appl. Cryst. **10**, 415 (1977)
58. Schmidt, P. W., Weil, C. G., Brill, O., in: X-ray and Electron Methods of Analysis. New York: Plenum, 1968, pp. 86–100
59. Mering, J., Tchoubar, D.: J. Appl. Cryst. **1**, 153 (1968)
60. Donati, J. R., Pascal, B., Renouprez, A. J.: Bull. Soc. Fr. Crist. **90**, 452 (1967)
61. Brusset, H., Donati, J. R.: J. Appl. Cryst. **2**, 55 (1969)
62. Renouprez, A. J., Imelik, B.: J. Appl. Cryst. **6**, 105 (1973)
63. Schelten, J., Hendricks, R. W.: J. Appl. Cryst. **11**, 297 (1978)
64. Renouprez, A. J., Dalmai, G., Weigel, D., Imelik, B.: J. Chim. Phys. **64**, 934 (1967)
65. Renouprez, A. J., Hoang-Van, C., Compagnon, P. A.: J. Catal. **34**, 411 (1974)
66. Goodisman, J., Brumberger, H., Cupelo, R.: J. Appl. Cryst. **14**, 305 (1981)
67. Gunn, E. L.: J. Phys. Chem. **62**, 928 (1958)
68. Benedetti, A., Cocco, G., Enzo, S., Piccaluga, G., Schiffini, L.: J. Chim. Phys. **78**, 961 (1981)
69. Cocco, G., Schiffini, L., Strukul, G., Carturan, G.: J. Catal. **65**, 348 (1980)
70. Carturan, G., Cocco, G., Schiffini, L., Strukul, G.: J. Catal. **65**, 359 (1980)
71. Gallezot, P., Alarcon Diaz, A., Dalmon, J. A., Renouprez, A. J., Imelik, B.: J. Catal. **39**, 334 (1975)
72. Breck, D. W.: Zeolite Molecular Sieves. New York: Wiley, 1974
73. Smith, J. V., in: Zeolite Chemistry and Catalysis, Rabo, J. A. (Ed.), Washington, D. C.: American Chemical Society No. 171, 1976, pp. 1–79
74. Sherman, J. D., Benett, J. M.: Advan. Chem. Ser. **121**, 52 (1973)
75. Smith, J. V.: Acta Crystallogr. **15**, 835 (1962)
76. Baerlocher, Ch., Meier, W. M.: Helv. Chim. Acta. **53**, 1285 (1970)
77. Wu, E. L., Lawton, S. L., Olson, D. H., Rohrman, A. C., Kokotailo, G. T.: J. Phys. Chem. **83**, 2777 (1979)
78. Maher, P. K., Hunter, F. D., Sherzer, J.: Advan. Chem. Ser. **101**, 266 (1981)
79. Gallezot, P., Beaumont, R., Barthomeuf, D.: J. Phys. Chem. **78**, 1550 (1974)
80. Mortier, W. J.: Compilation of Extra-Framework Sites in Zeolites; Guildford, U. K.: Butterworth Scientific Ltd., 1982
81. Gallezot, P., Ben Taarit, Y., Imelik, B.: J. Catal. **26**, 195 (1972)
82. Maxwell, I. E., De Boer, J.: J. Phys. Chem. **79**, 1874 (1975)
83. Gallezot, P., Imelik, B.: J. Phys. Chem. **77**, 652 (1973)
84. Gallezot, P., Ben Taarit, Y., Imelik, B.: J. Phys. Chem. **77**, 2556 (1973)
85. Olson, D. H.: J. Phys. Chem. **72**, 4366 (1968)
86. Pichat, P., Vedrine, J. C., Gallezot, P., Imelik, B.: J. Catal. **32**, 190 (1974)
87. Olson, D. H., Dempsey, E. J.: J. Catal. **13**, 221 (1969)
88. Mortier, W. J., Pluth, J. J., Smith, J. V.: J. Catal. **45**, 367 (1976)
89. Kim, Y., Seff, K.: J. Am. Chem. Soc. **99**, 7055 (1977)
90. Gellens, L. R., Mortier, W. J., Uytterhoeven, J. B.: Zeolites **1**, 11 (1981)
91. Gellens, L. R., Mortier, W. J., Uytterhoeven, J. P.: Zeolites **1**, 85 (1981)
92. Gallezot, P., Imelik, B.: Advan. Chem. Ser. **121**, 66 (1973)
93. Gallezot, P., Coudurier, G., Primet, M., Imelik, B., in: Molecular Sieves-II., Katzer, J. (Ed.). Washington, D. C.: American Chemical Society, 1977, p. 144
94. Riley, D. E., Seff, K.: Inorg. Chem. **14**, 714 (1975)
95. Mozzi, R. L., Warren, B. E.: J. Appl. Cryst. **2**, 164 (1969)
96. Debye, P.: Ann. Physik **46**, 809 (1915)
97. Ratnasamy, P., Leonard, A. J.: Catal. Rev. Sci.-Eng. **6**, 293 (1972)
98. Zernicke, F., Prins, J. A.: Zeit. Physik. **41**, 184 (1927)
99. Pings, C. J., Waser, J.: Chem. Phys. **48**, 3016 (1968)
100. Leadbetter, A. J., Wright, A. C.: J. Non-Cryst. Solds **7**, 141 (1972)
101. Kaplow, R., Strong, S. L., Averbach, B. L.: Phys. Rev. **A138**, 1336 (1965)

102. Gatineau, L.: J. Appl. Cryst. **5**, 255 (1972)
103. Konnert, J. H., Karle, J.: Acta Cryst. **A29**, 702 (1973)
104. Fripiat, J. J., Leonard, A. J., Baraké, N.: Bull. Soc. Chim. Fr. **30**, 122 (1963)
105. Leonard, A. J., Ratnasamy, P., Declerck, F. D., Fripiat, J. J.: Discuss. Faraday Soc. **52**, 98 (1971)
106. Leonard, A. J., Van Cauwelaert, F., Fripiat, J. J.: J. Phys. Chem. **71**, 695 (1967)
107. Ratnasamy, P., Rodrique, L., Leonard, A. J.: J. Phys. Chem. **77**, 2242 (1973)
108. Delvaux, G., Leonard, A., Genet, M., Grange, P., Delmon, B., in: Proc. 5th Ibero-American Symp. on Catalysis, Lisbon 1976, Portela, F., and Pulido, C. (Eds.), pp. 199–202
109. Gallezot, P., Bienenstock, A., Boudart, M.: Nouv. J. Chim. **2**, 263 (1978)
110. Moroz, E. M., Tripolskii, A. A., Ryndin, Y. A., Kuznetsov, B. N.: Kinet. Katal. **18**, 1531 (1977)
111. Ryndin, Y. A., Kuznetsov, B. N., Moroz, E. M., Tripolskii, A. A., Ermakov, Yu. I.: Kinet. Katal. **18**, 1536 (1977)
112. Moroz, E. M., Bogdanov, S. V., Ushakov, V. A., Levitskii, E. A.: Kinet. Katal. **20**, 173 (1979)
113. Ushakov, V. A., Moroz, E. M., Bursian, N. R., Kogan, S. B., Levitskii, E. A.: Kinet. Katal. **20**, 177 (1979)
114. Ratnasamy, P., Leonard, A. J., Rodrique, L., Fripiat, J. J.: J. Catal. **29**, 374 (1973)
115. Gallezot, P., Bergeret, G.: J. Catal. **72**, 294 (1981)
116. Gallezot, P.: Zeolites **2**, 103 (1982)
117. Gallezot, P., in: Proc. 5th Intl. Conf. on Zeolites, Naples 1980, Rees, L. V. C. (Ed.). London: Heyden, 1980, pp. 364–372.
118. Washeckeck, D., Wucherer, E. J., Dahl, L. F., Ceriotti, A., Longoni, G., Manassero, M., Sansoni, M., Chini, P.: J. Am. Chem. Soc. **101**, 6110 (1979)
119. Gallezot, P.: J. Chim. Phys. **78**, 881 (1981)
120. Lytle, F. W.: J. Catal. **43**, 376 (1976)
121. Yermakov, Yu. I., Kuznetsov, B. N., Ovsyannikova, I. A., Startsev, A. N., Erenburg, S. B., Sheromov, M. A., Mironenko, L. A.: React. Kinet. Catal. Letters **7**, 309 (1977)
122. Gallezot, P., Weber, R., Dalla Betta, R. A., Boudart, M.: Z. Naturforsch **A34**, 40 (1978)
123. Weber, R. S., Boudart, M., Gallezot, P., in: Growth and Properties of Metal Clusters, Bourdon, J. (Ed.). Amsterdam: Elsevier, 1980, pp. 415, 418
124. Tri, T. M., Massardier, J., Gallezot, P., Imelik, B., in: Proc. 7th Intl. Congr. on Catalysis, 1980, Seiyama, T., and Tanabe, K. (Eds.). Amsterdam: Elsevier, 1981, Vol. 1, p. 266–278
125. Fukushima, T., Katzer, J. R., Sayers, D. E., Cook, J. W., in: Proc. 7th Intl. Congr. on Catalysis, 1980, Seiyama, T., and Tanabe, K. (Eds.). Amsterdam: Elsevier, 1981, Vol. 1, pp. 79, 94
126. Renouprez, A. J., Moraweck, B., Imelik, B., Perrichon, B., Dominguez-Esquival, J. M., Jablonski, J., in: Proc. 7th Intl. Congr. on Catalysis, 1980, Seiyama, T., and Tanabe, K. (Eds.). Amsterdam: Elsevier, 1981, Vol. 1, pp. 173, 185
127. Brown, M., Peirls, R. E., Stern, E. A.: Phys. Rev. **B15**, 738 (1977)
128. Heald, S. M., Stern, E. A.: Phys. Rev. **B16**, 5549 (1977)
129. Natoli, C. R., Misemer, D. K., Doniack, S., Kutzler, F. W.: Phys. Rev. **A22**, 1104 (1980)
130. Honsley, J. A.: J. Chem. Phys. **76**, 1451 (1982)
131. Sayers, D. E., Stern, E. A., Lytle, F. W.: Phys. Rev. Lett. **27**, 1204 (1971)
132. Stern, E. A.: Phys. Rev. **B10**, 3027 (1974); Lytle, F. W., Sayers, D. E., Stern, E. A.: Phys. Rev. **B11**, 4825 (1975); Stern, E. A., Sayers, D. E., Lytle, F. W.: Phys. Rev. **B11**, 4836 (1975)
133. Ashley, C. A., Doniach, S.: Phys. Rev. **B11**, 1279 (1975)
134. Lee, P. A., Pendry, J. B.: Phys. Rev. **B11**, 2795 (1975)
135. Teo, B. K., Lee, P. A.: J. Am. Chem. Soc. **101**, 2815 (1979)
136. Bassi, I. W., Lytle, F. W., Parravano, G.: J. Catal. **42**, 139 (1976)
137. Lytle, F. W., Via, G. H., Sinfelt, J. H.: J. Chem. Phys. **67**, 3831 (1977)
138. Sinfelt, J. H., Via, G. H., Lytle, F. W.: J. Chem. Phys. **68**, 2009 (1978)

139. Friedman, R. M., Freeman, J. J., Lytle, F. W.: J. Catal. **55**, 10 (1978)
140. Moraweck, B., Clugnet, G., Renouprez, A. J.: Surface Sci. **81**, 631 (1979); Surface Sci. **106**, 35 (1981)
141. Joyner, R. W.: J. Chem. Soc. Faraday I **76**, 357 (1980)
142. Sinfelt, J. H., Via, G. H., Lytle, F. W.: J. Chem. Phys. **72**, 4832 (1980)
143. Short, D. R., Khalid, S. M., Katzer, J. R., Kelley, M. J.: J. Catal. **72**, 288 (1981)
144. Sinfelt, J. H., Via, G. H., Lytle, F. W., Greegor, R. B.: J. Chem. Phys. **75**, 5527 (1981)
145. Clausen, B. S., Topsøe, H., Candia, R., Madsen, J., Lengeler, B., Als-Nielsen, J., Christensen, F.: J. Phys. Chem. **85**, 3868 (1981)
146. Lyon, H. B., Somorjai, G. A.: J. Chem. Phys. **44**, 3707 (1966)
147. Reed, J., Eisenberg, P., Teo, B. K., Kincaid, B. M.: J. Am. Chem. Soc. **100**, 2375 (1978)
148. Greegor, R. B., Lytle, F. W.: J. Catal. **63**, 476 (1980)
149. Cargill, G. S., Tsuei, C. C., in: Rapidly Quenched Metal, Cantor, B., (Ed.). London: The Metals Society, 1978, Vol. 2, p. 337
150. Brown, G. S., Testardi, L. R., Wernick, J. H., Hallak, A. B., Geballe, T. H.: Solid State Commun. **23**, 875 (1977)
151. Cargill, G. S., in: Diffraction Studies on Non-Crystalline Substances, Hargittai, I., and Orville-Thomas, W. J. (Eds.). Amsterdam: Elsevier, 1981, pp. 733–780
152. Sinfelt, J. H., in: Catalysis — Science and Technology, Vol. 1. Anderson, J. R. and Boudart, M., (Eds.). Berlin: Springer, 1981, pp. 257–300

Subject Index

Author Index Volume 1–5

Bei Fragen zur Produktsicherheit wenden Sie sich bitte an:
If you have any questions regarding product safety,
please contact:

Walter de Gruyter GmbH
Genthiner Straße 13
10785 Berlin
productsafety@degruyterbrill.com